全栈技能修炼

使用Angular和Spring Boot打造全栈应用

王芃 著

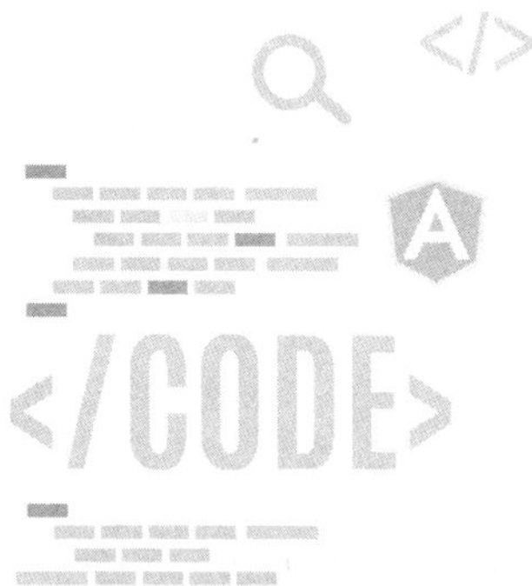

電子工業出版社
Publishing House of Electronics Industry
北京•BEIJING

内 容 简 介

本书涉及很多平台、框架和类库等，主要有前端使用的 Angular，后端使用的 Spring Boot 框架、Spring Security 安全框架，数据库涉及 MongoDB、Elasticsearch 和 Redis。此外，还会学习函数式编程、响应式编程（RxJS）、Redux 等理念，Swagger、JaVers 等工具及 Rest、WebSocket、微服务等概念。

一本书是无法深入这些技术细节的，这也不是本书的目标。希望通过本书，可以帮助读者开启一扇门，真正掌握这些让人眼花缭乱的编程语言、框架、平台、IDE 等技术背后的编程思想。

图书在版编目（CIP）数据

全栈技能修炼：使用 Angular 和 Spring Boot 打造全栈应用 / 王芃著. —北京：电子工业出版社，2019.9
ISBN 978-7-121-37183-7

Ⅰ. ①全… Ⅱ. ①王… Ⅲ. ①网页制作工具—程序设计 Ⅳ. ①TP393.092.2

中国版本图书馆 CIP 数据核字(2019)第 161845 号

责任编辑：石　倩
印　　刷：三河市良远印务有限公司
装　　订：三河市良远印务有限公司
出版发行：电子工业出版社
　　　　　北京市海淀区万寿路 173 信箱　邮编 100036
开　　本：787×980　1/16　印张：32　字数：732 千字
版　　次：2019 年 9 月第 1 版
印　　次：2019 年 9 月第 1 次印刷
定　　价：109.00 元

凡所购买电子工业出版社图书有缺损问题，请向购买书店调换。若书店售缺，请与本社发行部联系，联系及邮购电话：（010）88254888，88258888。

质量投诉请发邮件至 zlts@phei.com.cn，盗版侵权举报请发邮件至 dbqq@phei.com.cn。

本书咨询联系方式：010-51260888-819，faq@phei.com.cn。

推荐序

认识王艽是因为他的文章。

对我这样的惫懒中年人来说，要坚持不懈地写文章实在是有点难，每每提笔都难以成章——宁可靠写程序来麻醉自己。这，或许是另一种形式的“中年危机”？

但同样已届中年的王艽却能笔耕不辍，他不仅擅长写文章、写书，还经营着一家公司，让人叹服。

对于程序员来说，道路千万条，热忱第一条。无论你是希望像我一样在技术的道路上一条道走到黑，还是像王艽一样“技术、业务、管理”兼修，首先要具备的就是对技术的满腔热忱。

全栈，是技术领域的一条 Hard Way，选择它，你就要付出很多额外的努力——并非是 996，而是通过更多、更努力的思考（工作 955，思考 7×24）。其回报则是在多年之后，每当别人提起你时都会不禁赞叹：“瞧那位 Hardcore 的程序员！”没有付出就没有回报。相对于少数几条 Hard Way 来说，世界上还有无数条 Easy Way。如果你对 Hard Way 有着恐惧或犹疑，那么，请放下这本书，随便拿起一本“21 天精通×××”。

如果这都没有吓跑你，那么我要向你透露一个价值连城的秘密——“全栈并不难”。至少，对勤于思考的人来说，全栈是一个自然而然的结果，并不需要额外的代价。你之所以曾经觉得全栈难，是因为眼界受到限制。

现在，请听我说。

武侠小说里常常会说打通任督二脉之后功力大增，为什么打通任督二脉如此重要？因为它们统率着全身的经脉，它们是一切经脉的根本。相对于其他经脉来说，任督二脉很简单，但也最难打通。

全栈也是如此。经过多年的发展，技术体系变得非常庞杂——看看层出不穷的技术知识图就知道了，但是你是否曾注意到很多同样的思想被到处套用？不仅在同一个技术体系内如此，跨体系的套用其实更加明显。比如 RxJS 的事件流、后端的 MessageQueue 与现在大热的 Serverless 架构在深层思想上有着千丝万缕的联系。有了这种眼光，你不仅可以更轻松地跨领域学习，更重要的是，你会拥有极为难得的预见力。而技术预见力，能让你领先别人三年——哪怕半路在树下睡一觉都不用怕——果然是我这种惫懒中年人的上佳之选啊。当然，如果你是仍有着雄心壮志的青年，应该会更明白预见力的价值，有朝一日，你的预见力甚至不会再局限于技术。

与通常的理解不同，在我看来，全栈，不是特定技术的组合，而是一种思维方式，一种眼界。即使这本书讲的是 Angular + SpringBoot，即使我是在 Angular 领域的 Google 开发者专家，我仍然要郑重提醒你——阅读时请跳出具体的技术，努力从更高的层次上理解它。

全栈，就是技术的任督二脉。它把大量的技术思想贯穿在前后端这两条主线中。它们既有区别，又有联系。既不会抽象到让你找不到具体的例子去理解这些思想，也不会狭隘到让你只知其然而不知其所以然。所以，如果你曾是个前端开发者，那么请了解下 Java 的“注解”及其对 Spring 演化的影响，思考下“POJO + 注解”的模式为什么会流行起来。如果你曾是个后端开发者，那么请了解一下 RxJS 在前端的应用，及其与 MessageQueue 的共同点，了解下 Filter 和 Interceptor 的共同点及其背后的思想。而无论你是前端开发者还是后端开发者，思考下 Java 与 TypeScript 这两种语言背后的设计思想都会让你受益匪浅。

总之，不要囿于门户之见。没有前端工程师，也没有后端工程师，一个有技术追求的程序员，首先要是一个工程师。工程师的思维与热忱，才是你最宝贵的财富。

是为序。

雪狼——汪志成

ThoughtWorks 高级咨询师，Google Developer Expert

二十年码农，全栈工程师，儒生，Angular 中文文档译者

2019.05.14

前 言

全栈的目的并不是一个人搭建起一个复杂的商业应用，在现代软件大工程化的今天，这个目标既没有必要，也没有价值。全栈的概念对笔者来说是一个不断扩充视野、持续学习的过程，通过不同语言、不同框架、不同平台的学习，知道什么是好的软件设计，什么是好的 UI/UE 设计，什么是好的编程习惯；了解在不同的模型中，对同一个问题是怎样解决的，有什么优点和缺点。

让人眼花缭乱的编程语言、框架、平台、IDE 等，其实就是开发人员的工具箱。学会这些工具的使用不只是为了成为一个熟练的技术员，更重要的是，知道现实世界的问题是什么，对应的解决方案有哪些，然后才是选择合适的工具高效地解决问题。如今，在开源成为一个趋势的时代，对比其他行业来说，我们在使用这些工具的同时，还可以看到打造这些工具的开发者是如何思考问题、解决问题的。这个学习过程可以让我们打开眼界，也可以让我们在面对未知领域时有自信去探索。

如今，很多优秀的编程思想都会被各个平台吸收，比如约定优于配置、函数式编程、响应式编程，以及以注解方式为代表的元数据编程模型等。Java 这么成熟的平台也在不断地吸收新的元素，让自己更“酷”一些：Java 1.5 引入了注解，Java 8 引入了 Collectors、Stream 等函数式编程的元素。而一个好的框架也一定会得到多个平台、语言的支持，比如响应式编程的框架 Rx，最初只是微软为 .Net 平台开发的一套框架，但现在已经可以用 18 种编程语言实现了。

很多时候，随着接触的技术面的拓宽，你会发现在一个陌生平台或框架中，有太多令你会心一笑的东西，因为理念是相通的。在 Angular 中 Interceptor 的概念和 Spring 中 Interceptor 的概念如出一辙。如果你熟悉 RxJS，那么对于 Spring Reactor 就很快可以使用了，这是因为背后的思想是一样的。所以说，随着你见识的提高，对于新知识的掌握速度也会越来越快。

本书涉及很多平台、框架和类库等，主要有前端使用的 Angular，后端使用的 Spring Boot 框架、Spring Security 安全框架，数据库涉及 MongoDB、Elasticsearch 和 Redis。此外，还会学习函数式编程、响应式编程（RxJS）、Redux 等理念，Swagger、JaVers 等工具及 Rest、WebSocket、微服务等概念。

但一本书是无法深入这些技术细节的，这也不是笔者的目标。这本书想做的是希望帮助读者开启一扇门，读者读完本书后，如果觉得某个自己没有接触过的框架也不是那么可怕了，那么笔者的目的就达到了。不再对某个领域望而却步，其实就为你拓开了一片新天地。行万里路，读万卷书，这个道理在技术领域也是一样的，各种平台、框架、语言就像是路上的各种自然景观或人文景观，笔者愿意和你一起领略这多彩多姿的景色、升华自己的思想。

说句实话，虽然本书名字中有“全栈”，但是笔者以为“全栈”这个词还是有失偏颇的，给人的感觉好像样样通、样样松。确实，如果只是每方面都涉猎一些，确实无法做到各方面都精通，在某一个领域成为专家需要坚持不懈地进行学习和总结。但是任何一个领域学到一定程度时，就会发现在其他平台或编程语言上有类似的理念和思想，这些是相通的，已经超越语言和平台了。而这些恰恰又是在某个领域继续深入需要的一个台阶，持续学习的一个必要因素是不惧新的知识，而全栈学习就是要穿越这扇门，本书的目的就是想和你一起穿越这扇门。

作　者

目 录

第 1 章 技术的选型和环境搭建........1

1.1 技术选型........1

1.1.1 前端框架选型........2

1.1.2 后端框架选型........3

1.1.3 数据库选型........3

1.2 环境搭建........5

1.2.1 基础开发环境安装........5

1.2.2 IDE 的选择........7

1.2.3 字体的选择........8

1.2.4 定义通用的代码格式........9

1.3 工程项目的结构........10

1.3.1 前端项目........10

1.3.2 后端项目........13

1.3.3 整体项目工程的文件结构........14

第 2 章 使用 Angular 快速构造前端原型........17

2.1 Angular 基础概念........17

2.1.1 安装 Angular CLI........17

2.1.2 依赖性注入........22

2.1.3 组件........27

2.1.4 指令........28

2.1.5 管道........30

2.1.6 模块......31
2.1.7 模板驱动型表单......33
2.1.8 响应式表单......46
2.2 Angular Material 介绍......55
2.2.1 组件类别......55
2.2.2 布局控件：Sidenav......57
2.2.3 Flex 布局和 Angular Flex-layout......60
2.2.4 封装 Header/Footer/Sidebar......62
2.3 添加主题支持......78
2.3.1 Material Design 中对于主题的约束......79
2.3.2 主题的明与暗......79
2.3.3 Angular Material 中的主题......80
2.4 容器化 Angular 应用......83
2.4.1 什么是容器......83
2.4.2 安装 Docker......84
2.4.3 镜像仓库加速......85
2.4.4 创建 Angular 的 Docker 镜像......85
2.4.5 启动容器......87
2.4.6 使用 docker-compose 组织复杂的环境配置......88
2.4.7 使用 .dockerignore 文件......91
第 3 章 何谓后端......92
3.1 创建一个 Spring Boot 工程......93
3.1.1 通过 Gradle 创建......93
3.1.2 通过 Maven 创建......99
3.1.3 通过 IDE 创建......103
3.1.4 工程项目的组织......105
3.2 API 的构建可以如此简单......109
3.2.1 API 工程结构......109
3.2.2 领域对象......111
3.2.3 构造 Controller......112
3.2.4 启动服务......113
3.2.5 测试 API......115

3.3 MongoDB 支撑的 API 116
3.3.1 什么是 NoSQL 116
3.3.2 MongoDB 的集成 119
3.3.3 HATEOAS 124
3.3.4 “魔法”的背后 129
3.3.5 让后端也能热更新 132
3.4 容器化后端 135
3.4.1 手动创建镜像 135
3.4.2 使用 Gradle 自动化 Docker 任务 137
3.4.3 使用 docker-compose 组合服务 141
3.4.4 IDEA 中的 Gradle 支持 142
3.4.5 在容器中调试 143
第 4 章 登录鉴权功能的构建 145
4.1 模块化和组件化 145
4.1.1 登录的领域模型构建 145
4.1.2 前端页面设计 147
4.2 响应式编程初探 166
4.2.1 不同的视角 166
4.2.2 实现一个计数器 169
4.2.3 为什么要使用 Rx 177
4.2.4 Observable 的性质 178
4.2.5 RxJS 的调试 180
4.3 前端服务层 184
4.3.1 构建“伪”服务 184
4.3.2 构建“聪明组件” 188
4.3.3 路由处理 192
4.4 完成忘记密码前端设计 194
4.4.1 使用 RxJS 打造短信验证码控件 194
4.4.2 忘记密码向导“笨组件” 199
4.4.3 忘记密码的“聪明组件” 203

第 5 章 构建后端 API........205
5.1 HyperMedia API 与传统 API........205
5.1.1 领域对象........205
5.1.2 API 的可见控制........211
5.1.3 传统的 API 实现模式........218
5.2 Spring Data 中的查询........221
5.2.1 基础概念——Repository........221
5.2.2 查询方式........223
5.2.3 复杂类型查询........226
5.2.4 自定义查询........228
5.2.5 自定义 Repository........229
5.3 Controller 的构建........230
5.3.1 改造 TaskRepo 和 UserRepo........230
5.3.2 实现 Controller........231
5.3.3 登录........234
5.3.4 注册........235
5.3.5 忘记密码第一步：验证手机........236
5.3.6 忘记密码第二步：重置密码........244
5.3.7 API 的异常处理........246
5.4 构建安全的 API 接口........255
5.4.1 为什么要保护 API........256
5.4.2 什么是 JWT........256
5.4.3 JWT 的生成和解析........259
5.4.4 权限的设计........260
5.4.5 使用 Spring Security 规划角色安全........261
5.4.6 在 Spring Boot 中启用 Spring Security........265
5.4.7 改造用户对象........266
5.4.8 构建 JWT token 工具类........268
5.4.9 如何检查任何请求的授权信息........272
5.4.10 得到用户信息........275
5.4.11 配置 Spring Security........277
5.4.12 使用 JWT 进行 API 访问........292
5.5 跨域和 API 文档........305

5.5.1 跨域解决方案——CORS......305
5.5.2 API 文档......309

第 6 章 前端和 API 的配合......322

6.1 响应式的 HTTP API 处理......322
6.1.1 Angular 中的 HTTP 服务......322
6.1.2 Angular 的开发环境配置......323
6.1.3 在前端服务中使用 HttpClient......327
6.1.4 更改注册表单控件......331
6.2 RxJs 进阶......332
6.2.1 改造登录表单......332
6.2.2 RxJs 的高阶操作符......333
6.2.3 合并操作符......339
6.3 HTTP 拦截......342
6.3.1 实现一个简单的 HttpInterceptor......342
6.3.2 鉴权 HttpInterceptor......343
6.3.3 一个日志拦截器......345
6.4 Angular 路由......346
6.4.1 基准锚链接......346
6.4.2 Router 模块的简介......347
6.4.3 获取父路由的参数......354
6.4.4 获得前一个路由......354
6.4.5 Activated Route......354
6.5 安全守卫......355
6.5.1 激活守卫......356
6.5.2 激活子路由守卫......357
6.5.3 加载守卫......358
6.5.4 退出守卫......359
6.5.5 数据预获取守卫......359

第 7 章 后端不只是 API......361

7.1 缓存......361
7.1.1 配置 Cache......362

7.1.2 常用的缓存注解......364
7.1.3 测试缓存是否生效......366
7.2 Redis 作为缓存框架......368
7.2.1 Redis 的安装配置......369
7.2.2 在 Spring Boot 中集成 Redis......370
7.2.3 Redisson......373
7.3 使用 ElasticSearch 提升搜索性能......375
7.3.1 配置......375
7.3.2 构建用户查询 API......379
7.4 Spring Boot Actuator 和数据审计......390
7.4.1 初窥审计事件......390
7.4.2 实现应用的数据审计......394
7.4.3 JaVers 和 Spring Boot 集成......396
7.5 WebSocket 实时通信服务......401
7.5.1 HTTP 和 WebSocket 的区别和联系......402
7.5.2 何时使用 WebSocket......403
7.5.3 STOMP......403
7.5.4 WebSocket 配置......404
7.5.5 WebScoket 安全......406
7.5.6 建立一个实时消息 Controller......409
7.5.7 测试 WebSocket......409
7.6 Spring Boot 的自动化测试......411

第 8 章 前端的工程化......422

8.1 使用 Redux 管理状态......423
8.1.1 何时需要使用 Redux......423
8.1.2 Redux 的核心概念......424
8.1.3 在 Angular 中使用 Redux......429
8.1.4 Selector——状态选择器......434
8.2 使用 Effects 管理的副作用......437
8.3 使用 @ngrx/entity 提升生产效率......441
8.4 服务端渲染......446
8.4.1 Angular Universal 的工作机理......447

8.4.2 安装依赖......447
8.4.3 添加服务器端渲染模块......448
8.4.4 使用 Node.js Express 构建服务器......450
8.4.5 服务器端渲染中出现重复请求的处理......453

第 9 章 Spring Cloud 打造微服务......455

9.1 微服务的体系架构......455
9.1.1 服务即组件......456
9.1.2 微服务架构下的组织机构变化......456
9.1.3 产品化服务......458
9.1.4 持续集成和持续发布......458
9.1.5 监控和报警......458
9.1.6 Spring Cloud 项目依赖......459
9.2 配置服务和发现服务......461
9.2.1 配置中心是什么......461
9.2.2 发现服务......465
9.3 监控服务和路由服务......472
9.3.1 Spring Boot Admin......473
9.3.2 路由服务......478
9.4 微服务的远程调用......485
9.4.1 Feign Client......485
9.4.2 负载均衡......490

附录 A 常见云服务使用问题汇总......492

读者服务

轻松注册成为博文视点社区用户（www.broadview.com.cn），扫码直达本书页面。

- **下载资源**：本书如提供示例代码及资源文件，均可在 下载资源 处下载。
- **提交勘误**：您对书中内容的修改意见可在 提交勘误 处提交，若被采纳，将获赠博文视点社区积分（在您购买电子书时，积分可用来抵扣相应金额）。
- **交流互动**：在页面下方 读者评论 处留下您的疑问或观点，与我们和其他读者一同学习交流。

页面入口：http://www.broadview.com.cn/37183

第 1 章 技术的选型和环境搭建

本章主要集中讨论书中所采用的各种技术的选型，包括前端、后端、数据库、缓存等。

读者可以下载源码配合书籍阅读。[1]

1.1 技术选型

今天的技术领域真是让人眼花缭乱，除了前端各种层出不穷的框架，后端的技术又出现了容器、微服务等概念；数据库除了传统的关系型数据库，还有各种 NoSQL 数据库纷纷登场，分布式数据库的需求也越来越强；这还没算上大数据、人工智能等。这么多的概念，我们该怎么去选择，如何去学习？

其实，万变不离其宗，大家都认可的技术一般都会补充到现有技术中。比如注解本来是 .Net 平台先发展出来的，但由于这种方式确实可以减少大量的重复编码，而且更清晰易懂，所以 Java 平台也逐渐引入了这个特性。再比如“约定优于配置”（convention over configuration）这个设计范式本来是 Ruby on Rails 提倡的，由于确实省却了以前大量的配置带来的麻烦，所以现如今的大多数框架，不管是前端还是后端，都吸收了这个理念。

本书中采用的技术选型原则如下。

（1）成熟的框架：成熟的框架可能不会采用最时髦的一些概念，但是成熟本身意味着经历了考验。在我们选择技术方向时，前瞻的领域当然要关注，但对于商用系统来说，系统的稳定性是必须放到很高的优先级考虑的。因为我们肯定不想系统在上线后发生一些意想不到的事情。

---

1 本书的全部源码位于 https://github.com/wpcfan/gtm

（2）活跃的社区：在一个开源技术成为主流的时代，一个活跃的社区是考量这个项目是否有生命力的重要指标之一。活跃的社区意味着你可以很方便地找到一些常见问题的答案，而不是自己孤独地探索。活跃的社区还意味着更丰富的社区生态，社区会开发很多第三方的软件包来完善或补充现有框架中不太理想的地方，这可以极大地提升我们的开发效率，而不用“重造轮子”。

（3）工程化的支持：有一些小而美的框架在个人开发或者开发某个小功能时很好用，但是如果它缺乏工程化的支持，就无法推广到大团队中。工程化这个概念说起来比较宽泛，比如模块化的支持、自动化测试的支持、编译自动化的支持等。

（4）采用统一的编程思想：一个统一的编程思想或风格可以显著降低开发成本，比如在 Java 和 JavaScript 中都可以采用函数式编程风格，编程风格的统一会让团队协作起来更顺畅，也降低了团队人员调换的成本。如果团队能力强一些，则可以在前端、后端、客户端（Android、iOS）统一采用 Rx，也可以在多个平台和编程语言之上使用响应式编程作为交流的语言。

（5）快速开发的支持：如果框架中有一些诸如“约定优于配置”的特性，那么我们的开发会相对更方便、更容易一些；如果框架支持注解，就会让开发效率得到更大的提升。

1.1.1 前端框架选型

现在，前端框架多到让开发者眼花缭乱，但主流的三大框架还是逐渐浮出水面：React、Angular 和 Vue。从流行度来说，React 是目前全球最流行的前端框架，而 Vue 则是中国开发者最爱的前端框架。但本书中我们选用的是 Angular 6.x，为什么呢？有以下几个理由。

（1）无缝的 TypeScript 集成：TypeScript 在 React 和 Vue 上也可以使用，甚至官方也有较强的支持，但是都没有在 Angular 中使用顺畅。这主要是因为在 Angular 中，TypeScript 是首选语言，这一点在 React 和 Vue 中是做不到的。为什么 TypeScript 对于我们的技术选型这么重要呢？在大一点的公司里，基本都沉淀了很多面向对象的基础软件库、最佳实践及大量的熟练掌握面向对象概念的程序员。TypeScript 对于复用这些资源是非常必要的，而且如果使用原生的 JavaScript 开发大型项目，由于其缺乏强类型约束，在实际开发中则往往会出现难以调试的问题。可以这么说，对于有面向对象经验的开发者来说，TypeScript 有着近似于零的学习成本，而且能够提升代码质量。

（2）开发模型的良好体验：Angular 使用模块组织代码，基于组件进行开发，这个特性使得 Angular 开发在模式上非常接近后端 Java 开发或者 Android 和 iOS 应用开发。

（3）对于响应式编程良好的支持：RxJS 框架的开发者 Ben Lesh 已经入职 Google，入职前，其实 Angular 已经是前端框架中和 RxJS 配合度最好的了，后期 Angular 中的 RxJS 使用体验肯定越来越好。

1.1.2 后端框架选型

后端我们选择了 Spring Boot 2.x，Spring 已经有超过 15 年的开发时间了，从成熟度上来说都快“熟透”了。Spring 也基本成为 Java 开发中的事实标准，但是 Spring 一直在不断进化自身，吸收别的平台或框架的最佳实践经验。

Spring Boot 就是一个让 Spring 这棵老树发新芽的框架，Spring Boot 以“开箱即用”为基本理念，提供一套可以快速开发 Spring 应用的框架。Spring Boot 充分应用了注解、JavaConfig 等方式让我们摆脱传统 Spring 中的臃肿的 XML 配置文件，以相当简洁的方式专注于业务逻辑的开发。

Spring Boot 主要有以下特色。

- 创建一个单独可运行的 Spring 应用，无须依赖外部的 Web 容器。
- 直接嵌入 Tomcat、Jetty 或者 Undertow，无须发布 War 包。
- 提供各种 starter 依赖用以简化编译配置。
- 在可能的情况下，自动配置 Spring 和第三方类库。
- 提供生产环境的监控指标、健康检查及外部化的配置。
- 无代码生成，零 XML 配置。

1.1.3 数据库选型

1. MongoDB

在这么多 SQL 和 NoSQL 数据库喷薄而出的时代里，选择成了一件很痛苦的事情。在 SQL 这个层面上还好，因为 MySQL/MariaDB 基本一枝独秀（MariaDB 是在 Oracle 接手 MySQL 后，基于 MySQL 开发的社区维护的版本）。此外，如果原来比较习惯使用 Oracle，那么可以使用 PostgreSQL，两者比较像，迁移成本较低。

在 NoSQL 方向上，MongoDB 作为最早得到大规模应用的文档型数据库是一个非常稳妥的选择，而且 MongoDB 还在不断地提供可以和 SQL 数据库对标的功能，类似表连接的功能和事务等特性。本书中采用的数据存储的主要方式是 MongoDB。

目前来看，MongoDB 具备以下特点。

- 无 Schema，这一特性使 MongoDB 非常适合快速开发，因为开发过程中表结构可能经常变化，在 SQL 中，表结构的变化带来的维护问题经常制约开发人员的响应速度。

- 单一对象结构非常清晰。
- 没有复杂的连接（join）。
- 可以进行深度嵌套对象的查询。
- 支持使用类似 SQL 的查询语言。
- 易于优化。
- 易于拓展。
- 无须转换或映射 Java 对象到数据库。
- 更快的数据访问速度。
- 快速添加 Replica Set，提供高可用性。
- Spring Data 提供无缝集成。

2. Elasticsearch

Elasticsearch 是基于 Lucene 开发的搜索引擎，它提供了一个分布式的、支持多租户的全文搜索引擎，提供 JSON 形式的文档存储。和 MongoDB 类似的是，Elasticsearch 也是一个文档型数据库，但 Elasticsearch 的主要目的是解决搜索问题，也就是说使用 Elasticsearch 更多是使用其搜索特性，而不是其他的能力。

在本书中，我们使用 Elasticsearch 作为模糊搜索和条件搜索功能。Elasticsearch 一般是提供一整套方案，包括数据集合、日志分析引擎（logtash），以及数据分析和可视化工具（kibana），这个全家桶一般叫作“Elastic Stack”（你可能会经常看到 ELK 这个缩写，指的就是这个全家桶）。

Elasticsearch 的搜索能力强，不仅在于它可以支持搜索指定文档属性中的字符串，而且还可以分析文档，检查搜索的关键字是否存在于整个文档中，无论这个关键字是否中间有其他词的间隔；也会像 Google 那样评价所有结果的相关性；甚至包括更复杂的特性，比如同义词、多种拼写形式或者有一定拼写错误的容错率等。

本书采用 Elasticsearch 来为系统添加搜索功能，此外各个微服务系统的日志都会输出到 Elasticsearch，便于日志的统一管理和查询。

6. Redis

和 MongoDB 与 Elasticsearch 不同，Redis 是一个基于键值对的 NoSQL 内存数据库。它既可以当作数据库，又可以当作缓存或者消息代理。它支持多种数据结构：字符串、散列、列表、集合、排序集合、位图和地理位置索引等。Redis 有内建的复制机制，支持 Lua 脚本、LRU 回收、事务，以及不同程度的磁盘存储。

Redis 的优势如下所示。

- 性能极高：Redis 的读写速度极好，远远超过传统数据库。
- 丰富的数据类型：Redis 支持二进制案例的 String、List、Hash、Set 及 Ordered Set 数据类型操作。
- 事务：Redis 的所有操作都是原子性的，也就是要么成功执行，要么失败则完全不执行。
- 丰富的特性：Redis 还支持发布/订阅、通知、key 过期等特性。

在本书中，我们采用 Redis 作为缓存机制，而不是将其作为一个数据库。

1.2 环境搭建

对开发者来讲，一个友好的、强大的开发环境可以起到事半功倍的作用。本节中我们会把本书需要安装的开发环境和推荐的 IDE 进行配置。

1.2.1 基础开发环境安装

1. Ubuntu Linux

在 Ubuntu Linux 环境下，推荐使用 SDKMAN 来安装 Spring、Java 及 Grails 等依赖类库。安装 SDKMAN 非常简单，打开一个 terminal（命令行终端窗口），然后输入以下命令：

```
curl -s "https://get.sdkman.io" | bash
```

注意，在此过程中，安装脚本可能会提示缺少某些软件包，这时我们需要按屏幕提示进行软件包的安装，比如下面的这个提示，说的是我们的系统没有安装 zip 和 unzip。

```
Looking for a previous installation of SDKMAN...
Looking for unzip...
Looking for zip...
Not found.
=========================================================================================================
 Please install zip on your system using your favourite package manager.

 Restart after installing zip.
=========================================================================================================
```

那么，我们需要使用下面的命令进行安装：

```
apg-get install zip unzip
```

安装好软件包后，请再次执行上面的 SDKMAN 安装脚本并按照屏幕提示完成安装。如果你可以看到类似下面的输出，那么安装过程就结束了。

```
All done!


Please open a new terminal, or run the following in the existing one:

    source "/home/ubuntu/.sdkman/bin/sdkman-init.sh"

Then issue the following command:

    sdk help

Enjoy!!!
```

安装好之后，在 terminal 中再输入以下命令：

```
source "$HOME/.sdkman/bin/sdkman-init.sh"
```

这样就安装好了，你可以通过以下命令来验证安装是否成功：

```
sdk version
```

使用 sdk 这个命令可以很方便地安装依赖类库。

2. macOS

macOS 对开发者来讲是非常友好的，首先我们需要安装 brew，这是 macOS 上的一个包管理工具，类似于 Ubuntu Linux 中的 apt-get。在 terminal 中输入下面的命令即可完成安装。

```
/usr/bin/ruby -e "$(curl -fsSL
https://raw.githubusercontent.com/Homebrew/install/master/install)"
```

有了 brew 之后，再安装其他软件就简单多了，使用 brew install 命令就可以安装你需要的软件了。

macOS 原生的 terminal 其实还不错，但如果我们将默认的 shell 从 bash 换成 zsh，更准确地说，是用 OhMyZsh，那么整个 terminal 环境就会更美妙。

安装也只要一行命令而已：

```
sh -c "$(curl -fsSL https://raw.githubusercontent.com/robbyrussell/oh-
my-zsh/master/tools/install.sh)"
```

OhMyZsh 有很多既好看又好用的主题，这里推荐一个笔者非常喜欢的主题 spaceship，如果要安装，则可以在 terminal 中输入：

```
    curl -o - https://raw.githubusercontent.com/denysdovhan/spaceship-zsh-
theme/master/install.zsh | zsh
```

然后编辑：

```
~/.zshrc
ZSH_THEME="spaceship"
```

最后，重启 terminal 就可以享用这美味的命令行了。

3. Windows

Windows 作为使用最普及的操作系统，很多开发者在 Windows 上却过度地依赖图形化的 IDE，对于环境配置可能并不熟悉。其实在 Windows 系统中搭建一个好用的环境也不是很难，但首先是要有一个顺手一些的 terminal，这里推荐 cmder，其不但提供 bash 的体验到 Windows 上面，而且集成了很多好用的插件，比如类似 OhMyZsh 的 git 插件等，对于开发者来说是必备“神器”。

类似地，我们还需要一个类似 brew 这样的包管理工具，在 Windows 平台上，有很多类似的这种工具，这里推荐 Scoop，由于在设计上参考了 brew，体验是非常不错的。

1.2.2 IDE 的选择

在本书中，我们会分别涉及前端和后端的工程，一般来说，前端的工程笔者更喜欢轻量级编辑器，而对于后端，则需要一个完善的 IDE。

1. Visual Studio Code

轻量级编辑器，笔者推荐 Visual Studio Code，简称 VSC，但大家可不要被名字欺骗了，它和 Visual Studio 其实没有什么关系，除了都是微软团队开发的之外。

VSC 有很多好用的插件，有了这些插件的配合，无论是写代码还是写文档都特别畅快，下面具体介绍几个常用的插件。

- Angular 5 Snippets：Angular 开发的必备工具，有太多好用的代码模板了。
- Angular Language Service：为 Angular 提供代码自动完成、AOT 诊断信息、跳转到定义等实用功能。
- Debugger for Chrome：调试利器、设置断点、查看变量值等。
- Java Extension Pack：Java 开发插件集合，包括 Language Support for Java（TM）by Red Hat、Debugger for Java、Java Test Runner 和 Maven Project Explorer 四款插件，一般来说，如果使用 Maven 或 Gradle 作为管理脚本的 Spring 相关的 Web 开发还是不错的。当然，VSC 在 Java 方面的支持对比 IDEA 还是有较大差距的，因为 VSC 定位的是轻量级

编辑器，而不是完整的 IDE。

2. IntelliJ IDEA

这个老牌 IDE 始终是笔者心中的最佳大型 IDE，即便在 Eclipse 最火的年代，笔者也认为 IDEA 比 Eclipse 好用太多。

1.2.3 字体的选择

笔者日常使用的一款字体是 Fira Code，感觉无论是在代码编辑器中还是在终端窗口中看上去都极舒服，如图 1-1 所示。

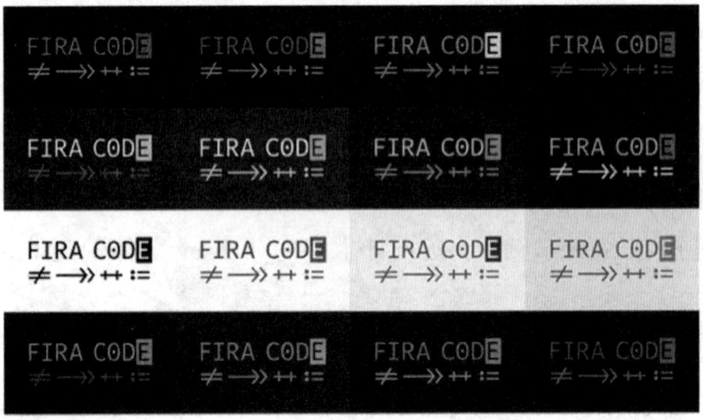

图 1-1

如果有兴趣的读者可以去 GitHub 网站下载。

如果需要在 VSC 中设置使用该字体，则需要到“首选项→设置”中添加如下配置：

```
{
    "editor.fontSize": 16,
    "editor.lineHeight": 24,
    "editor.fontLigatures": true,
    "editor.fontFamily": "Fira Code, 'Operator Mono', Menlo, Monaco,
'Courier New', monospace",
    "terminal.integrated.fontSize": 18,
    "terminal.integrated.fontFamily": "Fira Code, 'Operator Mono', Menlo,
Monaco, 'Courier New', monospace"
}
```

请记得要设置 "editor.fontLigatures": true。这样，我们就可以看到这个字体给我们带来的一些非常有趣的体验，比如 JavaScript 中常见的 === 变成了全等号，而 => 更像一个真正的箭头了，如图 1-2 所示。

```
export class NotFoundInterceptor implements HttpInterceptor {
  constructor(private _injector: Injector) {}

  intercept(
    req: HttpRequest<any>,
    next: HttpHandler
  ): Observable<HttpEvent<any>> {
    return next.handle(req).pipe(
      tap(
        event => {},
        err => {
          if (err.status === 404) {
            const router = this._injector.get(Router);
            router.navigate(['/404']);
          }
        }
      )
    );
  }
}
```

图 1-2

1.2.4 定义通用的代码格式

在大团队中，我们经常会遇到几个小组因为使用不同的 IDE 导致代码的格式显示得各式各样。也会因此在不同 IDE 中进行格式化代码时产生不必要的麻烦。所以统一定义一个不依赖于 IDE 的格式文件对于大工程和大团队来说是很有必要的。这里推荐 EditorConfig，一个被广泛支持的在众多 IDE 和编辑器中保持统一代码风格的开源项目。

EditorConfig 的配置非常简单，只需要建立一个 .editorconfig 文件，这个文件的格式很像 .ini 文件。

```
root = true

[*]
end_of_line = lf
insert_final_newline = true
charset = utf-8
max_line_length=140

[*.md]
max_line_length = off
trim_trailing_whitespace = false

[*.java]
```

```
indent_style = space
indent_size = 4
trim_trailing_whitespace = true

[*.yml]
indent_style = space
indent_size = 2
```

上面这个 EditorConfig 就是定义了一个“根”配置。是的，你可以在一个大工程的各个子项目或者子目录中使用各自的 EditorConfig 配置，但如果设置 root = true，就代表这个文件是根配置，应该放在项目的根目录下。

对于不同文件的后缀，我们可以单独为其定义格式，只需要在 [] 中指定后缀名，然后为其设置对应的格式定义即可。常用的配置项如下。

- end_of_line：指定换行符，可以是 LF 或者 CRLF，但这里使用小写。
- insert_final_newline：是否在文件末尾添加一个空行。
- charset：文件编码格式。
- max_line_length：每行最大字符数。
- trim_trailing_whitespace：是否删掉多余的空格。
- indent_style：缩进风格，是 tab 还是 space。
- indent_size：缩进大小。

1.3 工程项目的结构

1.3.1 前端项目

前端采用的工程结构类似下面的结构，第一层的几个文件夹如下。

- docker：用于容器的构建文件，比如 Dockerfile，以及在构建 Docker 时需要的一些文件。
- e2e：端到端测试文件，Web 自动化集成测试，基于 Protractor。
- src：源文件目录，我们平时主要接触的是这个目录。

其中第一层还包含若干文件。

- angular.json：Angular 工程配置文件。
- docker-compose.yml：容器集成脚本。
- karma.conf.js：自动化单元测试配置文件。
- package.json：项目依赖文件。

- protractor.conf.js：端到端测试配置文件。
- README.md：Markdown 格式的项目说明文档。
- tsconfig.json：TypeScript 的配置文件。
- tslint.json：TypeScript 的 lint 配置。
- yarn.lock：yarn 的依赖管理文件。

```
├── docker/
│   └── nginx/
│       ├── conf.d/
│       │   └── default.conf
│       └── Dockerfile
├── e2e/
│   ├── app.e2e-spec.ts
│   ├── app.po.ts
│   └── tsconfig.e2e.json
├── src/
├── angular.json
├── docker-compose.yml
├── karma.conf.js
├── package.json
├── protractor.conf.js
├── README.md
├── tsconfig.json
├── tslint.json
└── yarn.lock
```

src 目录下的文件结构是我们项目中主要使用的，所以展开讲一下。

- app：具体应用的源代码目录，src 下面可以有多个应用。
  - 在 app 目录下，我们一般按 Angular 的模块进行目录的规划，也就是说一般是每个模块一个目录，根模块除外。
  - 一个标准的应用，我们为它建立两个比较特殊的模块 CoreModule 和 SharedModule。前者每个应用启动时只初始化一次，后者的作用是共享一些常用模块和组件，避免产生在多个模块中导入类似模块、组件、指令等的麻烦操作。
  - 所有的领域对象我们存储在 domain 文件夹。
  - 工具类我们放在 utils 文件夹。
- assets：静态资源文件目录，比如图片等。
- environments：环境配置目录，在此可以建立多个环境配置文件，比如用于生产环境的 prod，用于开发环境的 dev 或者其他。

```
├── src/
│   ├── app/
│   │   ├── core/
│   │   │   ├── components/
│   │   │   │   ├── footer.component.ts
│   │   │   │   ├── header.component.ts
│   │   │   │   ├── page-not-found.component.spec.ts
│   │   │   │   ├── page-not-found.component.ts
│   │   │   │   └── sidebar.component.ts
│   │   │   ├── containers/
│   │   │   │   └── app/
│   │   │   │       ├── app.component.html
│   │   │   │       ├── app.component.scss
│   │   │   │       ├── app.component.spec.ts
│   │   │   │       └── app.component.ts
│   │   │   ├── app-routing.module.ts
│   │   │   ├── core.module.ts
│   │   │   └── material.module.ts
│   │   ├── domain/
│   │   │   ├── auth.ts
│   │   │   ├── menu.ts
│   │   │   ├── role.ts
│   │   │   └── user.ts
│   │   ├── profile/
│   │   │   ├── profile-routing.module.ts
│   │   │   └── profile.module.ts
│   │   ├── reducers/
│   │   │   └── index.ts
│   │   ├── shared/
│   │   │   ├── directives/
│   │   │   ├── permission.ts
│   │   │   └── shared.module.ts
│   │   ├── utils/
│   │   │   └── permission.util.ts
│   │   └── app.module.ts
│   ├── assets/
│   │   └── img/
│   │       └── 400_night_light.jpg
│   ├── environments/
│   │   ├── environment.prod.ts
│   │   └── environment.ts
│   ├── favicon.ico // 站点图标
│   ├── index.html // Angular 会编译成一个单页应用，这就是那个单页
│   ├── main.ts // 应用入口文件
```

```
│    ├── polyfills.ts // 浏览器兼容性适配文件
│    ├── styles.scss // 顶层样式文件
│    ├── test.ts // 测试入口文件
│    ├── theme.scss // 主题样式文件
│    ├── tsconfig.app.json // 开发时的 TypeScript 配置
│    ├── tsconfig.spec.json // 测试时的 TypeScript 配置
│    └── typings.d.ts
```

1.3.2 后端项目

我们的后端项目使用 Gradle 来处理多项目的构建，包括大项目和子项目的依赖管理及容器的建立等。对于 Gradle 项目来说，我们会有一个根项目，这个根项目下会建立若干子项目，具体文件结构如下：

```
|--backend（根项目）
|----common（共享子项目）
|------src（子项目源码目录）
|--------main（子项目开发源码目录）
|----------java（子项目开发 Java 类源码目录）
|----------resources（子项目资源类源码目录）
|------build.gradle（子项目 gradle 构建文件）
|----api（API 子项目）
|------src
|--------main
|----------java
|----------resources
|------build.gradle
|----report（报表子项目）
|------src
|--------main
|----------java
|----------resources
|------build.gradle
|--build.gradle（根项目构建文件）
|--settings.gradle（根项目设置文件）
```

其中，settings.gradle 可以包括各个子项目：

```
include 'subproject-1'
include 'subproject-2'
include 'subproject-2'
rootProject.name = 'gtm-backend'
```

1.3.3 整体项目工程的文件结构

```
├── docker/
│   ├── elasticsearch/
│   │   └── config/
│   ├── fluentd/
│   │   └── config/
│   ├── frontend/
│   │   ├── certs/
│   │   │   ├── letsencrypt/
│   │   │   └── self-signed/
│   │   ├── config/
│   │   └── three/
│   │       ├── libs/
│   │       └── textures/
│   ├── kibana/
│   │   └── config/
│   └── logstash/
│       ├── config/
│       └── pipeline/
├── frontend/
│   ├── docker/
│   │   └── nginx/
│   │       └── conf.d/
│   ├── e2e/
│   │   └── src/
│   └── src/
│       ├── app/
│       │   ├── actions/
│       │   ├── admin/
│       │   │   ├── actions/
│       │   │   ├── components/
│       │   │   ├── containers/
│       │   │   ├── data/
│       │   │   ├── effects/
│       │   │   ├── guards/
│       │   │   ├── reducers/
│       │   │   ├── services/
│       │   │   └── validators/
│       │   ├── auth/
│       │   │   ├── actions/
│       │   │   ├── components/
│       │   │   ├── containers/
│       │   │   ├── effects/
│       │   │   ├── guards/
```

```
│       │   │   ├── reducers/
│       │   │   ├── services/
│       │   │   └── validators/
│       │   ├── booking/
│       │   │   ├── actions/
│       │   │   ├── components/
│       │   │   ├── containers/
│       │   │   ├── effects/
│       │   │   └── reducers/
│       │   ├── core/
│       │   │   ├── components/
│       │   │   ├── containers/
│       │   │   ├── data/
│       │   │   ├── interceptors/
│       │   │   ├── mat-helpers/
│       │   │   └── services/
│       │   ├── domain/
│       │   ├── effects/
│       │   ├── profile/
│       │   ├── reducers/
│       │   ├── shared/
│       │   │   ├── components/
│       │   │   ├── containers/
│       │   │   ├── directives/
│       │   │   └── services/
│       │   ├── store/
│       │   └── utils/
│       ├── assets/
│       │   └── img/
│       └── environments/
├── gradle/
│   └── wrapper/
├── so-admin/
│   └── src/
│       ├── main/
│       │   ├── docker/
│       │   ├── java/
│       │   │   └── dev/
│       │   │       └── local/
│       │   │           └── smartoffice/
│       │   └── resources/
│       └── test/
├── so-api-service/
│   └── src/
│
```

```
│       ├── main/
│       │   ├── docker/
│       │   ├── java/
│       │   │   └── dev/
│       │   │       └── local/
│       │   │           └── smartoffice/
│       │   └── resources/
│       └── test/
│           ├── java/
│           │   └── dev/
│           │       └── local/
│           │           └── smartoffice/
│           │               └── api/
│           │                   ├── config/
│           │                   ├── rest/
│           │                   └── service/
│           └── resources/
└── so-discovery/
    ├── bin/
    │   └── main/
    │       └── dev/
    │           └── local/
    │               └── smartoffice/
    ├── build/
    │   ├── classes/
    │   │   └── java/
    │   │       └── main/
    │   │           └── dev/
    │   │               └── local/
    │   │                   └── smartoffice/
    │   ├── resources/
    │   │   └── main/
    │   └── tmp/
    │       └── compileJava/
    └── src/
        ├── main/
        │   ├── docker/
        │   ├── java/
        │   │   └── dev/
        │   │       └── local/
        │   │           └── smartoffice/
        │   └── resources/
        └── test/
```

第 2 章 使用 Angular 快速构造前端原型

本章会从 Angular 的核心概念出发，2.1 节以一系列小例子阐释这些概念的意义和使用方法。有 Angular 基础的读者可以跳过或者摘选自己感兴趣的内容看。在 2.2 节中，我们会一起来认识 Angular 的官方 UI 组件库 Angular Material，这是一套遵循谷歌 Material Design 风格的组件库。使用它的好处在于可以在组件标准化、动画、兼容性方面节省很大精力，即使你不熟悉 CSS 也可以做出很好看的 UI 效果。还会一起学习几个较常见的组件，当然只是最初的简单框架和页面，使用的是 Angular Material 组件库和 Angular FlexLayout 布局库。2.3 节我们会一起学习 Angular Material 的主题支持，学会如何定制化主题。2.4 节使用容器来构建应用，我们不会专门去讲关于容器的知识，但在书中需要使用容器的地方会有相应说明。使用容器的原因是它可以让整个开发部署的流程更加自动化，提高生产效率。

2.1 Angular 基础概念

Angular 中经常提到的几个概念：依赖性注入、模块、组件、指令、管道、模板驱动型表单和响应式表单。这些概念听上去令人头大，加上官网的解释又相对晦涩一些，导致很多人觉得 Angular 的学习门槛高。但其实这些概念并不复杂，下面我们逐一来揭开它们的面纱。

2.1.1 安装 Angular CLI

Angular CLI 是一套命令行工具，可以生成工程的脚手架，它隐藏了很多配置的烦琐细节，可以让我们更专注在逻辑代码的实现上。我们可以通过 npm 或者 yarn 来安装:

```
npm install -g @angular/cli
```

或者：

```
yarn add global @angular/cli
```

以上两种安装方式是笔者推荐的安装方式，但是有时可能会出现安装时间过长或者有些软件包无法下载的情况。如果实在无法安装成功，也可以采用淘宝团队提供的 cnpm，这个 cnpm 可以理解成使用淘宝的镜像软件仓库的 npm 中国加速版。

```
npm install -g cnpm --registry=https://registry.npm.taobao.org
```

安装 cnpm 之后，可以使用 cnpm 替代 yarn 或 npm。

可以通过如下命令测试 Angular CLI 是否安装成功，这个 ng 命令我们会经常用到，为什么叫 ng？因为 Angular 的简写就是 ng：

```
ng version
```

如果输出的是类似下面的样子，那么就代表安装成功了。

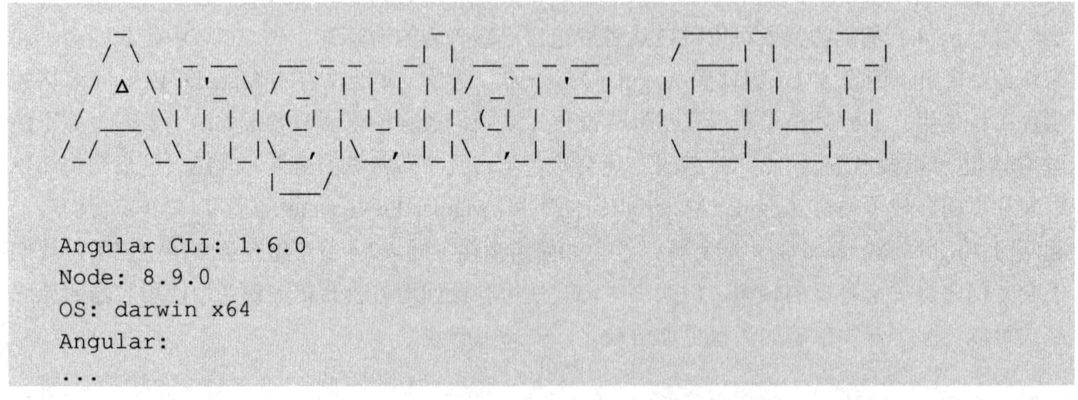

```
Angular CLI: 1.6.0
Node: 8.9.0
OS: darwin x64
Angular:
...
```

如果不是呢，那么很不幸，就是安装过程中有错误，你需要重复上面的安装步骤直至安装成功为止。

1. 搭建前端项目框架

下面，我们使用 Angular CLI 创建一个新的工程：

```
ng new frontend --style scss --skip-install
```

这里，注意 ng new <project name> 就是创建一个 Angular 工程，而后面的参数 --style scss 是告诉 Angular CLI 在创建工程时，我们会采用 scss 作为工程的样式工具，如果不加这个参数，那么工程默认的是 css。简单来说，scss 就是可编程的 css。

这个命令执行的过程可能会有点长，而且它默认使用了 yarn 进行安装，所以我们加了 --skip-install 这个参数跳过依赖的安装，这样后面如果你想使用 cnpm 就可以进入 client 目录手动安装依赖 cnpm install，如果使用推荐的安装命令，就是 yarn install 或 npm install。需要注意的一点是，frontend 目录才是工程目录，大部分的 Angular CLI 子命令都需要在此目录下执行才能生效。

```
> ng new frontend --style scss --skip-install
      Successfully initialized git.
CREATE frontend/README.md (1030 bytes)
CREATE frontend/angular.json (3841 bytes)
CREATE frontend/package.json (1404 bytes)
CREATE frontend/tsconfig.json (384 bytes)
CREATE frontend/tslint.json (2819 bytes)
CREATE frontend/.editorconfig (245 bytes)
CREATE frontend/.gitignore (503 bytes)
CREATE frontend/src/environments/environment.prod.ts (51 bytes)
CREATE frontend/src/environments/environment.ts (743 bytes)
CREATE frontend/src/favicon.ico (5430 bytes)
CREATE frontend/src/index.html (295 bytes)
CREATE frontend/src/main.ts (370 bytes)
CREATE frontend/src/polyfills.ts (3114 bytes)
CREATE frontend/src/test.ts (642 bytes)
CREATE frontend/src/assets/.gitkeep (0 bytes)
CREATE frontend/src/styles.scss (80 bytes)
CREATE frontend/src/browserslist (51 bytes)
CREATE frontend/src/karma.conf.js (1011 bytes)
CREATE frontend/src/tsconfig.app.json (194 bytes)
CREATE frontend/src/tsconfig.spec.json (282 bytes)
CREATE frontend/src/app/app.module.ts (314 bytes)
CREATE frontend/src/app/app.component.scss (0 bytes)
CREATE frontend/src/app/app.component.html (1141 bytes)
CREATE frontend/src/app/app.component.spec.ts (986 bytes)
CREATE frontend/src/app/app.component.ts (208 bytes)
CREATE frontend/e2e/protractor.conf.js (752 bytes)
CREATE frontend/e2e/src/app.e2e-spec.ts (299 bytes)
CREATE frontend/e2e/src/app.po.ts (208 bytes)
CREATE frontend/e2e/tsconfig.e2e.json (213 bytes)
```

如果看到类似上面的输出结果，工程就生成完毕了，我们进入 frontend 目录。

```
cd frontend
```

通过 yarn install 或 npm install 安装依赖，如果成功，则输出的结果和下面的类似：

```
> yarn install
yarn install v1.3.2
info No lockfile found.
[1/4] ? Resolving packages...
[2/4] ? Fetching packages...
```

```
    [3/4] ⁇  Linking dependencies...
    warning "@angular/cli > @angular-devkit/schematics >
@schematics/schematics@0.0.11" has unmet peer dependency "@angular-
devkit/core@0.0.22".
    [4/4] ⁇  Building fresh packages...
    success Saved lockfile.
    ⁇  Done in 70.30s.
```

我们的工程最后会与图 2-1 所示的组织形式一样。

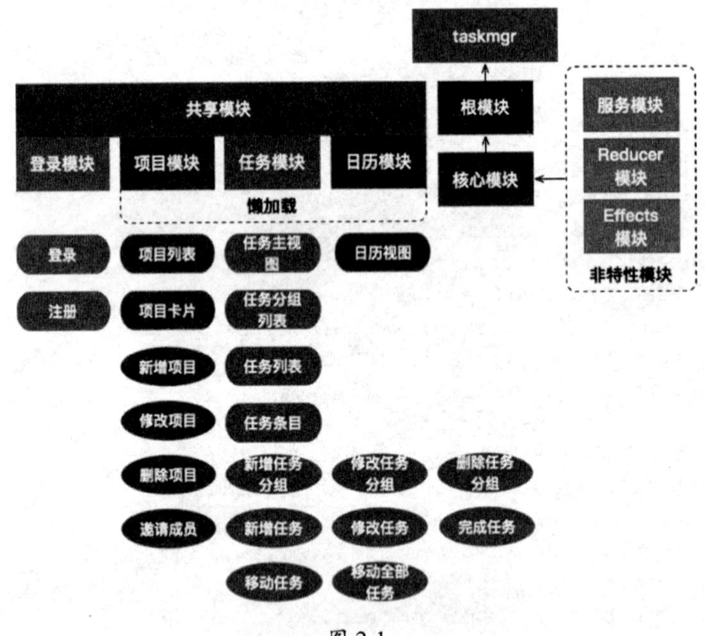

图 2-1

但在一开始，我们要先建立两个模块，一个叫核心模块（CoreModule），另一个叫共享模块（SharedModule）。核心模块的作用是初始化应用，以及加载必须要单例的功能，最常见的情况是，我们通常把 HTTP 的服务放到核心模块中，因为通常情况下，我们希望服务只被创建一次。

```
  > ng g m core
    create src/app/core/core.module.ts (188 bytes)
```

上面这个简单命令就是创建模块的命令，ng 后的那个 g 是 generate 的缩写，也就是生成的意思。而 m 自然就是 module 的缩写，就是模块的意思。如果你想生成组件，就写成 ng g c <component name>，生成指令就写成 ng g d <directive name>，具体的参数其实有很多，读者可以使用 ng help <subcommand name> 的形式查看。

回到我们的核心模块，将其改造成下面的样子：

```
import { NgModule, Optional, SkipSelf } from '@angular/core';

export class CoreModule {

  constructor(
    @Optional() @SkipSelf() parentModule: CoreModule) {
    if (parentModule) {
      throw new Error('CoreModule 已经装载，请仅在 AppModule 中引入该模块。');
    }
  }
}
```

这里改造的目的是让核心模块变成一个单例：应用生命周期内只加载一次。具体的逻辑如果看不懂，则请转到第 2 章 2.1.2 节去学习。

而共享模块通常包含需要在应用中多处使用的组件、指令、管道或其他共享代码，在共享模块中，我们经常会对模块进行导入和导出，这个看似吃力不讨好的操作是为了让所有导入共享模块的其他功能模块不用再去导入。

```
ng g m shared
  create src/app/shared/shared.module.ts (190 bytes)
```

同样地，我们需要改造一下这个共享模块，为什么把要导入、导出的模块都写成一个数组？这是为了后期我们可以方便地重用导入、导出的模块。

```
import { NgModule } from '@angular/core';
import { CommonModule } from '@angular/common';

const MODULES = [
  CommonModule,
];

const DECLARATIONS = [];
const EXPORT_COMPONENTS = [];
const ENTRYCOMPONENTS = [];

@NgModule({
  imports: MODULES,
  exports: [
    ...MODULES,
    ...EXPORT_COMPONENTS
  ],
  declarations: DECLARATIONS,
  entryComponents: [
    ENTRYCOMPONENTS
```

```
  ]
})
export class SharedModule {
}
```

2. 重新组织代码

默认情况下，新建的项目会自动生成根模块和根组件，这里我们需要做一个调整：在 src/app 下只保留 app.module.ts，而将根组件移动到 src/app/core/containers/app 中。但移动后，请注意调整相关文件中的引用路径的调整。这样调整的原因是我们希望以后的目录是按照模块或用途进行划分的，整体的目录结构应该如下所示：

```
app/
├── core/
│   ├── components/
│   ├── containers/
│   │   └── app/
│   │       ├── app.component.html
│   │       ├── app.component.scss
│   │       ├── app.component.spec.ts
│   │       └── app.component.ts
│   ├── app-routing.module.ts
│   ├── core.module.ts
│   └── material.module.ts
├── domain/
├── reducers/
├── shared/
│   └── shared.module.ts
├── utils/
└── app.module.ts
```

2.1.2 依赖性注入

1. 什么是依赖性注入

依赖性注入（Dependency Injection，DI）其实不是 Angular 独有的概念，这是一个已经存在很长时间的设计模式，也可以叫作控制反转（Inverse of Control）。我们从下面这个简单的代码片段入手来看看什么是依赖性注入，以及为什么要使用依赖性注入。

```
class Person {
  constructor() {
    this.address = new Address('北京', '北京', '朝阳区', 'xx 街 xx 号');
    this.id = Id.getInstance(ID_TYPES.IDCARD);
  }
}
```

在上面的代码中，我们在 Person 这个类的构造函数中初始化了构建 Person 所需要的依赖类：Address 和 ID，其中 Address 是个人的地址对象，而 ID 是个人身份对象。这段代码的问题在于除了引入内部所需的依赖之外，**它还知道了这些依赖创建的细节**，比如它知道 Address 的构造函数需要的参数（省、市、区和街道地址）和这些参数的顺序，它还知道 ID 的工厂方法和其参数（取得身份证类型的 ID）。

但这样做的问题究竟是什么呢？首先，这样的代码是非常难以进行单元测试的，因为在测试时我们往往需要构造一些不同的测试场景（比如我们想传入护照类型的 ID），但这种写法导致你没办法改变其行为。其次，我们在代码的可维护性和扩展性方面有了很大的障碍，设想一下，如果我们改变了 Address 的构造函数或 ID 的工厂方法，那么我们不得不去更改 Person 类。一个类还好，但如果几十个类都依赖 Address 或 Person，就会造成很大的麻烦。

那么解决的方法呢？也很简单，那就是我们把 Person 的构造改造一下：

```
class Person {
  constructor(address, id) {
    this.address = address;
    this.id = id;
  }
}
```

我们在构造中接受已经创建的 Address 和 ID 对象，这样在这段代码中就没有任何关于它们的具体实现了。换句话说，我们把创建这些依赖性的职责向上一级传递了出去（推卸责任啊）。现在我们在生产代码中可以这样构造 Person：

```
const person = new Person(
  new Address('北京', '北京', '朝阳区', 'xx 街 xx 号'),
  Id.getInstance(ID_TYPES.IDCARD)
);
```

而在测试时，可以方便地构造各种场景，比如我们将地区改为辽宁：

```
const person = new Person(
  new Address('辽宁', '沈阳', '和平区', 'xx 街 xx 号'),
  Id.getInstance(ID_TYPES.PASSPORT)
);
```

其实，这就是依赖性注入了，这个概念是不是很简单？但有的读者问了，那上一级如果是单元测试，那不还是有问题吗？是的，如果上一级需要测试，就得“推卸责任”到再上一级了。这样一级一级地“推卸”，最后会推到最终的入口函数，但这也不是办法啊，而且靠人工维护也很容易出错，这时候就需要有一个依赖性注入的框架来解决，这种框架一般叫作 DI 框架或者 IoC 框架。这种框架对于熟悉 Java 和 .Net 的读者不会陌生，鼎鼎大名的 Spring 最初就是一个这样的框架，当然现在它的功能丰富了，已经远不止这个功能了。

2. Angular 中的依赖性注入框架

Angular 中的依赖性注入框架主要包含下面几个角色（见图 2-2）。

- Injector（注入者）：使用 Injector 提供的 API 创建依赖的实例。
- Provider（提供者）：Provider 告诉 Injector 怎样创建实例（比如我们上面提到的是通过某个构造函数，还是工厂类创建等）。Provider 接受一个令牌，然后把令牌映射到一个用于构建目标对象的工厂函数。
- Dependency（依赖）：依赖是一种类型，这个类型就是我们要创建的对象的类型。

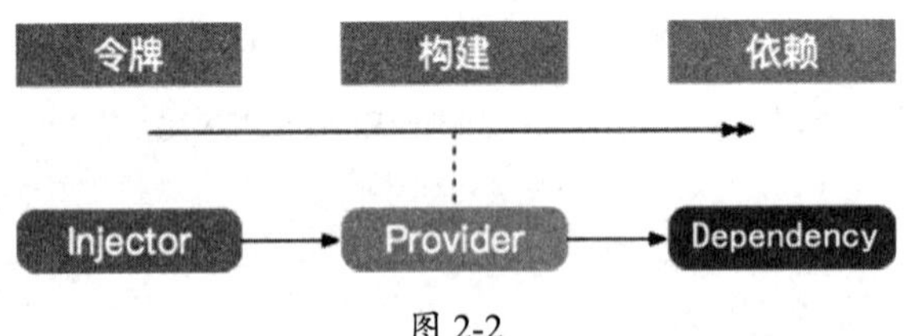

图 2-2

可能看到这里还是有些云里雾里，没关系，我们还是用例子来说明：

```
import { ReflectiveInjector } from '@angular/core';
const injector = RelfectiveInjector.resolveAndCreate([
  // provider 数组定义了多个提供者，provide 属性定义令牌
  // use××× 定义怎样创建的方法
  { provide: Person, useClass: Person },
  { provide: Address, useFactory: () => {
        if(env.testing)
            return new Address('辽宁', '沈阳', '和平区', 'xx 街 xx 号');
        return new Address('北京', '北京', '朝阳区', 'xx 街 xx 号');
    }
  },
  { provide: ID, useFactory: (type) => {
        if(type === ID_TYPES.PASSPORT)
            return Id.getInstance(ID_TYPES.PASSPORT, someparam);
        if(type === ID_TYPES.IDCARD)
            return Id.getInstance(ID_TYPES.IDCARD);
        return Id.getDefaultInstance();
    }
  }
]);

class Person {
  // 通过 @Inject 修饰器告诉 DI 这个参数需要什么样类型的对象
```

```
    // 请在 Injector 中帮我找到并注入对应参数中
    constructor(@Inject(Address) address, @Inject(Id) id) {
      // 省略
    }
}

// 通过 Injector 得到对象
const person = injector.get(Person);
```

在上述代码中，Angular 提供了 RelfectiveInjector 来解析和创建依赖的对象，你可以看到我们把这个应用中需要的 Person、ID 和 Address 都放在里面了。谁需要这些对象就可以向 Injector 请求，比如：injector.get(Person)，当然也可以是 injector.get(Address) 等。可以把它理解成一个依赖性的池子，想要什么来取就好了。

但是问题来了，首先，Injector 怎么知道如何创建你需要的对象呢？这个是靠 Provider 定义的，在刚刚的 RelfectiveInjector.resolveAndCreate() 中我们发现它是接受一个数组作为参数，这个数组就是一个 Provider 的数组。Provider 最常见的属性有两个：第一个是 provide，这个属性其实定义的是令牌，令牌的作用是让框架知道你要找的依赖是哪个，然后就可以在 use××× 这个属性定义的构建方式中将你需要的对象构建出来了。

那么 constructor(@Inject(Address) address, @Inject(Id) id) 这句怎么理解呢？由于我们想在 const person = injector.get(Person); 中取得 Person，但 Person 又需要两个依赖参数：Address 和 ID。@Inject(Address) address 是告诉框架我们需要的是一个令牌为 Address 的对象，这样框架就又到 Injector 中寻找令牌为 Address 对应的工厂函数，通过工厂函数构造好对象后又把对象赋值到 Address。

由于这里我们是用对象的类型来做令牌，所以上面的注入代码也可以写成下面的样子。利用 TypeScript 的类型定义，框架看到有依赖的参数就会去 Injector 中寻找令牌为该类型的工厂函数。

```
class Person {
  constructor(address: Address, id: Id) {
    // 省略
  }
}
```

而对于令牌为类型的并且是 useClass 的这种形式，由于前后都一样，对于这种 Provider 我们有一个语法糖：可以直接写成 { Person }，而不用完整地写成 { provide: Person, useClass: Person } 这种形式。当然还要注意 Token 不一定非得是某个类的类型，也可以是字符串，Angular 中还有 InjectionToken 用于创建一个可以避免重名的 Token。

那么，其实除了 useClass 和 useFactory，我们还可以使用 useValue 来提供一些简单的数据结构，比如我们可能希望把系统的 API 基础信息配置通过这种形式让所有想调用 API 的类都注入。如下面的例子中，基础配置就是一个简单的对象，里面有多个属性，这种情况用 useValue 即可。

```
{
  provide: 'BASE_CONFIG',
  useValue: {
    uri: 'https://dev.local/1.1',
    apiSecret: 'blablabla',
    apiKey: 'nahnahnah'
  }
}
```

3. 依赖性注入进阶

可能你会注意到，上面提到的依赖性注入有一个特点，即需要注入的参数如果在 Injector 中找不到对应的依赖，那么就会发生异常。但确实有些时候我们需要这样的特性：该依赖是可选的，如果有我们就这么做，如果没有就那样做。遇到这种情况怎么办呢？

Angular 提供了一个非常贴心的 @Optional 修饰器，这个修饰器用来告诉框架后面的参数需要一个可选的依赖。

```
constructor(@Optional(ThirdPartyLibrary) lib) {
    if (!lib) {
    // 如果该依赖不存在
    }
}
```

需要注意的是，Angular 的 DI 框架创建的对象都是单件（singleton）的，那么如果我们需要每次都创建一个新对象呢？有两个选择，第一种：在 Provider 中返回工厂而不是对象，像下面例子这样：

```
{
  provide: Address,
  useFactory: () => {
      // 注意：这里返回的是工厂，而不是对象
      return () => {
          if(env.testing)
              return new Address('辽宁', '沈阳', '和平区', 'xx 街 xx 号');
          return new Address('北京', '北京', '朝阳区', 'xx 街 xx 号');
      }
  }
}
```

第二种：创建一个 childInjector（子注入者）：injector.resolveAndCreateChild()

```
const injector = ReflectiveInjector.resolveAndCreate([Person]);
const childInjector = injector.resolveAndCreateChild([Person]);
// 此时父注入者和子注入者得到的 Person 对象是不同的
injector.get(Person) !== childInjector.get(Person);
```

子注入者还有一个特性：如果在 childInjector 中找不到令牌对应的工厂，它会去父注入者中寻找。换句话说，这父子关系（多重的）构成了一棵依赖树，框架会从最下面的子注入者开始寻找，一直找到最上面的父注入者。看到这里相信你就知道为什么父组件声明的 Provider 对于子组件是可见的，因为子组件在自己的 constructor 中如果发现有找不到的依赖就会到父组件中去找。

在实际的 Angular 应用中我们其实很少会直接显式使用 Injector 去完成注入，而是在对应的模块、组件等的元数据中提供 Provider 即可，这是由于 Angular 框架帮我们完成了这部分代码，它们其实在元数据配置后由框架放入 Injector 中了。

2.1.3 组件

组件是 Angular 应用中最基础的 UI 构建元素，从 UI 界面上可以暂时这么理解，一个组件就是渲染后的 HTML 页面中的某一块区域的元素、样式及事件的集合。一个 Angular 的应用其实就是一棵组件树。组件其实是指令的一个子集，但不同的是，组件拥有自己的模板。任何一个组件都必须从属于一个模块，即在模块的公告（declaration）中声明。

@Component 修饰符用于标识一个类是 Angular 组件，和模块类似的也有很多元数据属性，一个典型的组件如下面代码所示。

```
@Component({
  // 其他组件调用时使用 <app-login></app-login> 标签
  // selector 可以想象成一个自定义的 HTML 标签
  selector: 'app-login',
  changeDetection: ChangeDetectionStrategy.OnPush,
  template: `
    <input type="text" placeholder="您的 Email">
    <input type="password" placeholder="您的密码">
  `,
  styleUrls: ['./login.component.scss']
})
export class LoginComponent implements OnInit {
    constructor(){
        // 省略
    }

    ngOnInit(){
        // 省略
```

```
        }
    }
```

组件中支持的常见元数据如下所示。

- animations：动画定义。
- changeDetection：设置脏值检测的策略。
- encapsulation：样式的封装策略。
- entryComponents：在此视图中需要动态加载的组件列表。
- providers：对此组件及其子组件可见的 Provider 列表。
- selector：用于定位此组件的选择器。
- styleUrls：外部样式 URL 列表。
- styles：内联的样式。
- template：内联的模板。
- templateUrl：指向外部的模板文件 URL。
- viewProviders：仅为此组件及其视图上的子元素可见的 Provider 列表。

组件有生命周期的概念：Angular 创建、渲染控件；创建、渲染子控件；当数据绑定属性改变时做检查；在把控件移除 DOM 之前销毁控件，等等。Angular 提供生命周期的"钩子"（Hook）以便于开发者可以得到这些关键过程的数据，以及在这些过程中做出响应的能力。这些函数和顺序如图 2-3 所示。上面的例子中我们监听组件的 OnInit 事件。

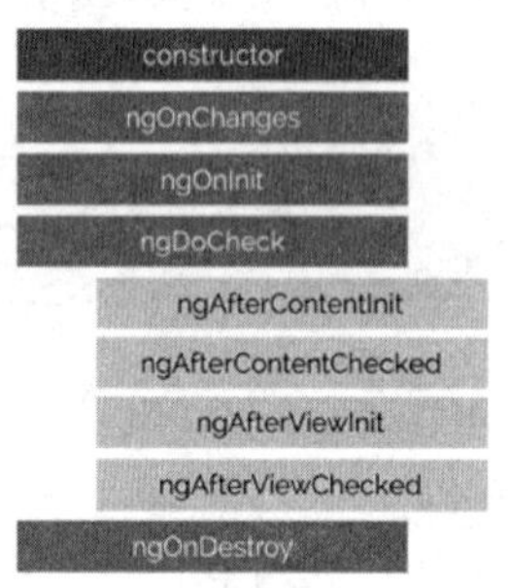

图 2-3

2.1.4 指令

Angular 中的指令分成三种：结构型（structural）指令和属性型（attribute）指令，还有一种是什么呢？就是组件（component），组件本身就是一个带模板的，有更多生命周期的 Hook 的指令。组件这里就不做讨论了，我们主要看前两种指令，它们大致的形式可以理解成类似于 HTML 中元素的属性，比如 <input type="text">" 中的 type。

下面的 ngIf 就是一个典型的结构型指令，结构型指令一般前方都有一个 *，用于控制 HTML 的布局，它们一般会生成或调整一些 DOM 元素，结构型指令没有绑定的概念，也就不会在指令外面发现 () 或 []。

```
<a *ngIf="user.login">退出</a>
```

那么，为什么会有这个看起来这么奇怪的 * 呢？它的作用是什么呢？加了这个 *，指令会把自己所在的元素包装在一对 <ng-template></ng-template> 的模板之中，也就是说上面的代码相当于下面的形式，注意此时 * 没有了，所以其实这个 * 也是一个语法糖，写一个 * 省去了写嵌套的 <ng-template></ng-template>。

```
<ng-template ngIf="user.login">
  <a>退出</a>
</ng-template>
```

把指令所在的元素包含在模板中的好处就是，你可以取得完全的控制权，我们可以根据需要将此模板内的元素显示、隐藏、增加或删除等。对于这个模板，我们在指令内部可以通过 TemplateRef 得到其引用。知道了这些，我们可以一起来看看 ngIf 这个指令的源码，你就会明白为什么根据表达式的真假，当前元素就可以显示或不显示了：

```
import {Directive, Input, TemplateRef, ViewContainerRef} from
'@angular/core';
@Directive({selector: '[ngIf]'})
export class NgIf {
  private _hasView = false;

  constructor(private _viewContainer: ViewContainerRef, private
_template: TemplateRef<Object>) {}

  @Input()
  set ngIf(condition: any) {
    if (condition && !this._hasView) {
      this._hasView = true;
      // 如果条件为真，则创建该模板元素
      this._viewContainer.createEmbeddedView(this._template);
    } else if (!condition && this._hasView) {
      this._hasView = false;
      // 否则清空视图
      this._viewContainer.clear();
    }
  }
}
```

常见的内建结构型指令除了 ngIf，还有 ngFor 和 ngSwitchCase 等。

属性型指令是改变 DOM 元素的外观或行为的指令，而常见的内建属性型指令有 ngModel 和 ngClass，我们会在后面的模板驱动型表单中讲解 ngModel，这里主要看一下 ngClass。很多时候，我们希望动态地改变元素或组件的样式，这时 ngClass 就派上用场了。ngClass 可以接收很多种类型的参数，包括以空格分隔的字符串、数组，以及像下面这样的对象。对象的 key 是 CSS 的类名，value 是一个布尔值。笔者认为这种对象的形式最灵活，因为你可以在程序中动态控制各个 CSS 类的 true 或 false，即是否使用。

```
<some-element [ngClass]="{'first': true, 'second': true, 'third':
false}">...</some-element>
```

属性型指令经常会在 DOM 元素属性的外面套上 [] 或 ()，比如上面的 [ngClass]="×××"，[] 是说 "×××" 是一个表达式（或者对象、变量），请将这个表达式的值赋值给 ngClass。但如果不加 []，那么 Angular 会认为你要把 "×××" 这个字符串赋值给 ngClass 了。那么 () 呢？这个是用来绑定事件的，有时候指令或组件会有事件（也就是输出型参数），这个时候就需要用 () 绑定这个事件，用来监听处理。

2.1.5 管道

管道是一种在模板中使用的快速方便地进行数据变换的快捷方式，从使用上来说是“| 管道名称: "参数" ”这样的形式。这个特性可以让我们很快将数据在界面上以想要的格式输出。看一个小例子：

```
import { Component, OnDestroy } from '@angular/core';

@Component({
    selector: 'app-playground',
    template: `
    <p>Without Pipe: Today is {{ birthday }} </p>
    <p>With Pipe: Today is {{ birthday | date:"MM/dd/yy" }} </p>
    `,
    styleUrls: ['./playground.component.css']
})
export class PlaygroundComponent {
    birthday = new Date();
    constructor() { }
}
```

输出结果如图 2-4 所示。

```
Without Pipe: Today is Mon Dec 26 2016 01:02:39 GMT+0800 (CST)

With Pipe: Today is 12/26/16
```

图 2-4

2.1.6 模块

什么是模块？简单来说就是把一些功能组织到一起完成一个相对独立的事情。打个比方，常见的应用都有登录、注册、忘记密码这些功能，但这些功能在应用中其他的地方使用的场景多吗？好像联系不大，这块功能是相对独立的，那么也就是说我们可以把这些登录注册相关的功能封装在一个模块中（这里我们给这个模块起个名字，叫 AuthModule），然后提供一些必要的接口和组件给外部，系统的其他部分想调用时可以导入这个模块，不必关心这些功能是怎么实现的。比如说，我们在登录模块中增加一个微信登录的方式，就只需要在这个 AuthModule 中改动即可，其他模块不需要进行更改。

可能有些熟悉 JavaScript 的读者还会问，JavaScript 中已经有模块的概念了，为什么还要再搞出一个呢？这是由于 JavaScript 中的模块概念是每一个文件一个模块，但是当我们有很多文件时，比如我们有 10 个文件，而这时需要增加 5 个文件，这两个文件是需要依赖这 10 个文件的，那么就需要在这 5 个文件中导入那 10 个文件。当工程大起来之后，不仅工作量增加，代码维护也很麻烦。所以用模块形式组织代码是大工程的必要步骤。

在 Angular 中，我们会使用一个 @NgModule 的修饰符来标识一个类是模块，在这个修饰符中，定义一个模块的元数据，如下所示。

- declarations：用于列出属于此模块的组件、指令和管道。
- imports：用于导入该模块依赖的其他模块。
- exports：如果有想让其他模块使用的组件、指令和管道，则在此列出。
- providers：列出提供给模块内部用于注入的资源，通常是服务。
- entryComponents：列出进入模块就要实例化的组件，用于动态加载。
- bootstrap：只有根模块有此项，用于指定应用启动后的根组件。

一个典型的模块在 Angular 中的定义如下所示：

```
@NgModule({
    declarations: [ // 此模块包含的组件/指令/管道都列在这个数组中
        HomeComponent,
        AComponent, BComponent,
        CDirective,
        DPipe,
    ],
    imports: [ // 该模块所依赖的其他模块在这个数组中列出并进行导入
        CommonModule,
        ThridPartyModule,
    ],
    exports: [ // 导出给外部模块可以使用的组件/指令/管道等
        AComponent
```

```
    ]
    providers: [ // 该模块要提供给模块内部注入使用的类库，一般是服务
        MyOwnService,
    ],
    entryComponents: [ // 对于有些组件需要在进入模块后就创建出来，比如对话框等
        HomeComponent
    ]
})
export class MyModule { }
```

任何一个 Angular 应用都至少有一个模块：根模块。习惯上我们把根模块命名为 AppModule。另外需要注意一点，模块是一个类，尽管很多时候我们都没有为这个类写具体的方法，但不代表这个类不能写具体的内部的成员方法。事实上，有一些情况是一定要写方法才可以达成目标的，比如，如果你希望模块被导入后要进行一些初始化工作：

```
@NgModule({
    // 省略
})
export class MyModule {
    constructor(){
        doInitWork();
    }
    void doInitWork(){
      // 省略
    }
}
```

我们看到内建的模块，比如 HttpModule 或者很多第三方模块，在导入的同时，其提供的 providers 也就可以调用了，并不用显式地在我们的模块的 providers 数组中列出。这其实也是利用了模块的构造或实例方法完成的。像下面的例子中，我们在模块类中提供一个静态方法返回模块和 providers，这样只要在根模块的 imports 中写入 MyModule.forRoot()，即可完成在导入模块的同时将模块提供的 providers 也提供给根模块。注意一点，命名上习惯性地将只在根模块使用的方法命名为 forRoot。

```
@NgModule({
  // 省略
})
class MyModule {
  static forRoot() {
    return {
      ngModule: MyModule,
      providers: [ AuthService ]
    }
```

```
  }
}
```

2.1.7 模板驱动型表单

在企业应用开发时，表单是必须的，和面向消费者的应用不同，企业领域的开发中，表单的使用量是惊人的。这些表单的处理其实是一件挺复杂的事情，比如有的是涉及多个标签的表单，有的是向导形式多个步骤的，各种复杂的验证逻辑和时不时需要弹出的对话框等。

Angular 中提供两种类型的表单处理机制，一种叫模板驱动型（Template Driven）的表单，另一种叫模型驱动型表单（Model Driven），这后一种也叫响应式表单（Reactive Form），由于模板驱动中有一个 ngModel 的指令容易与这里说的模型驱动型表单混淆，所以本书中使用后一种说法：响应式表单。

模板驱动型表单和 AngularJS 对于表单的处理类似，即把一些指令（比如 ngModel）、数据值和行为约束（比如 require、minlength 等）绑定到模板中（模板就是组件元数据 @Component 中定义的那个 template），这也是模板驱动这个叫法的来源。总体来说，这种类型的表单通过绑定把很多工作交给了模板。

1. 模板驱动型表单的例子

还是用例子来说话，比如我们有一个用户注册的表单，用户名就是 email，还需要填的信息有：住址、密码和重复密码。这应该是比较常见的注册时需要的信息。那么我们第一步来建立领域模型：

```
// src/app/domain/index.ts
export interface User {
  // 新的用户 ID 一般由服务器自动生成，所以可以为空，用 ? 标示
  id?: string;
  email: string;
  password: string;
  repeat: string;
  address: Address;
}

export interface Address {
  province: string; // 省份
  city: string; // 城市
  area: string; // 区县
  addr: string; // 详细地址
}
```

接下来建立模板文件，下面是一个最简单的 HTML 模板，先不增加任何的绑定或事件处理：

```
<!-- template-driven.component.html -->
<form novalidate>
  <label>
    <span>电子邮件地址</span>
    <input
      type="text"
      name="email"
      placeholder="请输入您的 email 地址">
  </label>
  <div>
    <label>
      <span>密码</span>
      <input
        type="password"
        name="password"
        placeholder="请输入您的密码">
    </label>
    <label>
      <span>确认密码</span>
      <input
        type="password"
        name="repeat"
        placeholder="请再次输入密码">
    </label>
  </div>
  <div >
    <label>
      <span>省份</span>
      <select name="province">
        <option value="">请选择省份</option>
      </select>
    </label>
    <label>
      <span>城市</span>
      <select name="city">
        <option value="">请选择城市</option>
      </select>
    </label>
    <label>
      <span>区县</span>
      <select name="area">
        <option value="">请选择区县</option>
      </select>
    </label>
```

```
      <label>
        <span>地址</span>
        <input type="text" name="addr">
      </label>
    </div>
    <button type="submit">注册</button>
  </form>
```

渲染之后的效果如图 2-5 所示。

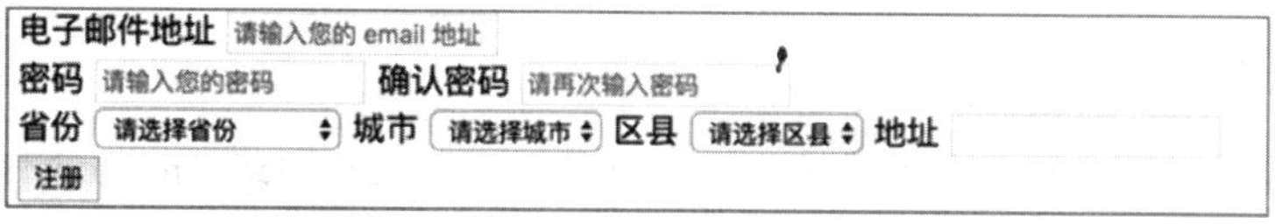

图 2-5

2. 数据绑定

对于模板驱动型表单的处理，我们首先需要在对应的模块中引入 FormsModule，这一点千万不要忘记了。

```
// 省略导入

@NgModule({
  imports: [
    CommonModule,
    FormsModule
  ],
  exports: [TemplateDrivenComponent],
  declarations: [TemplateDrivenComponent]
})
export class FormDemoModule { }
```

进行模板驱动类型的表单处理的一个必要步骤就是建立数据的双向绑定，那么我们需要在组件中建立一个类型为 User 的成员变量并赋初始值，代码如下所示。

```
// template-driven.component.ts
// 省略元数据和导入的类库信息
export class TemplateDrivenComponent implements OnInit {

  user: User = {
    email: '',
    password: '',
    repeat: '',
    address: {
```

```
        province: '',
        city: '',
        area: '',
        addr: ''
      }
    };
    // 省略其他部分
  }
```

有了这样一个成员变量之后，我们在组件模板中就可以使用 ngModel 进行绑定了。

3. 令人困惑的 ngModel

在 Angular 中可以使用三种形式的 ngModel 表达式：ngModel、[ngModel] 和 [(ngModel)]。但无论哪种形式，如果你要使用 ngModel 就必须为该控件（比如下面的 input）指定一个 name 属性，如果你忘记添加 name，则多半会看到下面这样的错误：

```
ERROR Error: Uncaught (in promise): Error: If ngModel is used within a form tag, either the name attribute must be set or the form control must be defined as 'standalone' in ngModelOptions.
```

假如我们使用的是 ngModel，没有任何中括号、小括号，则代表着我们创建了一个 FormControl 的实例，这个实例将会跟踪值的变化、用户的交互、验证状态，以及保持视图和领域对象的同步等工作。

```
<input
  type="text"
  name="email"
  placeholder="请输入您的 email 地址"
  ngModel>
```

如果将这个控件放在一个 Form（表单）中，那么 ngModel 会自动将这个 FormControl 注册为 Form 的子控件。下面的例子中，我们在 <form> 中加上 ngForm 指令，声明这是一个 Angular 可识别的表单，而 ngModel 会将 <input> 注册成表单的子控件，这个子控件的名字就是 email，而且 ngModel 会基于这个子控件的值去绑定表单的值，这也是为什么需要显式声明 name 的原因。

其实，在导入 FormsModule 时，所有的 <form> 标签都会默认地被认为是一个 ngForm，因此并不需要显式地在标签中写 ngForm 这个指令。

```
<!-- ngForm 并不需要显示声明，任何 <form> 标签默认都是 ngForm -->
<form novalidate ngForm>
  <input
    type="text"
    name="email"
```

```
    placeholder="请输入您的 email 地址"
    ngModel>
</form>
```

这一切现在都是不可见的，所以读者可能还是有些困惑，那么下面我们将其“可视化”，这需要引用一下表单对象，所以使用 #f="ngForm" 以便可以在模板中输出表单的一些特性。

```
<!-- 使用 # 把表单对象导出到 f 这个可引用变量中 -->
<form novalidate #f="ngForm">
  ...
</form>
<!-- 将表单的值以 JSON 形式输出 -->
{{f.value | json}}
```

这时，如果在电子邮件地址中输入 sss，可以看到图 2-6 所示的以 JSON（一种轻量级的数据交换格式）形式出现的表单值。

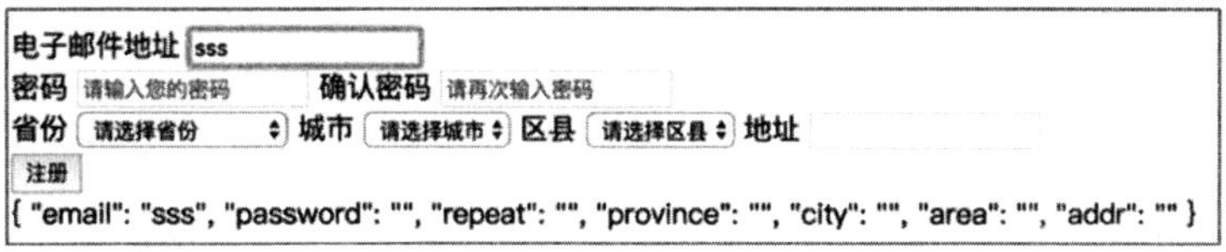

电子邮件地址 sss
密码 请输入您的密码 确认密码 请再次输入密码
省份 请选择省份 城市 请选择城市 区县 请选择区县 地址
注册
{ "email": "sss", "password": "", "repeat": "", "province": "", "city": "", "area": "", "addr": "" }

图 2-6

接下来，我们看看 [ngModel] 有什么用？如果想给控件设置一个初始值怎么办呢，这时就需要进行单向绑定，方向是从组件到视图。我们可以做的是在初始化 User 时，将 email 的属性设置成 wang@163.com，如下所示。

```
user: User = {
    email: 'wang@163.com',
    ...
  };
```

而且，在模板中使用 [ngModel]="user.email" 进行单向绑定，这个语法其实和普通的属性绑定是一样的，用中括号表示这是一个要进行数据绑定的属性，等号右边的是需要绑定的值（这里是 user.email）。我们就可以得到如图 2-7 所示的输出了，email 的初始值被绑定成功！

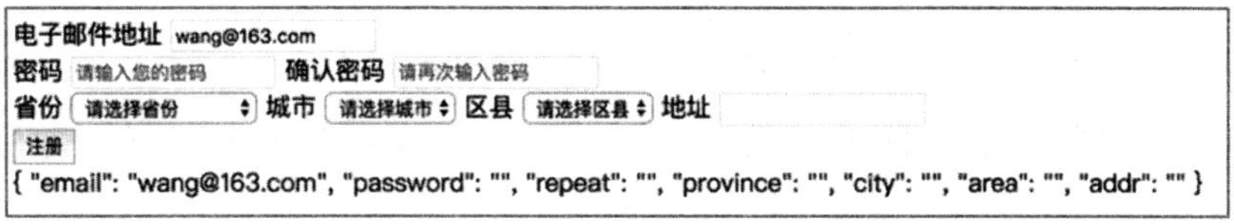

电子邮件地址 wang@163.com
密码 请输入您的密码 确认密码 请再次输入密码
省份 请选择省份 城市 请选择城市 区县 请选择区县 地址
注册
{ "email": "wang@163.com", "password": "", "repeat": "", "province": "", "city": "", "area": "", "addr": "" }

图 2-7

但上面的例子存在一个问题，数据的绑定是单向的。也就是说，在输入框进行输入的时候，user 的值不会随之改变。为了更好地说明，我们将 user 和 表单的值同时输出。

```
<div>
  <span>user: </span> {{user | json}}
</div>
<div>
  <span>表单：</span> {{f.value | json}}
</div>
```

此时，我们将默认的电子邮件改成 wang@gmail.com，表单的值改变，但 user 并未改变，如图 2-8 所示。

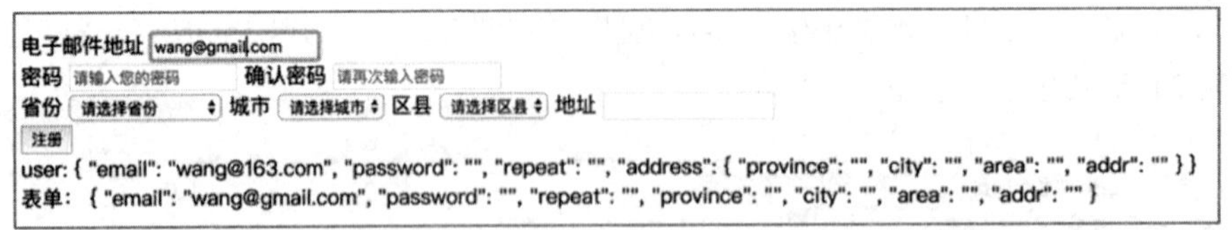

图 2-8

如果我们希望在输入时，这个输入的值也会反向地影响 user 对象的值，那就需要用到双向绑定，也就是 [(ngModel)] 需要上场了，如图 2-9 所示。

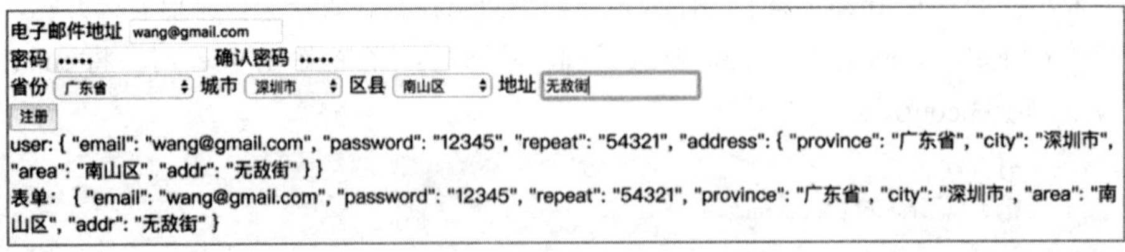

图 2-9

无论如何，这个 [()] 的表达很奇怪，其实这个表达是一个“语法糖”。只要我们知道下面的两种写法是等价的，就可以清楚地理解：用这个语法糖就不用既写数据绑定，又写事件绑定了。

```
<input [(ngModel)]="user.email">
<input [ngModel]="user.email"` (ngModelChange)="user.email = $event">
```

如果我们仔细观察上面的输出，就会发现一个问题：user 中有一个嵌套对象 address，而表单中是没有嵌套对象的。如果要使表单中的结构和领域对象的结构一致，就需要使用 ngModelGroup。ngModelGroup 会创建并绑定一个 FormGroup 到该 DOM 元素。FormGroup 又是什么呢？简单来说，是一组 FormControl。

```
    <!-- 使用 ngModelGroup 来创建并绑定 FormGroup  -->
    <div ngModelGroup="address">
      <label>
        <span>省份</span>
        <select name="province" (change)="onProvinceChange()"
[(ngModel)]="user.address.province">
          <option value="">请选择省份</option>
          <option [value]="province" *ngFor="let province of
provinces">{{province}}</option>
        </select>
      </label>
      <!-- 省略其他部分 -->
    </div>
```

我们来看一下如图 2-10 所示的输出结果，现在已经完全一致了。

电子邮件地址 wang@163.com
密码 请输入您的密码 确认密码 请再次输入密码
省份 请选择省份 城市 请选择城市 区县 请选择区县 地址
注册
user: { "email": "wang@163.com", "password": "", "repeat": "", "address": { "province": "", "city": "", "area": "", "addr": "" } }
表单: { "email": "wang@163.com", "password": "", "repeat": "", "address": { "province": "", "city": "", "area": "", "addr": "" } }

图 2-10

4．数据验证

模板驱动型的表单验证主要由模板来处理，在使用之前，需要界定一下验证规则。

- 三个必填项：email、password 和 repeat。
- email 的形式需要符合电子邮件的标准。
- password 和 repeat 必须一致。

当然，除了这几个规则，我们还希望在表单未验证通过时提交按钮是不可用的。

```
  <form novalidate #f="ngForm">
    <label>
      <span>电子邮件地址</span>
      <input
        type="text"
        name="email"
        placeholder="请输入您的 email 地址"
        [ngModel]="user.email"
        required
        pattern="([a-zA-Z0-9]+[_|_|.]?)*[a-zA-Z0-9]+@([a-zA-Z0-
9]+[_|_|.]?)*[a-zA-Z0-9]+.[a-zA-Z]{2,4}">
    </label>
```

```
  <div>
    <label>
      <span>密码</span>
      <input
        type="password"
        name="password"
        placeholder="请输入您的密码"
        [(ngModel)]="user.password"
        required
        minlength="8">
    </label>
    <label>
      <span>确认密码</span>
      <input
        type="password"
        name="repeat"
        placeholder="请再次输入密码"
        [(ngModel)]="user.repeat"
        required
        minlength="8">
    </label>
  </div>
  <!-- 省略其他部分 -->
  <button type="submit" [disabled]="f.invalid">注册</button>
</form>
<div>
```

Angular 中有几种内建支持的验证器（validator）：

- required：需要 FormControl 有非空值
- minlength：需要 FormControl 有最小长度的值
- maxlength：需要 FormControl 有最大长度的值
- pattern：需要 FormControl 的值可以匹配正则表达式
- email：内建的电子邮件的验证器

如果我们想看到结果，那么可以在模板中加上下面的代码，将错误以 JSON 形式输出即可。

```
<div>
  <span>email 验证：</span> {{f.controls.email?.errors | json}}
</div>
```

我们看到，如果不填电子邮件，那么错误的 JSON 是 {"required": true}，这告诉我们目前有一个 required 的规则没有被满足，如图 2-11 所示。

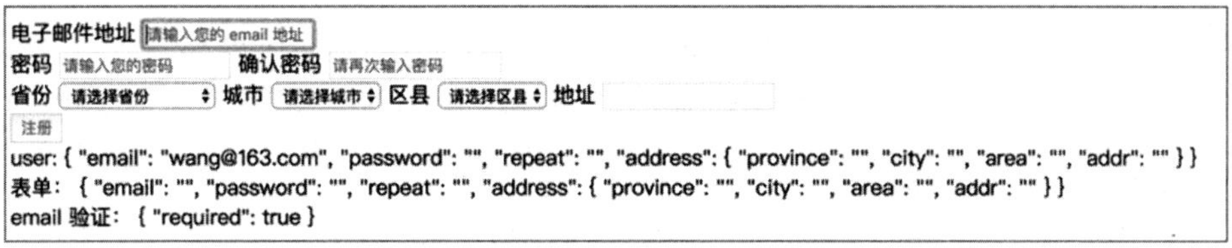

电子邮件地址 请输入您的 email 地址
密码 请输入您的密码 确认密码 请再次输入密码
省份 请选择省份 城市 请选择城市 区县 请选择区县 地址
注册
user: { "email": "wang@163.com", "password": "", "repeat": "", "address": { "province": "", "city": "", "area": "", "addr": "" } }
表单： { "email": "", "password": "", "repeat": "", "address": { "province": "", "city": "", "area": "", "addr": "" } }
email 验证： { "required": true }

图 2-11

当我们输入一个字母 w 之后，就会发现错误变成了下面的样子。这是因为我们对于 email 应用了多个规则，当必填项被满足后，系统会继续检查其他验证结果。

```
    {
    "pattern":
        {
            "requiredPattern": "^([a-zA-Z0-9]+[_|_|.]?)*[a-zA-Z0-9]+@([a-zA-
Z0-9]+[_|_|.]?)*[a-zA-Z0-9]+.[a-zA-Z]{2,4}$",
            "actualValue": "w"
        }
    }
```

通过几次实验，我们应该可以得出结论，当验证未通过时，验证器返回的是一个对象，key 为验证规则（比如 required、minlength 等），value 为验证结果。如果验证通过，则返回的是一个 null。

知道这一点后，其实就可以做出验证出错的提示了。为了方便引用，我们先导出 ngModel 到一个 email 引用，然后访问这个 FormControl 的各个属性：验证的状态（valid/invalid）、控件的状态（是否获得过焦点—— touched/untouched，是否更改过内容—— pristine/dirty）。

```
<label>
  <span>电子邮件地址</span>
  <input
    ...
    [ngModel]="user.email"
    #email="ngModel">
</label>
<div *ngIf="email.errors?.required && email.touched" class="error">
  email 是必填项
</div>
<div *ngIf="email.errors?.pattern && email.touched" class="error">
  email 格式不正确
</div>
```

内建的验证器不适用两个密码的比较，我们需要自己定义一个验证器。对于响应式表单来说，会比较简单一些，但对模板驱动型表单，则需要实现一个指令来使验证器更通用和更一致。

因为我们希望实现的是和 required、minlength 等差不多的形式，比如 validateEqual ="repeat"。

```
<div>
    <label>
      <span>密码</span>
      <input
        type="password"
        name="password"
        placeholder="请输入您的密码"
        [(ngModel)]="user.password"
        required
        minlength="8"
        validateEqual="repeat">
    </label>
    <label>
      <span>确认密码</span>
      <input
        type="password"
        name="repeat"
        placeholder="请再次输入密码"
        [(ngModel)]="user.repeat"
        required
        minlength="8">
    </label>
  </div>
```

如果要实现这种形式的验证器，就要建立一个指令，而且这个指令应该实现 Validator 接口。一个基础的框架如下：

```
import { Directive, forwardRef } from '@angular/core';
import { NG_VALIDATORS, Validator, AbstractControl } from
'@angular/forms';

@Directive({
  selector: '[validateEqual][ngModel]',
  providers: [
    {
      provide: NG_VALIDATORS,
      useExisting: forwardRef(()=>RepeatValidatorDirective),
      multi: true
    }
  ]
})
export class RepeatValidatorDirective implements Validator{
  constructor() { }
```

```
  validate(c: AbstractControl): { [key: string]: any } {
    return null;
  }
}
```

我们还没有开始正式写验证逻辑，但上面的框架已经出现了几个有意思的点。

- Validator 接口要求必须实现的一个方法是 validate(c: AbstractControl): ValidationErrors | null;。这也是我们前面提到的若验证正确则返回 null，否则返回一个对象，虽然没有严格的约束，但其 key 一般用于表示这个验证器的名字或者验证规则的名字，value 一般是失败的原因或验证结果。
- 与组件类似，指令也有 selector 这个元数据，用于选择哪个元素应用该指令，那么我们这里除了要求 DOM 元素应用 validateEqual，还需要它是一个 ngModel 元素，这样它才是一个 FormControl，在使用 validate 方法时才是合法的。
- Angular 中，在 FormControl 上执行验证器有一个内部机制：Angular 维护一个令牌叫作 NG_VALIDATORS 的 multi provider（简单来说，Angular 为一个单一令牌注入多个值的这种形式叫 multi provider）。所有的内建验证器都是加到这个 NG_VALIDATORS 的令牌上的，因此在做验证时，Angular 注入了 NG_VALIDATORS 的依赖，也就是所有的验证器，然后一个个地按顺序执行。因此我们这里也把自己加到这个 NG_VALIDATORS 中。
- 如果我们直接写成 useExisting: RepeatValidatorDirective，则会出现一个问题，RepeatValidatorDirective 还没有生成，你怎么能在元数据中使用呢？这就需要使用 forwardRef 来解决这个问题，它接受返回一个类的函数作为参数，但这个函数不会立即被调用，而是在该类声明后被调用，也就避免了 undefined 的状况。

下面我们就来实现这个验证逻辑，由于密码和确认密码有主从关系，并非完全的平行关系。也就是说，密码是一个基准对比对象，当密码改变时，我们不应该提示密码和确认密码不符，而是应该将错误放在确认密码中，所以我们给出另一个属性 reverse。

```
export class RepeatValidatorDirective implements Validator{
  constructor(
    @Attribute('validateEqual') public validateEqual: string,
    @Attribute('reverse') public reverse: string) { }

  private get isReverse() {
    if (!this.reverse) return false;
    return this.reverse === 'true' ? true: false;
  }

  validate(c: AbstractControl): { [key: string]: any } {
```

```
    // 控件自身值
    let self = c.value;

    // 要对比的值，也就是在 validateEqual="ctrlname" 的那个控件的值
    let target = c.root.get(this.validateEqual);

    // 不反向查询且值不相等
    if (target && self !== target.value && !this.isReverse) {
      return {
        validateEqual: true
      }
    }

    // 反向查询且值相等
    if (target && self === target.value && this.isReverse) {
        delete target.errors['validateEqual'];
        if (!Object.keys(target.errors).length) target.setErrors(null);
    }

    // 反向查询且值不相等
    if (target && self !== target.value && this.isReverse) {
        target.setErrors({
            validateEqual: true
        })
    }

    return null;
  }
}
```

这样改造后，模板文件中对密码和确认密码的验证器如下：

```
<input
    type="password"
    name="password"
    placeholder="请输入您的密码"
    [(ngModel)]="user.password"
    #password="ngModel"
    required
    minlength="8"
    validateEqual="repeat"
    reverse="true">
<!-- 省略其他部分 -->
<input
    type="password"
```

```
    name="repeat"
    placeholder="请再次输入密码"
    [(ngModel)]="user.repeat"
    #repeat="ngModel"
    required
    minlength="8"
    validateEqual="password"
    reverse="false">
```

输出结果如图 2-12 所示。

图 2-12

5．表单的提交

表单的提交比较简单，绑定表单的 ngSubmit 事件即可：

```
<form novalidate #f="ngForm" (ngSubmit)="onSubmit(f, $event)">
```

需要注意的一点是，如果不给 button 指定类型，则会被当作 type="submit"，所以当按钮不是提交表单时，则需要显式指定 type="button"。而且如果遇到点击“提交”按钮刷新页面的情况，则意味着默认的表单提交事件引起了浏览器的刷新，这种时候需要阻止事件冒泡。

```
onSubmit({value, valid}, event: Event){
  if(valid){
    console.log(value);
  }
  event.preventDefault();
}
```

2.1.8 响应式表单

响应式表单乍一看很像模板驱动型表单，但响应式表单需要引入一个不同的模块：是

ReactiveFormsModule，而不是 FormsModule。

```
import {ReactiveFormsModule} from "@angular/forms";
@NgModule({
  // 省略其他
    imports: [..., ReactiveFormsModule],
  // 省略其他
})
// 省略其他
```

1. 与模板驱动型表单的区别

接下来，我们还是利用前面的例子，用响应式表单的要求改写一下：

```
<form [formGroup]="user" (ngSubmit)="onSubmit(user)">
  <label>
    <span>电子邮件地址</span>
    <input type="text" formControlName="email" placeholder="请输入您的
email 地址">
  </label>
  <div *ngIf="user.get('email').hasError('required') &&
user.get('email').touched" class="error">
    email 是必填项
  </div>
  <div *ngIf="user.get('email').hasError('pattern') &&
user.get('email').touched" class="error">
    email 格式不正确
  </div>
  <div>
    <label>
      <span>密码</span>
      <input type="password" formControlName="password" placeholder="请输
入您的密码">
    </label>
    <div *ngIf="user.get('password').hasError('required') &&
user.get('password').touched" class="error">
      密码是必填项
    </div>
    <label>
      <span>确认密码</span>
      <input type="password" formControlName="repeat" placeholder="请再次
输入密码">
    </label>
    <div *ngIf="user.get('repeat').hasError('required') &&
user.get('repeat').touched" class="error">
      确认密码是必填项
```

```
        </div>
        <div *ngIf="user.hasError('validateEqual') &&
user.get('repeat').touched" class="error">
          确认密码和密码不一致
        </div>
      </div>
      <div formGroupName="address">
        <label>
          <span>省份</span>
          <select formControlName="province">
            <option value="">请选择省份</option>
            <option [value]="province" *ngFor="let province of
provinces">{{province}}</option>
          </select>
        </label>
        <label>
          <span>城市</span>
          <select formControlName="city">
            <option value="">请选择城市</option>
            <option [value]="city" *ngFor="let city of (cities$ |
async)">{{city}}</option>
          </select>
        </label>
        <label>
          <span>区县</span>
          <select formControlName="area">
            <option value="">请选择区县</option>
            <option [value]="area" *ngFor="let area of (areas$ |
async)">{{area}}</option>
          </select>
        </label>
        <label>
          <span>地址</span>
          <input type="text" formControlName="addr">
        </label>
      </div>
      <button type="submit" [disabled]="user.invalid">注册</button>
    </form>
```

这段代码和模板驱动型表单的那段看起来差不多，但是有几个区别：

- 表单多了一个指令 [formGroup]="user"。
- 去掉了对表单的引用 #f="ngForm"。
- 每个控件多了一个 formControlName。

- 同时每个控件也去掉了验证条件，比如 required、minlength 等。
- 在地址分组中用 formGroupName="address" 替代了 ngModelGroup="address"。

模板上的区别大概就这样了，接下来我们来看看组件的区别：

```
import { Component, OnInit } from '@angular/core';
import { FormControl, FormGroup, Validators } from "@angular/forms";
@Component({
  selector: 'app-model-driven',
  templateUrl: './model-driven.component.html',
  styleUrls: ['./model-driven.component.css']
})
export class ModelDrivenComponent implements OnInit {

  user: FormGroup;

  ngOnInit() {
    // 初始化表单
    this.user = new FormGroup({
      email: new FormControl('', [Validators.required,
Validators.pattern(/([a-zA-Z0-9]+[_|_|.]?)*[a-zA-Z0-9]+@([a-zA-Z0-
9]+[_|_|.]?)*[a-zA-Z0-9]+.[a-zA-Z]{2,4}/)]),
      password: new FormControl('', [Validators.required]),
      repeat: new FormControl('', [Validators.required]),
      address: new FormGroup({
        province: new FormControl(''),
        city: new FormControl(''),
        area: new FormControl(''),
        addr: new FormControl('')
      })
    });
  }
  onSubmit({value, valid}){
    if(!valid) return;
    console.log(JSON.stringify(value));
  }
}
```

从上面的代码中我们可以看到，这里的表单（FormGroup）是由一系列的表单控件（FormControl）构成的。其实 FormGroup 的构造函数有三个参数：controls（表单控件『数组』，其实不是数组，是一个类似字典的对象）、validator（验证器）和 asyncValidator（异步验证器），其中只有 controls 数组是必须的参数，后两个都是可选参数。

```
// FormGroup 的构造函数
```

```
constructor(
  controls: {
    [key: string]: AbstractControl;
  },
  validator?: ValidatorFn,
  asyncValidator?: AsyncValidatorFn
)
```

上述代码没有使用验证器和异步验证器的可选参数，而且我们提供 controls 的方式是，一个 key 对应一个 FormControl。比如下面的 key 是 password，对应的值是 new FormControl('', [Validators.required])。这个 key 对应的就是模板中的 formControlName 的值，模板代码中设置了 formControlName="password"，而表单控件会根据这个 password 的控件名来跟踪实际渲染出的表单页面上的控件（比如 <input formcontrolname="password">）的值和验证状态。

```
password: new FormControl('', [Validators.required])
```

那么，可以看出，这个表单控件的构造函数同样也有三个可选参数，分别是：控件初始值（formState）、控件验证器或验证器数组（validator）、控件异步验证器或异步验证器数组（asyncValidator）。在上面的那行代码中，初始值为空字符串，验证器是必选，而异步验证器我们没有提供。

```
// FormControl 的构造函数
constructor(
  formState?: any, // 控件初始值
  validator?: ValidatorFn | ValidatorFn[], // 控件验证器或验证器数组
  asyncValidator?: AsyncValidatorFn | AsyncValidatorFn[] // 控件异步验证器
或异步验证器数组
)
```

由此可以看出，响应式表单区别于模板驱动型表单的主要特点在于，其是由组件类去创建、维护和跟踪表单的变化，而不是依赖模板。

那么，我们是否在响应式表单中还可以使用 ngModel 呢？当然可以，但如果这样，那么表单的值会在两个不同的位置存储 ngModel 绑定的对象和 FormGroup，设计上我们一般要避免这种情况，也就是说尽管可以这么做，但不建议这么做。

2．使用 FormBuilder 快速构建表单

上面的表单构造方法虽然也不算太麻烦，但是若表单项目逐渐多起来，就会比较烦索，所以 Angular 提供了一种快捷构造表单的方式，即使用 FormBuilder。

```
import { Component, OnInit } from '@angular/core';
import { FormBuilder, FormGroup, Validators } from "@angular/forms";
@Component({
```

```
  selector: 'app-model-driven',
  templateUrl: './model-driven.component.html',
  styleUrls: ['./model-driven.component.css']
})
export class ModelDrivenComponent implements OnInit {

  user: FormGroup;

  constructor(private fb: FormBuilder) {
  }

  ngOnInit() {
    // 初始化表单
    this.user = this.fb.group({
      email: ['', [Validators.required, Validators.email]],
      password: ['', Validators.required],
      repeat: ['', Validators.required],
      address: this.fb.group({
        province: [],
        city: [],
        area: [],
        addr: []
      })
    });
  }
  // 省略其他部分
}
```

使用 FormBuilder，可以无须显式声明 FormControl 或 FormGroup。FormBuilder 提供三种类型的快速构造：control、group 和 array，分别对应 FormControl、FormGroup 和 FormArray。我们在表单中最常见的一种是通过 group 来初始化整个表单。在上面的例子中，group 接受一个字典对象作为参数，这个字典中的 key 就是这个 FormGroup 中 FormControl 的名字，值是一个数组，数组中的第一个值是控件的初始值，第二个值是同步验证器的数组，第三个值是异步验证器数组（第三个值并未出现在我们的例子中）。这其实已经在隐性地使用 FormBuilder.control 了，可以参看下面的 FormBuilder 中的 control 函数定义，其实 FormBuilder 利用我们给出的值构造了相对应的 control：

```
control(
    formState: Object,
    validator?: ValidatorFn | ValidatorFn[],
    asyncValidator?: AsyncValidatorFn | AsyncValidatorFn[]
    ): FormControl;
```

此外，还值得注意的一点是 address 的处理，我们可以清晰地看到 FormBuilder 支持嵌套，遇到 FormGroup 时仅仅需要再次使用 this.fb.group({...}) 即可。这样我们的表单在拥有大量的表单项时，构造起来就方便多了。

3．自定义验证

对于响应式表单来说，构造一个自定义验证器是非常简单的，比如我们上面提到过的验证密码和重复输入密码是否相同的需求，我们在响应式表单中来试一下。

```
    validateEqual(passwordKey: string, confirmPasswordKey: string):
ValidatorFn {
      return (group: FormGroup): {[key: string]: any} => {
        const password = group.controls[passwordKey];
        const confirmPassword = group.controls[confirmPasswordKey];
        if (password.value !== confirmPassword.value) {
          return { validateEqual: true };
        }
        return null;
      }
    }
```

这个函数的逻辑比较简单：首先接受两个字符串（是 FormControl 的名字），然后返回一个 ValidatorFn。但是这个函数里面比较奇怪，比如 (group: FormGroup): {[key: string]: any} => {...} 是什么意思呢？还有，这个 ValidatorFn 是什么？我们来看一下定义：

```
export interface ValidatorFn {
    (c: AbstractControl): ValidationErrors | null;
}
```

这样就清楚了，ValidatorFn 是一个对象定义，这个对象中有一个方法，此方法接受一个 AbstractControl 类型的参数（其实也就是我们的 FormControl，而 AbstractControl 为其父类），而这个方法还要返回 ValidationErrors，ValidationErrors 的定义如下：

```
export declare type ValidationErrors = {
    [key: string]: any;
};
```

回过头来再看这句 (group: FormGroup): {[key: string]: any} => {...}，大家就应该明白为什么这么写了，这其实就是在返回一个 ValidatorFn 类型的对象。只不过我们利用 javascript/typescript 对象展开的特性把 ValidationErrors 写成了 {[key: string]: any}。

弄清楚这个函数的逻辑后，我们怎么使用呢？非常简单，先看代码：

```
    this.user = this.fb.group({
      email: ['', [Validators.required, Validators.email]],
```

```
      password: ['', Validators.required],
      repeat: ['', Validators.required],
      address: this.fb.group({
        province: [],
        city: [],
        area: [],
        addr: []
      })
    }, {validator: this.validateEqual('password', 'repeat')});
```

上述代码和最初的代码相比，多了一个参数，那就是 {validator: this.validateEqual('password', 'repeat')}。FormBuilder 的 group 函数接受两个参数，第一个就是那串长长的，我们叫它 controlsConfig，用于表单控件的构造，以及每个表单控件的验证器。但是如果一个验证器是要计算多个 field，那么我们可以把它作为整个 group 的验证器。所以 FormBuilder 的 group 函数还接受第二个参数，这个参数中可以提供同步验证器或异步验证器。同样还是一个字典对象，若是同步验证器，key 则写成 validator，若是异步验证，则写成 asyncValidator。

现在我们可以保存代码，启动 ng serve，到浏览器中看一下结果，如图 2-13 所示。

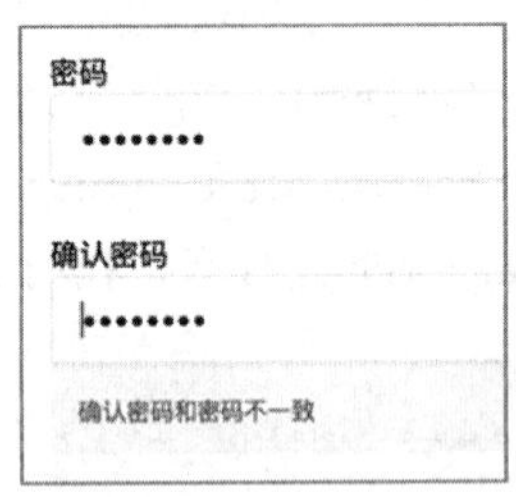

图 2-13

4. FormArray 有什么用

我们在购物网站中经常遇到需要维护多个地址的情况，因为有些商品需要送到公司，有些需要送到家里，还有些给父母采购的需要送到父母那里。这就是一个典型的 FormArray 可以派上用场的场景。所有的这些地址的结构都是一样的，有省、市、区县和街道地址，那么我们来看看在响应式表单中怎么做。

首先，需要把 HTML 模板改造一下，现在的地址是多项了，所以需要在原来的地址部分的外面再套一层，并且声明成 formArrayName="addrs"。FormArray 顾名思义是一个数组，所以我们要对这个控件数组做一个循环，然后让每个数组元素是 FormGroup，只不过 [formGroupName]="i" 是让 formGroupName 等于该数组元素的索引。

```
  <div formArrayName="addrs">
      <button (click)="addAddr()">Add</button>
      <div *ngFor="let item of user.controls['addrs'].controls; let i =
index;">
        <div [formGroupName]="i">
          <label>
            <span>省份</span>
            <select formControlName="province">
              <option value="">请选择省份</option>
              <option [value]="province" *ngFor="let province of
provinces">{{province}}</option>
            </select>
          </label>
          <label>
            <span>城市</span>
            <select formControlName="city">
              <option value="">请选择城市</option>
              <option [value]="city" *ngFor="let city of (cities$ |
async)">{{city}}</option>
            </select>
          </label>
          <label>
            <span>区县</span>
            <select formControlName="area">
              <option value="">请选择区县</option>
              <option [value]="area" *ngFor="let area of (areas$ |
async)">{{area}}</option>
            </select>
          </label>
          <label>
            <span>地址</span>
            <input type="text" formControlName="street">
          </label>
        </div>
      </div>
    </div>
```

改造好模板后，我们需要在类文件中也进行对应处理，去掉原来的 address: this.fb.group({...})，换成 addrs: this.fb.array([])。

```
  this.user = this.fb.group({
    email: ['', [Validators.required, Validators.email]],
    password: ['', Validators.required],
    repeat: ['', Validators.required],
    addrs: this.fb.array([])
```

```
}, {validator: this.validateEqual('password', 'repeat')});
```

但这样我们看不到，也增加不了新的地址，因为我们还没有处理添加的逻辑。下面我们就添加一下：其实就是建立一个新的 FormGroup，然后将其加入 FormArray 数组中。

```
addAddr(): void {
  (<FormArray>this.user.controls['addrs']).push(this.createAddrItem());
}

private createAddrItem(): FormGroup {
  return this.fb.group({
    province: [],
    city: [],
    area: [],
    street: []
  })
}
```

到这里我们的结构就建好了，保存后，到浏览器中去试试添加多个地址吧，如图 2-14 所示。

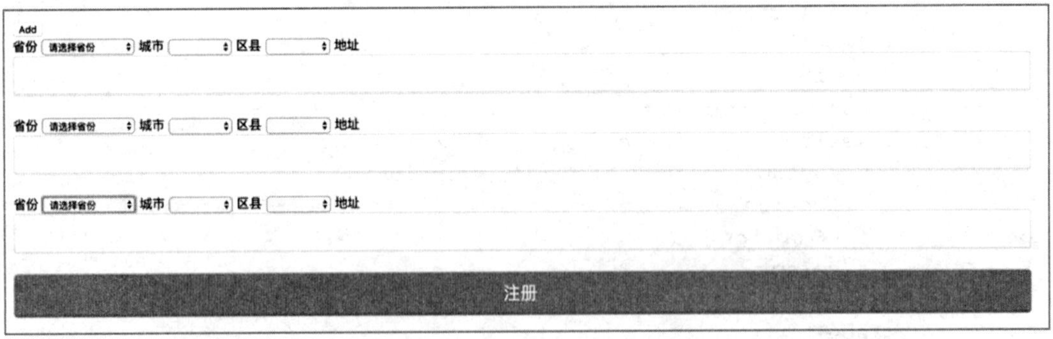

Add
省份 请选择省份 城市 区县 地址

省份 请选择省份 城市 区县 地址

省份 请选择省份 城市 区县 地址

注册

图 2-14

5. 响应式表单的优势

响应式表单的优势是可测试能力。模板驱动型表单进行单元测试是比较困难的，因为验证逻辑是写在模板中的。但验证器逻辑的单元测试对于响应式表单来说非常简单，因为验证器只是一个函数。

当然，除了这个优点，我们对表单可以有完全的掌控：从初始化表单控件的值、更新和获取表单值的变化到表单的验证和提交，这一系列的流程都在程序逻辑的控制之下。

更重要的是，我们可以使用函数响应式编程的风格来处理各种表单操作，因为响应式表单提供了一系列支持 Observable 的接口 API。那么这又能说明什么呢？有什么用呢？

无论是表单本身还是控件都可以看作是一系列的基于时间维度的数据流，这个数据流可以被多个观察者订阅和处理，由于 valueChanges 本身是个 Observable，所以我们就可以利用 RxJS 提供的丰富操作符，将一个对数据验证、处理等的完整逻辑清晰地表达出来。当然现在我们不会对 RxJS 进行深入的讨论，后面有专门针对 RxJS 进行讲解的章节。

```
this.form.valueChanges
        .filter((value) => this.user.valid)
        .subscribe((value) => {
           console.log("现在时刻表单的值为 ",JSON.stringify(value));
        });
```

在上面的例子中，我们首先取得表单值的变化，然后过滤掉表单存在非法值的情况，再输出表单的值。这只是一个非常简单的 Rx 应用，随着逻辑复杂度的增加，后面会见证 Rx 卓越的处理能力。

2.2 Angular Material 介绍

Angular Material 是 Angular 团队官方开发的一套符合 Google Material 风格的 Angular UI 组件库。这套组件的使用方式常常让笔者联想起 Android 的开发，笔者感觉这应该也是 Google 努力的方向之一——让 Web 开发更像 App 开发或者后端开发。

安装 @angular/material 可以在 terminal 中输入：

```
yarn add @angular/material @angular/cdk
```

如果看到类似下面的输出，那就安装成功了。

```
•100% ➜ yarn add @angular/material @angular/cdk
yarn add v1.3.2
[1/4] ?  Resolving packages...
[2/4] ?  Fetching packages...
[3/4] ?  Linking dependencies...
[4/4] ?  Building fresh packages...
success Saved lockfile.
success Saved 2 new dependencies.
├─ @angular/cdk@5.0.2
└─ @angular/material@5.0.2
?  Done in 29.78s.
```

2.2.1 组件类别

@angular/material 在 2.0.0-beta.8 之前是单独的一个包，但后来团队把其中的一些公用功能及组件抽离出来放到了一个单独的 @angular/cdk 包中。这个 cdk（component dev kit，组件开发工具包）以后可以作为你开发自己风格组件库的基础，因为它封装了很多公共特性的支持，你不需要从零开始。cdk 中提供的主要功能如图 2-15 所示。

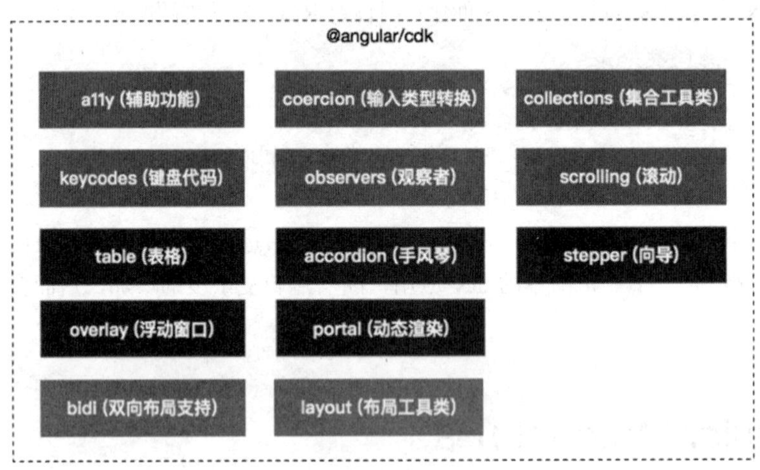

图 2-15

总体来说，@angular/material 提供了 30 多个组件，以及主题、字体的支持，并通过 @angular/flex-layout 提供了 flexbox 布局的 Angular 封装。

（1）组件。

表单控件类：Autocomplete、Checkbox、Datepicker、Form Field、Input、Radio Button、Select、Slider、Slide Toggle

导航类：Menu、Sidenav、Toolbar

布局类：List、Grid List、Card、Stepper、Tabs、Expansion Panel

按钮和提示类：Button、Button Toggle、Icon、Progress Spinner、Progress Bar

弹出类：Dialog、Tooltip、Snackbar、

数据表格类：Table、Sort Header、Paginator

（2）布局支持：通过提供独立的 @angular/flex-layout，这个软件包不仅提供 flex 的封装，也提供响应式页面设计需要的各种 API 和指令。

（3）主题支持：主题的支持主要由框架提供的一系列 scss 函数来实现，因此如果希望有主题的自定义时，则需要以 scss 形式提供样式。

2.2.2 布局控件：Sidenav

在 Google 的 Material 设计语言中，对于一个应用的布局经常采用的形式是一个内容区块加上一个侧面菜单，这个菜单既可以是滑出的也可以是固定在侧面的，如图 2-16 所示。

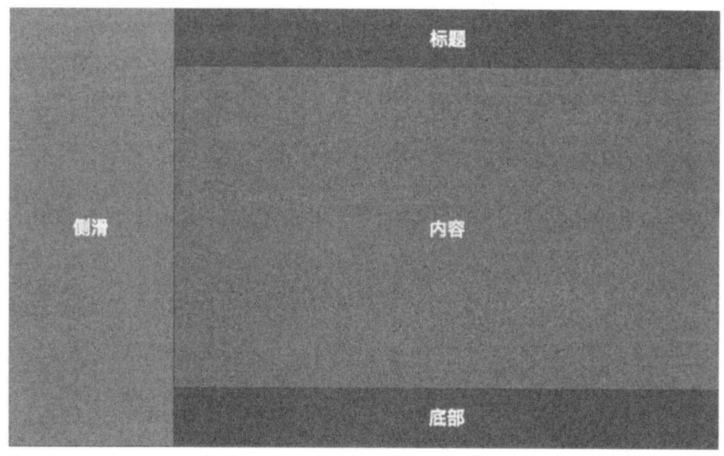

图 2-16

所以，对于这种常见的布局，Google 提供了一种布局控件——Sidenav——来帮助开发者很方便地实现。

```
<mat-sidenav-container fullscreen>
  <mat-sidenav #sidenav mode="over">
    <app-sidebar></app-sidebar>
  </mat-sidenav>
  <div class="site">
    <header>
      ...
    </header>
    <main>
      <router-outlet></router-outlet>
    </main>
    <footer>
      ...
    </footer>
  </div>
</mat-sidenav-container>
```

如上述代码所示，这种布局一般需要两个组件相互配合。

第一个是 <mat-sidenav-container>，作为侧滑控件的容器，一般也可以用作整个 App 的容器。再在这个容器中使用 <mat-sidenav> 构建可侧滑的内容。

第二个 <mat-sidenav-content>，也就是我们可以把 <sidenav> 对应的内容放入<mat-sidenav-content> 中，但如果不写，Angular 也会默认创建一个元素把其他部分封装在里面，所以不写这个也是可以的。

这个组件需要封装在一个叫作 MatSidenavModule 的模块中，所以如果要使用，则需要导入这个模块，我们现在把它放到共享模块当中：

```
import { NgModule } from '@angular/core';
import { CommonModule } from '@angular/common';
import { MatSidenavModule } from '@angular/material';

const MATERIAL_MODULES = [
  MatSidenavModule,
];

const MODULES = [
  ...MATERIAL_MODULES,
  CommonModule,
];

@NgModule({
  declarations: [],
  imports: MODULES,
  exports: [
    ...MODULES,
  ]
})
export class SharedModule {}
```

这个模块除了提供 Sidenav ，还提供了一个类似功能的控件 Drawer，那么，Drawer 和 Sidenav 的区别是什么？答案是区域的大小。如果是整个页面级（也就是全屏）的侧滑，则使用 Sidenav；对于页面上的某个小区域，如果也要实现类似的效果，则使用 Drawer。

```
<mat-drawer-container class="container">
  <mat-drawer mode="side" opened="true">抽屉内容</mat-drawer>
  <mat-drawer-content>主要内容</mat-drawer-content>
</mat-drawer-container>
```

如果你看得足够仔细，则会发现这个控件 Sidenav 有一系列属性。下面我们介绍其中最常用的几个属性，第一个是 mode，如表 2-1 所示。

表 2-1

mode	说　明
over	这个值是默认值，效果是侧边浮在主要内容之上
push	侧边会向右或向左挤走主要内容的部分区域
side	侧边回合主要内容并列

如果是抽屉，那么自然会有打开和关闭的状态，针对这两种状态，Sidenav 提供了若干方法、属性和事件，如表 2-2 所示。

表 2-2

名　称	类　型	描　述	示　例
open	方法	打开侧边	sidenav.open()
close	方法	关闭侧边	sidenav.close()
toggle	方法	反转当前状态	sidenav.toggle()
opened	属性	打开的状态	[opened]="status"
closed	属性	关闭的状态	[closed]="status"
closedStart	事件	开始关闭的事件	(closedStart)="handleClose()"
openedStart	事件	开始打开的事件	(openedStart)="handleOpen()"

下面以 toggle 为例讲解，我们给 Sidenav 起一个引用名字 sidenav，然后在 button 的点击事件处理中调用 toggle()。

```
<mat-sidenav-container fullscreen>
  <mat-sidenav #sidenav mode="over">
    <app-sidebar></app-sidebar>
  </mat-sidenav>
  <div class="site">
    <header>
      <button (click)="sidenav.toggle()"> 切换开关状态 </button>
    </header>
    <main>
      <router-outlet></router-outlet>
    </main>
    <footer>
      ...
    </footer>
  </div>
</mat-sidenav-container>
```

2.2.3 Flex 布局和 Angular Flex-layout

在 Flex 布局之前，若想使用 HTML/CSS/JavaScript 去进行一些特殊布局，一般都需要进行一些修改。可以说 CSS 引入 Flex 布局给前端开发注入了一股“清流”。当然之后还有 grid 布局等现代布局方式，但由于支持 grid 布局的浏览器还不是很普遍，而各主流浏览器对 Flex 的支持则已经比较成熟了。我们这里同样不会对 Flex 做详细介绍。

@angular/flex-layout 是 Angular 团队给出的一个基于 Flex 布局的 Angular 类库，那么为什么不直接使用 Flex 布局而要使用这个封装类库呢？其实答案并不是非此即彼的，很多时候我们可以混用，但 @angular/flex-layout 提供了方便的指令用来自动化 Flex 布局和媒体查询，而且其最大的优势是提供了一个强大的 Responsive API，让开发者可以方便地开发适合多种屏幕布局的响应式应用。

安装 @angular/flex-layout，方法如下：

```
yarn add @angular/flex-layout
```

1．常见的指令和用法

- fxLayout：标识一个元素为 Flex 容器，值分别为 row 和 column，指明容器方向。
- fxLayoutAlign：指定子元素按容器方向和交叉轴（cross axis）的排布方式，相当于 CSS 中的 justify-content 和 align-content
- fxFlex：相当于 CSS 中的 Flex 布局，可以接受三个值——flex-grow、flex-shrink、flex-basis

这些指令可以和 @angular/flex-layout 的媒体查询断点结合使用，如表 2-3 所示。

表 2-3

断　点	媒体查询
xs	'screen and (max-width: 599px)'
sm	'screen and (min-width: 600px) and (max-width: 959px)'
md	'screen and (min-width: 960px) and (max-width: 1279px)'
lg	'screen and (min-width: 1280px) and (max-width: 1919px)'
xl	'screen and (min-width: 1920px) and (max-width: 5000px)'
lt-sm	'screen and (max-width: 599px)'
lt-md	'screen and (max-width: 959px)'
lt-lg	'screen and (max-width: 1279px)'
lt-xl	'screen and (max-width: 1919px)'

续表

断　点	媒体查询
gt-xs	'screen and (min-width: 600px)'
gt-sm	'screen and (min-width: 960px)'
gt-md	'screen and (min-width: 1280px)'
gt-lg	'screen and (min-width: 1920px)'

也就是，我们可以这样来写一个响应式布局：

```
<div fxFlex="50%" fxFlex.gt-sm="100%">
...
</div>
```

2．完成首页布局

我们要对首页进行的布局初看上去比较简单，但当内容较少时，footer 一般会上移，这就比较难看了，我们希望的是，无论内容多少，footer 始终在页尾，下面我们看看怎样使用 Flex 布局达成这个效果：

```
// styles.scss

html, body, app-root {
  margin: 0;
  width: 100%;
  height: 100%;
}

.site {
  width: 100%;
  min-height: 100%;
}

.full-width {
  // 这个类和布局无关，后面会使用它来使某些组件撑满空间
  width: 100%;
}

.fill-remaining-space {
  // 使用 flexbox 填充剩余空间
  // @angular/material 中的很多控件使用了 Flex 布局
  flex: 1 1 auto;
}
```

首先我们在 styles.scss 中将几个顶级元素（html、body、app-root）的边距设为 0，并且让

它们充满整个空间。fullscreen 指令起的作用和前面的 CSS 差不多，就是让 mat-sidenav-container 也变成边距为 0、长宽比例为 100%。然后我们需要把主要内容区域设置成一个垂直方向的 flexbox（<div class="site" fxLayout="column">），这样它的子元素（header、main、footer）会按照纵向排列。而我们对于 main 又设置了 fxFlex="1"，使得这个元素会尽可能占据剩余空间，这样它就会把 header 和 footer 分别挤到页首和页尾，也就达成了我们的目的。

```
<mat-sidenav-container fullscreen>
  <mat-sidenav #sidenav mode="over">
    ...
  </mat-sidenav>
  <div class="site" fxLayout="column">
    <header>
      ...
    </header>
    <main fxFlex="1" fxLayout="column" fxLayoutAlign="center strech">
      <router-outlet></router-outlet>
    </main>
    <footer>
      ...
    </footer>
  </div>
</mat-sidenav-container>
```

2.2.4 封装 Header/Footer/Sidebar

接下来，我们会创建 3 个组件，分别渲染页面头部、尾部和侧边栏。

1. Header（页首）

这个页首按照如下需求实现。

- 纯色背景。
- 左侧有一个控制侧边栏显示的图标按钮，点击它可以切换侧边栏的显示与隐藏。
- 右侧有一个普通主题和黑夜模式的切换控件。
- 最右侧有一个退出按钮。
- 这些按钮应该可以控制显示和隐藏，因为后面我们会应对登录前后的状态。比如登录前侧边栏菜单是否显示、退出按钮不应显示等。

我们采用 Angular Material 的 toolbar，工具栏一般适合有一排按钮或者多排按钮，也可以用作页首或页尾。

如果要使用 Angular Material toolbar，则需要导入 MatToolbarModule。

```
import { MatToolbarModule } from '@angular/material/toolbar';
```

这个模块提供了两个组件：MatToolbar 和 MatToolbarRow。前者是定义一个 toolbar，后者是如果希望 toolbar 是多行，则采用类似下面的写法。此外，MatToolbar 提供了一个 color 属性，用来设置背景色，由于遵循 Material 标准，所以可选的值有 primary、accent 和 warn。

```
<mat-toolbar color="primary">
  <mat-toolbar-row>
    <span>第一行</span>
  </mat-toolbar-row>

  <mat-toolbar-row>
    <span>第二行</span>
  </mat-toolbar-row>
</mat-toolbar>
```

我们的项目中为了集中管理 Material 组件，可以创建一个新的模块 MyMaterialModule，首先，把我们需要的 Material 组件所需要的模块都统一做一次导出。

```
// 省略导入

/**
 * NgModule that includes all Material modules.
 */
@NgModule({
  exports: [
    MatButtonModule,
    MatIconModule,
    MatListModule,
    MatMenuModule,
    MatSidenavModule,
    MatSlideToggleModule,
    FlexLayoutModule,
  ]
})
export class MyMaterialModule {}
```

然后在 CoreModule 中导入这个模块，这样在全局我们只导入了一次 Material。

```
imports: [
  CommonModule,
  MyMaterialModule,
  HttpClientModule, //如果有 mat-icon 或 mat-icon-button 是必须要导入的
  BrowserAnimationsModule // material 的动画效果必须导入这个模块
]
```

接下来，我们要实现 HeaderComponent，这里我们把它设计成一个“笨组件”。所谓“笨组件”就是指这个组件并不知道业务逻辑，它的全部行为都由若干属性定义，也就是说它就像一个木偶组件，外部使用时设置它做什么，它就做什么。这么笨的组件能有什么用呢？其实我们使用的大部分组件都是“笨组件”，比如 HTML 的 <button>、<link>、<div> 等。由于“笨组件”不了解业务逻辑反倒使它具有更灵活的发挥空间，复用性也更好。一般来说，一个系统中“笨组件”更多，而“聪明组件”较少为好，这样需求发生变更时，系统修改起来会比较快。

对 HeaderComponent，我们定义两个输入型属性和 3 个输出型属性（在外部看来是事件）。

- title：站点名称，在页首以较大字号显示。
- hideForGuest：是否对未登录客户隐藏。
- toggleMenuEvent：点击左侧按钮。
- toggleDarkModeEvent：切换黑夜模式。
- logoutEvent：点击“退出”按钮。

```
// 省略导入

@Component({
  selector: 'app-header',
  template: `
  <mat-toolbar color="primary">
    <button mat-icon-button (click)="toggleMenu()" *ngIf="!hideForGuest">
      <mat-icon>menu</mat-icon>
    </button>
    <span>{{ title }}</span>
    <span class="fill-remaining-space"></span>
    <mat-slide-toggle (change)="toggleDarkMode($event.checked)"
*ngIf="!hideForGuest">黑夜模式</mat-slide-toggle>
    <button *ngIf="!hideForGuest" mat-button (click)="handleLogout()">退
出</button>
  </mat-toolbar>
  `,
  changeDetection: ChangeDetectionStrategy.OnPush
})
export class HeaderComponent {
  @Input() title = '企业协作平台';
  @Input() hideForGuest = false;
  @Output() toggleMenuEvent = new EventEmitter<void>();
  @Output() toggleDarkModeEvent = new EventEmitter<boolean>();
  @Output() logoutEvent = new EventEmitter();

  toggleMenu() {
```

```
    this.toggleMenuEvent.emit();
  }

  handleLogout() {
    this.logoutEvent.emit();
  }

  toggleDarkMode(checked: boolean) {
    this.toggleDarkModeEvent.emit(checked);
  }
}
```

这个组件中除用到 mat-toolbar，还用到了 mat-icon-button、mat-buton、mat-icon 和 mat-slide-toggle。

（1）按钮

按钮这个组件是最常用的 Material 组件之一，它有很多种类型，如表 2-4 所示。

表 2-4

指　令	描　述
mat-button	标准的长方形按钮，没有凸起的视觉效果
mat-raised-button	标准的长方形按钮，有凸起的视觉效果
mat-icon-button	图标按钮，透明背景的圆形按钮，一般会包含一个 mat-icon
mat-fab	浮动按钮，它是有凸起效果的圆形按钮，默认背景色为 accent
mat-mini-fab	和 mat-fab 类似，但是更小

需要注意的是，除了 <button> 可以应用这个指令，<a> 也可以。按钮提供了如表 2-5 所示的属性和方法。

表 2-5

属性/方法	描　述
@Input() color: ThemePalette	颜色属性
@Input() disableRipple: boolean	禁用水波纹效果，默认是点击时有水波纹效果
@Input() disabled: boolean	是否禁用该按钮
ripple: MatRipple	设置按钮的水波纹实例
focus()	聚焦

（2）图标

mat-icon 是一个图标组件，支持 Google 为 Material Design 设计的图标字体 Material Icon Font。使用 mat-icon 需导入 MatIconModule。

```
import { MatIconModule } from '@angular/material/icon';
```

这个字体中含有哪些图标读者可以访问 https://material.io/icons 查看。在应用中使用图标字体有几个好处。

- 字体体积较小，比图片的体积小很多。
- 字体可以无损缩放，所以可以适配多种分辨率。

当然，除了支持自己的图标字体，它也支持矢量 SVG 格式的图标和其他字体图标，如 FontAwesome 等。

我们首先来看第一种方式与 Material Icon Font 的配合使用，我们将字体图标的名称嵌入一对闭合的 mat-icon 标签中，就可以显示图标了。

```
<mat-icon>home</mat-icon>
```

由于 FontAwesome 字体是采用 CSS 中的:before 选择器来显示图标的。如果要和类似 FontAwesome 的字体配合，那么我们需要设置 mat-icon 中的 fontSet 属性为其 CSS 类名，然后在 fontIcon 属性设置其图标名称。比如，在 FontAwesome 中的 HTML 代码如果是下面的样子：

```
<i class="fas fa-birthday-cake"></i>
```

那么，我们使用 mat-icon 可以写成：

```
<mat-icon fontSet="fa" fontIcon="fa-birthday-cake"></mat-icon>
```

当然也可以通过 MatIconRegistry 设置一个别名：

```
constructor(iconRegistry: MatIconRegistry, sanitizer: DomSanitizer) {
    iconRegistry
        .registerFontClassAlias('fontawesome', 'fa');
  }
```

然后写成下面的样子：

```
<mat-icon fontSet="fontawesome" fontIcon="fa-birthday-cake"></mat-icon>
```

MatIconRegistry 是由 MatIconModule 提供的一个服务，它用于注册和显示 mat-icon 使用的图标。几个常用的方法如表 2-6 所示。

表 2-6

方　法	描　述
addSvgIcon(iconName: string, url: SafeResourceUrl)	通过 URL 注册一个图标
addSvgIconInNamespace(namespace: string, iconName: string, url: SafeResourceUrl)	使用指定的命名空间通过 URL 注册一个图标
addSvgIconSet(url: SafeResourceUrl)	使用默认命名控件通过 URL 注册一组图标
addSvgIconSetInNamespace(url: SafeResourceUrl)	使用指定的命名空间通过 URL 注册一组图标
registerFontClassAlias(alias: string, className: string)	为图标字体使用的 CSS 类名称定义一个别名

如果要使用 SVG 图标，那么我们需要注入 MatIconRegistry 和 DomSanitizer，然后将 SVG 图标注册到 MatIconRegistry。

```
constructor(iconRegistry: MatIconRegistry, sanitizer: DomSanitizer) {
    iconRegistry.addSvgIcon(
        'thumbs-up',
        sanitizer.bypassSecurityTrustResourceUrl('assets/img/examples/
thumbup-icon.svg'));
  }
```

在模板中就可以通过设置 mat-icon 的 SVG 图标为刚才注册的名称，这样 SVG 图标就会显示出来了。

```
<mat-icon svgIcon="project"></mat-icon>
```

当然，在很多时候，我们不希望在某个组件加载时才逐一地加载图标，因为如果那样，每个使用图标的组件可能就都得注入重复的 MatIconRegistry 和 DomSanitizer 和重复的初始化代码。不仅代码冗余，而且性能很差。所以我们把 SVG 图标的载入等操作放入 CoreModule 的构造函数中，这样就在应用启动后只初始化一次，而且以后添加或修改资源时，位置也很集中。其他组件都是单纯的消费者，无须关心这些 SVG 图标或者字体图标的注册。

```
import { MatIconRegistry } from '@angular/material';
import { DomSanitizer } from '@angular/platform-browser';

/**
 * 加载图标，包括 SVG 图标和 FontAwesome 字体图标等
 *
 * @param ir a MatIconRegistry 实例，用于注册图标资源
 * @param ds a DomSanitizer 实例，用于忽略安全检查返回一个 URL
```

```
 */
export const loadIconResources = (ir: MatIconRegistry, ds: DomSanitizer)
=> {
  const imgDir = 'assets/img';
  const avatarDir = `${imgDir}/avatar`;
  const sidebarDir = `${imgDir}/sidebar`;
  const iconDir = `${imgDir}/icons`;
  const dayDir = `${imgDir}/days`;
  ir
    .addSvgIconSetInNamespace(
      'avatars',
      ds.bypassSecurityTrustResourceUrl(`${avatarDir}/avatars.svg`)
    )
    .addSvgIcon(
      'unassigned',
      ds.bypassSecurityTrustResourceUrl(`${avatarDir}/unassigned.svg`)
    )
    .addSvgIcon(
      'project',
      ds.bypassSecurityTrustResourceUrl(`${sidebarDir}/project.svg`)
    )
    .addSvgIcon(
      'projects',
      ds.bypassSecurityTrustResourceUrl(`${sidebarDir}/projects.svg`)
    )
    .addSvgIcon(
      'month',
      ds.bypassSecurityTrustResourceUrl(`${sidebarDir}/month.svg`)
    )
    .addSvgIcon(
      'week',
      ds.bypassSecurityTrustResourceUrl(`${sidebarDir}/week.svg`)
    )
    .addSvgIcon(
      'day',
      ds.bypassSecurityTrustResourceUrl(`${sidebarDir}/day.svg`)
    )
    .addSvgIcon(
      'move',
      ds.bypassSecurityTrustResourceUrl(`${iconDir}/move.svg`)
    )
    .registerFontClassAlias('fontawesome', 'fa');
};
```

然后，在 CoreModule 的构造函数中调用工具函数：

```
// 略去注解和导入部分
export class CoreModule {
  constructor(
    @Optional()
    @SkipSelf()
    parentModule: CoreModule,
    ir: MatIconRegistry,
    ds: DomSanitizer
  ) {
    if (parentModule) {
      throw new Error('CoreModule 已经装载，请仅在 AppModule 中引入该模块。');
    }
    // 加载刚刚创建的工具函数
    loadIconResources(ir, ds);
  }
}
```

为了支持这些图标字体，还需要引入一些 CSS，请将 src/index.html 更改成下面的样子：

```
<!doctype html>
<html lang="en">

<head>
  <meta charset="utf-8">
  <title>Gtm</title>
  <base href="/">

  <meta name="viewport" content="width=device-width, initial-scale=1">
  <link rel="icon" type="image/x-icon" href="favicon.ico">
  <link href="https://fonts.googleapis.com/icon?family=Material+Icons"
rel="stylesheet">
  <link href="https://fonts.googleapis.com/css?family=Roboto:300,400,
500" rel="stylesheet">
  <link href="http://lib.baomitu.com/font-awesome/5.0.8/web-fonts-with-
css/css/fontawesome-all.min.css" rel="stylesheet">
</head>

<body>
  <app-root></app-root>
</body>

</html>
```

（3）滑动开关

滑动开关需要导入 MatSlideToggleModule，提供了一种可以通过点击或拖曳切换开/关状态的交互方式。

```
<mat-slide-toggle (change)="toggleDarkMode($event.checked)"
*ngIf="!hideForGuest">黑夜模式</mat-slide-toggle>
```

在 HeaderComponent 中，我们可以看到这个控件提供 change 事件，这个事件（输出型属性）的定义是 @Output() change: EventEmitter<MatSlideToggleChange>，我们可以看到它的参数是 MatSlideToggleChange，有以下两个属性。

- checked：布尔型，代表最新的 MatSlideToggle 状态。
- source：产生事件的源控件。

所以，我们在事件处理中可以使用 $event.checked 将最新的状态传入处理函数。

2. Footer（页尾）

页尾在这里就非常简单了，只是居中显示版权信息。

```
import { Component, ChangeDetectionStrategy } from '@angular/core';

@Component({
  selector: 'app-footer',
  template: `
  <mat-toolbar color="primary">
    <span class="fill-remaining-space"></span>
    <span>&copy; 2018 版权所有：接灰的电子产品</span>
    <span class="fill-remaining-space"></span>
  </mat-toolbar>
  `,
  changeDetection: ChangeDetectionStrategy.OnPush
})
export class FooterComponent {}
```

3. Sidebar（侧边栏）

我们在 src/app/core/components 中建立 sidebar.component.ts。在侧边栏中，采用了另一个控件 mat-nav-list，是列表控件的一种，这种控件一般用来展现一系列结构类似的数据。Angular Material 提供了这种控件的多种变化类型：

- 简单列表——<mat-list> 用于展现列表数据。
- 导航列表——<mat-nav-list> 用于点击后导航到不同的视图。
- 选择列表——<mat-selection-list> 用于选择多项数据。

此外，还提供了一系列指令，可以配合列表类型实现更多效果：

- matLine——用于实现多行的列表条目。
- matListIcon——用于给条目添加图标。
- matListAvatar——给列表条目增加头像支持。
- dense——形成紧凑排布的列表。

下面的两个指令不是专门用于 List 的：

- mat-divider——一条横线，用于分隔内容。我们在这个例子里面用于菜单分组。默认粗度 1dp，依据主题的明暗，透明度分别是 12% 黑色或 12% 白色。
- matSubheader——用于列表分组标题的显示，一般用于 List、Grid 和 Menu 中，默认 48dp。

```
// 省略导入

@Component({
  selector: 'app-sidebar',
  template: `
  <mat-nav-list>
    <ng-container *ngFor="let group of menuGroups">
      <h3 matSubheader> {{ group.name }} </h3>
      <ng-container *ngFor="let item of group.items">
        <mat-list-item
          [routerLink]="item.routerLink"
          (click)="menuClick(item.emitData)" [ngSwitch]="item.iconType">
          <mat-icon matListIcon *ngSwitchCase="iconType.SVG"
[svgIcon]="item.iconName"></mat-icon>
          <mat-icon matListIcon *ngSwitchCase="iconType.FONT_AWESOME"
[fontSet]="item?.fontSet" [fontIcon]="item.iconName"></mat-icon>
          <mat-icon matListIcon
*ngSwitchDefault>{{ item.iconName }}</mat-icon>
          <h4 matLine> {{ item.title }} </h4>
          <p matLine> {{ item.subtitle }} </p>
        </mat-list-item>
      </ng-container>
      <mat-divider></mat-divider>
    </ng-container>
  </mat-nav-list>
  `,
  styles: [
    `
  .day-num {
    font-size: 48px;
    width: 48px;
```

```
    height: 48px;
  }
  `
  ],
  changeDetection: ChangeDetectionStrategy.OnPush
})
export class SidebarComponent {
  @Input() menuGroups: MenuGroup[] = [];
  @Output() menuClickEvent = new EventEmitter<any>();

  menuClick(data: any) {
    if (data) {
      this.menuClickEvent.emit(data);
    } else {
      this.menuClickEvent.emit();
    }
  }
  // 前面的 get 关键字表示这个方法是一个属性方法，可以当成属性使用
  get iconType() {
    return IconType;
  }
}
```

这个侧边栏组件的内容不是很多，但里面有一些值得注意的知识点。

- get 关键字。TypeScript 提供 get 和 set 关键字用于标识属性方法，一般如果是方法，则在模板中调用，需要写上括号和参数等，比如上面的 iconType ，如果按照方法写，则应该写成 iconType()。但属性方法可以直接当成属性使用，就可以直接写成 iconType。模板中无法直接访问 IconType 这个类型，这也是我们为什么用一个属性方法去得到它。
- ngSwitch 的使用。枚举中有三个值，这时候使用 ngSwitch 这种分支条件指令就很方便。
- 这又是一个“笨组件”，我们可以看到这个组件不关心具体的业务，只是渲染菜单，这种形式的可扩展性和可维护性都非常好。我们只需要在使用 <app-sidebar> 的地方去构造一个 MenuGroup 数组，并赋值给组件的 menuGroups 属性即可。

由于菜单的数据类型比较相似，所以构造了 MenuGroup 和 MenuItem 来组织菜单的数据结构。这个文件位于 src/app/domain 中，我们创建一个 menu.ts 用来定义菜单相关的数据结构。一个菜单条目包含菜单的标题、副标题、图标名称和要导航的链接。而一个菜单分组会包含多个菜单项。每个菜单项可以是 Material、FontAwesome 或者是 SVG 类型的图标，所以我们定义了一个枚举 IconType 用来标识类别。而特殊的对于 FontAwesome 的图标，有一个可选属性 fontSet，定义中的 ? 表示这个属性设置成可选。

```
export enum IconType {
  MATERIAL_ICON,
  FONT_AWESOME,
  SVG
}

export interface MenuItem {
  routerLink: string[];
  iconName: string;
  iconType: IconType;
  fontSet?: string;
  title: string;
  subtitle: string;
  emitData: any;
}

export interface MenuGroup {
  name: string;
  items: MenuItem[];
}
```

4. 实验效果

要查看效果，就需要在 AppComponent 中使用这些组件，并为其设置必要的属性或处理产生的事件。在模板中，使用 app-header、app-footer 以及 app-sidebar。

```
<mat-sidenav-container fullscreen>
  <mat-sidenav #sidenav mode="over">
    <app-sidebar [menuGroups]="sidebarMenu"></app-sidebar>
  </mat-sidenav>
  <div class="site" fxLayout="column">
    <app-header (toggleMenuEvent)="sidenav.toggle()"></app-header>
    <main fxFlex="1" fxLayout="column" fxLayoutAlign="start center"
fxLayoutGap="10px">
      <router-outlet></router-outlet>
    </main>
    <app-footer></app-footer>
  </div>
</mat-sidenav-container>
```

在 app-header 中需要处理图标按钮点击事件，这个事件在 app-header 中定义成了 toggleMenuEvent，这个事件发生时需要切换 sidenav 的显示和隐藏。我们是给 mat-sidenav 定义了一个引用名 #sidenav，这样后面就可以使用 sidenav 引用这个组件了。(toggleMenuEvent)="sidenav.toggle()" 这个语句的意思就是当 toggleMenuEvent 发生时，我们就调用 sidenav.toggle()

切换侧边栏的显示和隐藏。

而在 app-sidebar 中，设置了 [menuGroups]="sidebarMenu"，这个 sidebarMenu 是什么意思？别急，我们还没有定义这个成员变量，接下来看一下它的定义：

```
import { Component } from '@angular/core';
import { MenuGroup, IconType } from '../../../domain/menu';

@Component({
  selector: 'app-root',
  templateUrl: './app.component.html',
  styleUrls: ['./app.component.scss']
})
export class AppComponent {
  isCollapsed = true;
  sidebarMenu: MenuGroup[] = [];
  constructor() {
    this.sidebarMenu = [
      {
        name: '项目列表',
        items: [
          {
            title: '项目一',
            subtitle: '项目一的描述',
            iconName: 'project',
            iconType: IconType.SVG,
            emitData: 'abc123',
            routerLink: ['/projects/abc123']
          },
          {
            title: '项目二',
            subtitle: '项目二的描述',
            iconName: 'home',
            iconType: IconType.MATERIAL_ICON,
            emitData: 'abc234',
            routerLink: ['/projects/abc234']
          },
          {
            title: '项目三',
            subtitle: '项目三的描述',
            iconName: 'fa-bell',
            iconType: IconType.FONT_AWESOME,
            fontSet: 'fontawesome',
            emitData: 'abc345',
            routerLink: ['/projects/abc345']
```

```
            }
          ]
        }
      ];
    }
}
```

在查看效果之前，还需要做几件事，现在还没有做路由模块，也就是说我们并没有告诉 Angular 什么链接应该显示什么组件。所以还需要在 core 中建立一个根路由模块 app-routing.module.ts，这个根路由模块决定了顶层的路由设计，后面无论是采用预加载方式还是懒加载方式，一级的路由定义都会在这个模块中。

```
// 省略导入

const routes: Routes = [
  {
    path: '**', component: PageNotFoundComponent
  }
];

@NgModule({
  imports: [RouterModule.forRoot(routes)],
  exports: [RouterModule]
})
export class AppRoutingModule {
}
```

读者可能会注意到上面我们定义了一个路由，对所有链接都让一个 PageNotFoundComponent 进行处理。理论上用户通过浏览器访问的任何不存在的链接都会指向这个页面。但暂时我们先让所有链接都指向这个页面。在 src/app/core/components 中新建一个 page-not-found.component.ts。

```
import { Component, OnInit, ChangeDetectionStrategy } from
'@angular/core';

@Component({
  template: `
  <div fxLayout="row" fxLayoutAlign="center center">
    <mat-card>
      <mat-card-header>
        <mat-card-title> 又迷路了... </mat-card-title>
        <mat-card-subtitle> 404 - 没有找到页面 </mat-card-subtitle>
      </mat-card-header>
      <img mat-card-image [src]="notFoundImgSrc">
      <mat-card-actions>
```

```
        <button mat-button color="primary"> 回首页 </button>
      </mat-card-actions>
    </mat-card>
  </div>
  `,
  styles: [
    `
    :host {
      display: flex;
      flex: 1 1 auto;
    }
    mat-card {
      width: 70%;
    }
    `
  ],
  changeDetection: ChangeDetectionStrategy.OnPush
})
export class PageNotFoundComponent {
  notFoundImgSrc = 'assets/img/400_night_light.jpg';
}
```

这个页面非常简单，就是建立了一个带图片的卡片，在卡片的标题和副标题中写上让访问者理解目前处境的语言。在 Material Design 中，卡片是基本的显示内容的一种形式，如表 2-7 所示。

表 2-7

组　件	功能描述
<mat-card-title>	卡片标题
<mat-card-subtitle>	卡片副标题
<mat-card-content>	卡片内容，最好是文本内容
<img mat-card-image>	卡片图片，默认拉伸图片到容器宽度
<mat-card-actions>	卡片下方的按钮容器
<mat-card-footer>	卡片的尾部
<mat-card-header>	对于更复杂的卡片头部，可以采用这个组件定制化，里面可以含有 <mat-card-title>、<mat-card-subtitle> 和 <img mat-card-avatar>
<mat-card-title-group>	可以把标题（<mat-card-title>）、副标题（<mat-card-subtitle>）和图片（<img mat-card-sm-image> 或 <img mat-card-md-image> 或 <img mat-card-lg-image>）合并进一个区块

使用卡片，我们需要导入 MatCardModule，所以要更改 src/app/core/material.module.ts，加上这个模块。这样一个一个添加很麻烦，我们这里就偷点懒，索性一次性地把所有的 Material 组件都加上，但请在正式项目中只包含你实际用到的模块。

```
// 省略导入

/**
 * NgModule that includes all Material modules.
 */
@NgModule({
  exports: [
    MatAutocompleteModule,
    MatButtonModule,
    MatButtonToggleModule,
    MatCardModule,
    MatCheckboxModule,
    MatChipsModule,
    MatTableModule,
    MatDatepickerModule,
    MatDialogModule,
    MatExpansionModule,
    MatFormFieldModule,
    MatGridListModule,
    MatIconModule,
    MatInputModule,
    MatListModule,
    MatMenuModule,
    MatPaginatorModule,
    MatProgressBarModule,
    MatProgressSpinnerModule,
    MatRadioModule,
    MatRippleModule,
    MatSelectModule,
    MatSidenavModule,
    MatSlideToggleModule,
    MatSliderModule,
    MatSnackBarModule,
    MatSortModule,
    MatStepperModule,
    MatTabsModule,
    MatToolbarModule,
    MatTooltipModule,
    MatNativeDateModule,
    CdkTableModule,
```

```
    A11yModule,
    BidiModule,
    CdkAccordionModule,
    ObserversModule,
    OverlayModule,
    PlatformModule,
    PortalModule,
    MatTreeModule,
    MatBadgeModule,
    MatBottomSheetModule,
    MatMomentDateModule,
    FlexLayoutModule
  ]
})
export class MyMaterialModule {}
```

接下来就可以启动应用了。在应用根目录 ng serve、yarn start 或者 npm start 下，你应该可以看到如图 2-17 所示的页面了。

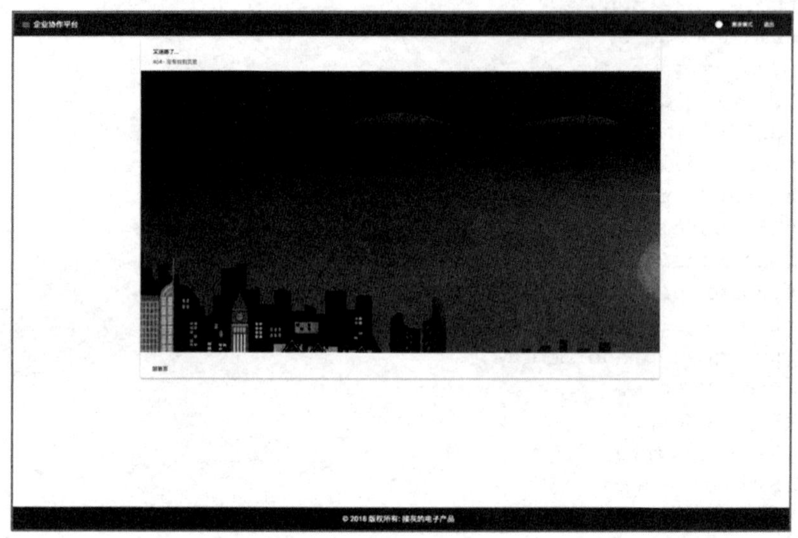

图 2-17

2.3 添加主题支持

Angular Material 是一个从诞生起就有主题支持的类库，既然叫 Material，它的主题也是要受 Material Design 设计风格约束的。

2.3.1 Material Design 中对于主题的约束

Material Design 中的颜色主要由调色板（palette）来提供，值得注意的是，Material Design 并不鼓励使用太多颜色，而提倡对颜色进行限制。调色板以一些基础色为基准，通过填充光谱来为 Android、Web 和 iOS 环境提供一套完整可用的颜色，基础色的饱和度是 500。

构成一个主题的颜色需要在众多基础色中选出同一颜色系的三个色度及一个强调色（可区别与前一种颜色），强调色用于某些元素的强调。

Google 的 Material Design 团队提供了一个在线调色工具，大家可以自行前往去体会一下，如图 2-18 所示。

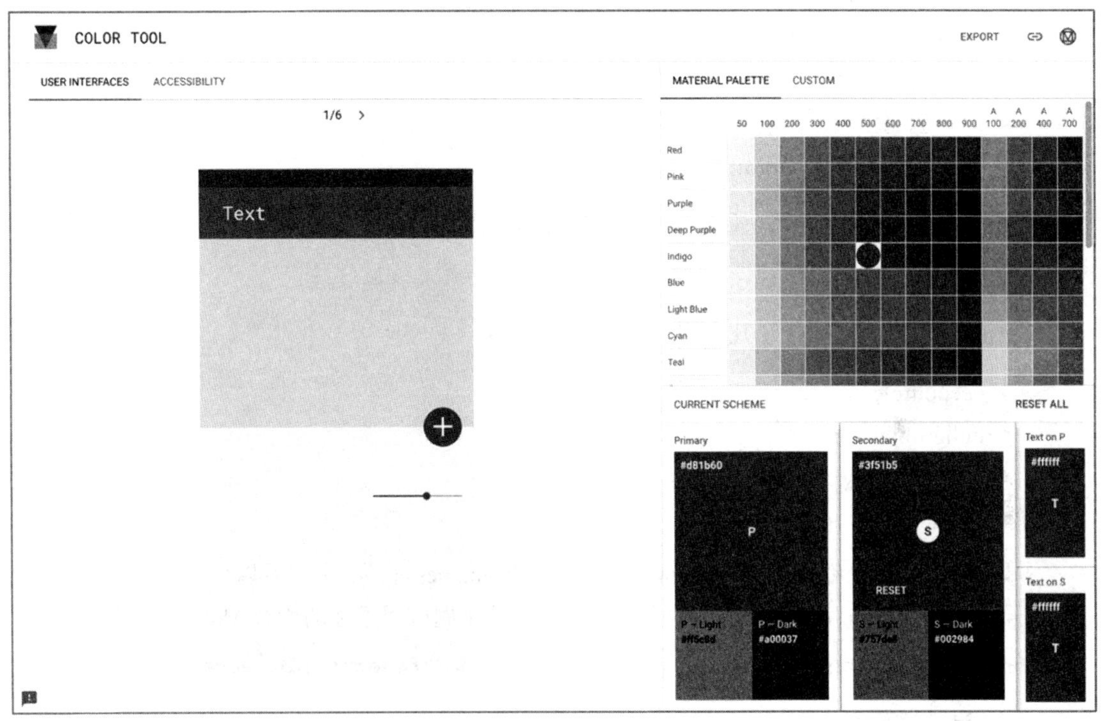

图 2-18

如果我们严格地执行这种约束，那么即使自定义了颜色，看上去也一样是 Material Design 的风格。

2.3.2 主题的明与暗

Material Design 认为主题是对应用提供一致性色调的方法。样式指定了表面的亮度、阴影的层次和字体元素的适当不透明度。为了提高应用间的一致性，提供两种明暗的选择：浅色和深色两种（即 light 和 dark）。

换句话说，即使选择了同样的基础色和强调色，你的主题也会有偏亮和偏暗两种选择。比如想实现白天模式和黑夜模式，那么这两种模式就有明暗的区别。

2.3.3 Angular Material 中的主题

知道了以上的基础概念后，就比较好理解 Angular Material 中的主题了。Angular Material 的主题由以下几部分构成。

- 基础色调色板：用于大部分的组件。
- 强调色调色板：用于浮动按钮和交互元素。
- 警告色调色板：用于显示错误状态，一般为红色。
- 前景色调色板：用于文字和图标的颜色。
- 背景色调色板：用于元素背景的颜色。

在 Angular Material 中，主题在编译时被静态地输出生成。

1. 内建主题

Angular Material 提供了几个官方配置好的主题：

- deeppurple-amber.css
- indigo-pink.css
- pink-bluegrey.css
- purple-green.css

以上内建主题都在 @angular/material/prebuilt-themes 目录下，所以一般来说，在你的 styles.css 或者 styles.scss 中导入对应主题，就可以将主题应用于你的所有 Material 组件了。

```
@import '~@angular/material/prebuilt-themes/deeppurple-amber.css';
```

2. 自定义主题

在很多情况下，内建主题是不能满足需求的，那么 Angular Material 也是允许自定义主题。

我们为主题风格单独建立一个文件 theme.scss，在 style.scss 中导入 theme.scss：

```
// styles.scss
@import 'theme.scss';
```

下面我们来看看如何写一个主题：

```
// theme.scss
@import '~@angular/material/theming';

// mat-core 提供了 Angular Material 的公共风格
// 这个 mat-core 只能在整个应用中包含一次
@include mat-core();

// 使用 mat-palette 添加基础色和强调色调色板
// 对于每种调色板，你可以手动指定三种饱和度，分别对应默认、亮和暗
// $mat-indigo 和$mat-pink 是 Material 定义的基本色中的两种，具体都有哪些颜色可
// 以查看 node_modules/@angular/material/_theming.scss
// 如果还是不满足你的需求，则可以自己定义颜色变量
/**
$mat-custom: (
  50: #fce4ec,
  100: #f8bbd0,
  200: #f48fb1,
  300: #f06292,
  400: #ec407a,
  500: #e91e63,
  600: #d81b60,
  700: #c2185b,
  800: #ad1457,
  900: #880e4f,
  A100: #ff80ab,
  A200: #ff4081,
  A400: #f50057,
  A700: #c51162,
  contrast: (
    50: $dark-primary-text,
    100: $dark-primary-text,
    200: $dark-primary-text,
    300: $dark-primary-text,
    400: $dark-primary-text,
    500: $light-primary-text,
    600: $light-primary-text,
    700: $light-primary-text,
    800: $light-primary-text,
    900: $light-primary-text,
    A100: $dark-primary-text,
    A200: $light-primary-text,
    A400: $light-primary-text,
    A700: $light-primary-text,
  )
```

```
);
**/
$app-primary: mat-palette($mat-indigo);
$app-accent: mat-palette($mat-pink, A200, A100, A400);

// 错误调色板，默认为红色
$app-warn: mat-palette($mat-red);

// 通过 mat-light-theme 将上面定制化的调色板组合成一个主题，注意主题分为明和暗两种
// mat-light-theme：明
// mat-dark-theme：暗
$app-theme: mat-light-theme($app-primary, $app-accent, $app-warn);

// 接下来这句代码就是把主题风格包含，以便组件可以使用
@include angular-material-theme($app-theme);
```

必须注意的是，一般情况下，主题的样式只需导入一次，不要在多个文件中导入，以免样式风格被重复生成。

3. 如何包含多个主题

有时候，我们可能希望有多个主题，方便用户切换，或者不同模块使用不同主题。这种情况下，可以使用多个 CSS 类多次包含 angular-material-theme 来处理。

```
.app-dark-theme {
  @include angular-material-theme($dark-theme);
}

.app-other-theme {
  @include angular-material-theme($other-theme);
}
```

在对应组件的样式中应用该样式即可：

```
mat-sidenav-container.myapp-dark-theme {
  background: black;
}
```

4. 对于特殊组件的支持

在 Angular Material 中有一些组件是比较特殊的，比如对话框、弹出式菜单等，这一类组件是动态创建的，而且一般是要覆盖在其他组件之上显示的，这一类组件在 Angular Material 中都是 OverlayContainer。对于这类组件，如果要支持主题，则需要多做一点工作。

- 在构造函数中注入 OverlayContainer。
- 在对应的切换主题或者应用主题的函数中，使用上面得到的对象的 getContainer

Element().classList.add 和 getContainerElement().classList.remove 方法来添加或删除主题的支持。

```
constructor(private oc: OverlayContainer) {}

// 切换主题
switchDarkTheme(dark: boolean) {
    if (dark) {
      // 将定义的主题类应用到 OverlayContainer
      this.oc.getContainerElement().classList.add('myapp-dark-theme');
    } else {
      // 将定义的主题类从 OverlayContainer 的风格列表中移除
      this.oc.getContainerElement().classList.remove('myapp-dark-theme');
    }
  }
```

2.4 容器化 Angular 应用

2.4.1 什么是容器

容器其实可以理解成一个硬件虚拟化后的运行镜像，如果认为虚拟机是模拟运行的一套操作系统（提供了运行态环境和其他系统环境）和跑在上面的应用。那么 Docker 容器就是独立运行的一个或一组应用以及它们的运行环境，如图 2-19 所示。

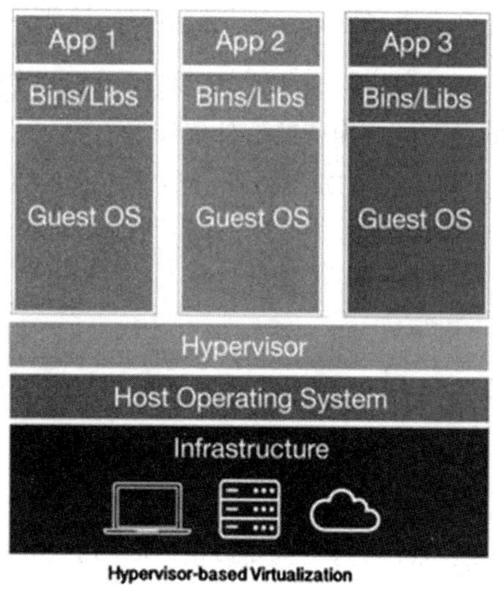

图 2-19

图 2-18 左侧是虚拟机的架构，图中的例子有 4 个操作系统（Operating System）。它们是 1 个 Host Operating System 和 3 个 Guest Operating System，每个虚拟机中都有一个独立的 Kernel。虚拟机可以让你能充分利用硬件资源。你可以购买性能超强的物理机器，然后在上面运行多个虚拟机。你可以有一个数据库虚拟机及很多运行相同版本的定制应用程序的虚拟机所构成的集群。你可以在有限的硬件资源获得很多的扩展能力。如果觉得还需要更多的虚拟机，而且宿主硬件还有容量，则可以添加任何你需要的虚拟机。如果不再需要一个虚拟机，则可以关闭该虚拟机甚至删除虚拟机镜像。但虚拟机的缺陷在于所有分配给一个虚拟机的资源是专有的。假如一台物理主机有 8 GB 的内存，如果建立一个拥有 1GB 内存的虚拟机，那么它只会使用这 1GB 进行内测，即使内存不足，它也不会动用剩余的 7 GB 内存。没有资源的动态分配，每台虚拟机拥有且只能使用分配给它的资源。

图 2-18 右侧为容器架构，容器其实是一个进程，操作系统认为它只是一个运行中的进程。另外，该容器进程也分配了自己的 IP 地址。一旦有了一个 IP 地址，该进程就是宿主网络中可识别的资源。然后，可以在容器管理器上运行命令，使容器 IP 映射到主机中能访问公网的 IP 地址。建立了该映射，容器就是网络上一个可访问的独立机器，从概念上类似于虚拟机。

因为容器是一个进程，所以它可以动态地共享主机上的资源。如果容器只需要 1GB 内存，则只会使用 1GB。如果它需要 4GB，则使用 4GB。CPU 和硬盘等也是如此。和典型虚拟机的静态方式不同，所有这些资源的共享都由容器管理器来管理的。而且，由于摆脱了硬件虚拟化，容器可以非常快速地启动。

容器的好处是：既有虚拟机独立和封装的优点，又屏蔽了静态资源专有的缺陷，可以共享资源。另外，由于容器能快速加载到内存，在扩展到多个容器时你能获得更好的性能。

那么对开发者来说，由于容器既快速又方便，就解决了部署的痛苦。有经验的开发者都知道开发环境好用的应用到了生产环境会出现各种问题，因为现代的软件依赖很多第三方类库和其他环境支持，稍微有点变化就会导致程序无法正确运行。Docker 给开发者带来的好处就是可以让开发环境和生产环境一致，无缝部署到生产环境。

2.4.2 安装 Docker

一般来说，在 Linux 系统中可以使用官方的安装脚本，国内的各种云服务购买的 Linux 虚拟机上都可以这么安装。

```
curl -sSL https://get.docker.com/ | sh
```

在 macOS 或者 Windows 系统上面，需要访问 Docker 官网下载安装包，如图 2-20 所示。

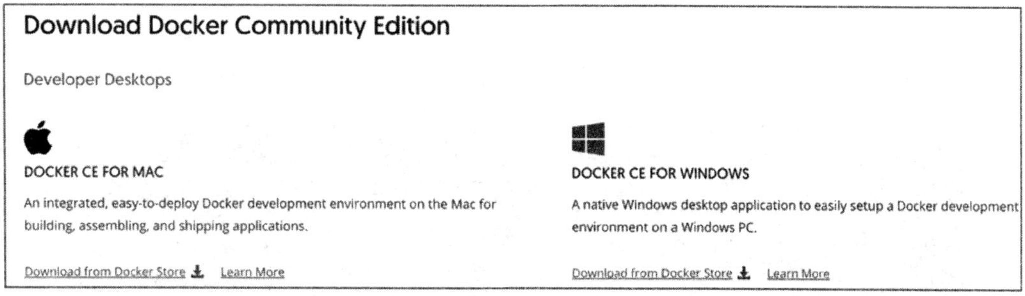

Download Docker Community Edition

Developer Desktops

DOCKER CE FOR MAC

An integrated, easy-to-deploy Docker development environment on the Mac for building, assembling, and shipping applications.

Download from Docker Store　Learn More

DOCKER CE FOR WINDOWS

A native Windows desktop application to easily setup a Docker development environment on a Windows PC.

Download from Docker Store　Learn More

图 2-20

2.4.3 镜像仓库加速

很多开源的开发工具或软件仓库在国内访问速度欠佳，这时候一般采用的方法是使用国内的镜像进行加速。Docker 也不例外，这里提供一种使用腾讯云提供的 Docker 软件镜像仓库，可以加速镜像下载。

在 Linux 系统中可以采用下面脚本进行设置。

```
sudo mkdir -p /etc/docker
sudo tee /etc/docker/daemon.json <<-'EOF'
{
  "registry-mirrors": ["https://mirror.ccs.tencentyun.com"]
}
EOF
sudo systemctl daemon-reload
sudo systemctl restart docker
```

2.4.4 创建 Angular 的 Docker 镜像

将容器应用到我们的 Angular 项目中是比较简单的，首先在项目工程的根目录建立一个叫 docker 的文件夹，然后在 docker 文件夹下面建立 nginx 文件夹，并在此文件夹中创建一个叫 Dockerfile 的文件。

我们在这个 Dockerfile 文件中要做两件事：

- 编译 Angular 应用，编译后的文件会存在 dist 目录下。
- 将 dist 目录下的 HTML、CSS 和 JavaScript 文件发布到 Nginx。

为了更好地理解后面的容器化过程，手动完成这样的部署，先编译项目，在项目根目录执行下面的命令：

```
ng build --prod
```

如果编译成功，那么你会看到类似图 2-21 所示的输出效果，成功编译后会产生几个 HTML、CSS 和 JavaScript 文件，这些文件位于项目根目录下的一个叫 dist 的子目录。

```
➜ ng build -prod
Your global Angular CLI version (1.7.2) is greater than your local
version (1.6.8). The local Angular CLI version is used.

To disable this warning use "ng set --global warnings.versionMismatch=false".
Date: 2018-03-09T06:57:12.154Z
Hash: 5d5e369d15f97d37c167
Time: 53311ms
chunk {0} main.dac2b5969f09d6776a67.bundle.js (main) 872 kB [initial] [rendered]
chunk {1} polyfills.abdf53ca655716e505e0.bundle.js (polyfills) 59.4 kB [initial] [rendered]
chunk {2} styles.a6f88da6763efc59b51c.bundle.css (styles) 89.5 kB [initial] [rendered]
chunk {3} inline.aec7a2ee582a00ef0ace.bundle.js (inline) 1.45 kB [entry] [rendered]
```

图 2-21

如果我们把这些文件复制到一个 HTTP 服务器的 Web 目录中，就可以访问我们生成的页面了。如果计算机上已经安装了 Nginx，则可以把 dist 目录下的内容复制到 /usr/share/nginx/html 目录下，然后看看效果。

其实 Docker 要做的工作和上述手动的过程类似，在编译过程中，分别需要 node 镜像，就相当于安装 node.js 环境，编译之后的发布其实只需要把内容复制到 Nginx 镜像的 Web 目录下。

```
## 第一阶段：构建 Angular 应用

## 编译这个阶段需要的是 node 镜像，所以我们以 'node:9-alpine' 为基础
## 另外给这一阶段起个友好的名称，叫 builder，以便在第二阶段可以方便地引用第一阶段的成
果
FROM node:9-alpine as builder

## 单独复制 'package.json'，在下一层安装相关依赖
COPY package.json ./

## 把 node_modules 保存在单独的一层以避免以后每次构建时都重复做 npm install
## 采用阿里团队提供的镜像
RUN npm config set registry https://registry.npm.taobao.org && npm i &&
mkdir /ng-app && cp -R ./node_modules ./ng-app

## 指定工作目录
WORKDIR /ng-app

COPY . .

## 在生产模式下编译 Angular 应用
ARG env=production
RUN npm run build -- --prod --configuration $env
```

```
## 第二阶段：设置 Nginx 服务器

FROM nginx:1.13.8-alpine

## 将我们的 Nginx 配置文件复制到镜像中的 /etc/nginx/conf.d/ 目录
COPY ./docker/nginx/conf.d/default.conf /etc/nginx/conf.d/

## 删除默认站点
RUN rm -rf /usr/share/nginx/html/*

## 从'builder'阶段将编译后的文件复制到 Nginx 默认的站点目录
'/usr/share/nginx/html'
COPY --from=builder /ng-app/dist /usr/share/nginx/html

## 执行命令启动 Nginx
CMD ["nginx", "-g", "daemon off;"]
```

从上面的文件可以看出来，如果想构建一个镜像，那么首先可以先以现有的镜像为基础更改。比如第一阶段中我们需要在 node.js 环境中编译 Angular 应用，所以使用 node:9-alpine 镜像，在 Dockerfile 中很多操作就是在把手动的环境配置写进去，比如 COPY、RUN 命令等。而且很棒的一点是 Dockerfile 还支持多阶段镜像，也就是说我们可以基于多个镜像，这个例子中，我们第一阶段使用了 node，而第二个阶段使用了 nginx。

然后我们就可以用这个 Dockerfile 编译自己的镜像了，执行下面的命令生成镜像文件。

```
## '-f' 开关指定使用的 Dockerfile 文件
## '-t' 开关指定容器的名字，也就是下面的 ng-app
## 'ng-app' 后面的 '.' 是指定上下文路径，Dockerfile 中的 COPY 等命令会基于这个路径
来操作
docker build -t ng-app . -f ./docker/nginx/Dockerfile
```

2.4.5 启动容器

启动这个容器的方法很简单，做好端口映射，-p <本地端口>:<容器端口> 用于指定本地和容器的端口映射，注意前面的端口是本地端口。如果没有 SSL 要求或者没有配置 HTTPS，那么其实是不用加-p 443:443 的。命令中的--rm 表示这是一个临时容器，如果容器停止，就会被删除。命令中的 -d 表示在后台运行，而最后的 gtm:dev 就是我们上一步生成的镜像名称。

```
docker run --rm -d -p 443:443 -p 80:80 gtm:dev
```

容器启动后，你可以访问 http://localhost 来查看目前的成果。

2.4.6 使用 docker-compose 组织复杂的环境配置

真正的开发环境或者生产环境一般不是由单纯的某一个服务构成的，而是由一组服务构成的，比如有负责前端渲染的服务器，有后端的 API 服务、数据库服务等。所以当我们构建了一个容器后，面临的一个问题就是如何组织项目依赖的各个服务。

我们可以通过在工程中建立一个 docker-compose.yml 文件来实现上述目标。将这个文件建立在工程根目录，.yml 格式的文件可读性很强，简单来说就是以缩进体现结构，以键/值对体现属性设置。

在 services 下面可以列出项目需要的所有服务，目前只有一个 Nginx，所以只能看到一个，但随着后面的学习，会逐渐增加进去的。

```
version: '3'
services:
  nginx:
    build:
      context: .
      dockerfile: ./docker/nginx/Dockerfile
      args:
        - env=dev
    container_name: nginx
    ports:
      - 80:80
```

有了 docker-compose.yml 文件，我们怎么使用呢？

使用下面的命令进行整个 docker-compose 中需要的镜像的构建，并在构建后启动所有服务。一般在更新了镜像所需的文件时，采用以下命令重新构建。

```
docker-compose up --build
```

上面的命令如果没有产生错误，则输出应该是类似下面的样子：

```
Building nginx
Step 1/12 : FROM node:8-alpine as builder
 ---> 7c2983dfbf98
Step 2/12 : COPY package.json ./
 ---> Using cache
 ---> bcb0d43e7df7
Step 3/12 : RUN npm i && mkdir /ng-app && cp -R ./node_modules ./ng-app
 ---> Running in 789733c11030
npm WARN deprecated nodemailer@2.7.2: All versions below 4.0.1 of
Nodemailer are deprecated. See https://nodemailer.com/status/
npm WARN deprecated mailcomposer@4.0.1: This project is unmaintained
```

```
npm WARN deprecated socks@1.1.9: If using 2.x branch, please upgrade to
at least 2.1.6 to avoid a serious bug with socket data flow and an import
issue introduced in 2.1.0
npm WARN deprecated node-uuid@1.4.8: Use uuid module instead
npm WARN deprecated buildmail@4.0.1: This project is unmaintained
npm WARN deprecated socks@1.1.10: If using 2.x branch, please upgrade to
at least 2.1.6 to avoid a serious bug with socket data flow and an import
issue introduced in 2.1.0

> uws@9.14.0 install /node_modules/uws
> node-gyp rebuild > build_log.txt 2>&1 || exit 0


> node-sass@4.8.3 install /node_modules/node-sass
> node scripts/install.js

Downloading binary from https://github.com/sass/node-
sass/releases/download/v4.8.3/linux_musl-x64-57_binding.node
Download complete
Binary saved to /node_modules/node-sass/vendor/linux_musl-x64-
57/binding.node
Caching binary to /root/.npm/node-sass/4.8.3/linux_musl-x64-
57_binding.node

> uglifyjs-webpack-plugin@0.4.6 postinstall
/node_modules/webpack/node_modules/uglifyjs-webpack-plugin
> node lib/post_install.js


> node-sass@4.8.3 postinstall /node_modules/node-sass
> node scripts/build.js

Binary found at /node_modules/node-sass/vendor/linux_musl-x64-
57/binding.node
Testing binary
Binary is fine
npm notice created a lockfile as package-lock.json. You should commit
this file.
npm WARN ajv-keywords@3.1.0 requires a peer of ajv@^6.0.0 but none is
installed. You must install peer dependencies yourself.
npm WARN optional SKIPPING OPTIONAL DEPENDENCY: fsevents@1.1.3
(node_modules/fsevents):
npm WARN notsup SKIPPING OPTIONAL DEPENDENCY: Unsupported platform for
fsevents@1.1.3: wanted {"os":"darwin","arch":"any"} (current:
```

```
{"os":"linux","arch":"x64"})

   added 1558 packages in 233.974s
   Removing intermediate container 789733c11030
    ---> 54a94bef793a
   Step 4/12 : WORKDIR /ng-app
   Removing intermediate container 6f9b336dc0e8
    ---> 9d6e2a9c8deb
   Step 5/12 : COPY . .
    ---> 1b9cc29a0831
   Step 6/12 : ARG env=prod
    ---> Running in 0eb37fa6530c
   Removing intermediate container 0eb37fa6530c
    ---> 2234f5373e66
   Step 7/12 : RUN npm run build -- --prod --configuration $env
    ---> Running in 3d26db2ca897

   > gtm@0.0.0 build /ng-app
   > ng build "--prod" "--configuration" "dev"

   Date: 2018-03-31T14:09:08.208Z
   Hash: 2e6fc11c769df2c5d8e8
   Time: 51145ms
   chunk {0} main.d46b2e467d69bc70f0dd.bundle.js (main) 915 kB [initial]
[rendered]
   chunk {1} polyfills.cc16b94dcfd6baec7194.bundle.js (polyfills) 60.1 kB
[initial] [rendered]
   chunk {2} styles.77df39064a08875d96f3.bundle.css (styles) 89.3 kB
[initial] [rendered]
   chunk {3} inline.8460b3901605c8e449fe.bundle.js (inline) 1.45 kB [entry]
[rendered]
   Removing intermediate container 3d26db2ca897
    ---> dfe339b68abf
   Step 8/12 : FROM nginx:1.13.8-alpine
    ---> bb00c21b4edf
    ---> bb00c21b4edf
   Step 9/12 : COPY ./docker/nginx/conf.d/default.conf /etc/nginx/conf.d/
    ---> adf15ab968f4
   Step 10/12 : RUN rm -rf /usr/share/nginx/html/*
    ---> Running in c6a67abeb18c
   Removing intermediate container c6a67abeb18c
    ---> 2cb90aad97a1
   Step 11/12 : COPY --from=builder /ng-app/dist /usr/share/nginx/html
    ---> 4f2d2de05dc1
```

```
Step 12/12 : CMD ["nginx", "-g", "daemon off;"]
 ---> Running in 437cb9c00d5b
Removing intermediate container 437cb9c00d5b
 ---> 522c49bf937c
Successfully built 522c49bf937c
Successfully tagged gtm_nginx:latest
Creating nginx ...
Creating nginx ... done
Attaching to nginx
```

此时，可以访问 http://localhost 来看看我们的成果。

启动全部服务，可以采用下面的命令，其中 -d 表示后台运行。

```
docker-compose up -d
```

停止全部服务，可以采用下面的命令：

```
docker-compose down
```

重新启动 docker-compose 中的某一项服务：

```
docker-compose restart nginx
```

查看某一项服务的日志：

```
docker-compose logs nginx
```

2.4.7 使用 .dockerignore 文件

和 git 很像的一点是，容器中也有类似 .gitignore 的文件，叫作 .dockerignore。它们起的作用也非常相似，都是要过滤掉一些文件，也就是在这个文件中列出的文件都不会参与到容器的构建动作中，对于容器来说，这些文件是不可见的。

所以务必注意，如果你在 Dockerfile 中写了类似 COPY 这样的命令，那么一定要记得检查一下是否有你想使用的目录，却在 .dockerignore 文件中列出了。

```
## dependencies
node_modules/
## compiled output
dist/
dist-server/
tmp/
out-tsc/
.git/
.vscode/
```

第 3 章 何谓后端

有前端就会有后端，其实这两端原本是在一起的。十几年前，当时的热门词汇叫 Web Framework（框架），各种 Web 框架层出不穷，就知今天前端各种框架百花齐放。很多现在大家很熟悉的模式都是当时普及的，比如 MVC、IoC 等。

任何系统中界面始终是一个绕不过去的大“坑”，因为需求变化在 UI 上体现得最明显，所以这些框架多用于解决如何快速开发 Web 界面，并且易于维护。这些框架竞争后胜出的，或者叫普及率最高的 Web 框架是 Structs。伴随着 Web 框架，还有数据库 ORM 框架及中间件框架等逐渐出现，在 Java 领域经常听到的 SSH 指的就是 Structs、Spring、Hibernate， SSM 指的就是 Structs、Spring、MyBatis。

发展到这个时候，其实开发效率已经不错了，但是上述框架有一个问题，那就是后端人员要懂得 HTML、CSS、JavaScript，这个对开发人员要求还是蛮高的，而在当时，后端开发人员对于这些前端技术的掌握程度普遍不是特别好。随着互联网的发展，Web 的展现要求越来越高，这就决定了我们必须要有专业的人员来去应对前端，而不是像原来的 Web 框架一样把前后端耦合在一起，这也就是前后端分离的一个重要意义——进一步专业分工了 Web 开发。

一旦开始进行了前后端的分离，生产效率飞速提升，后端没有了界面的考虑后架构更轻、责任更单一、目标更专注，开发也变得更简单了。Spring Boot 并非是要替代 Spring，而是在这样一个时代背景下诞生的快速配置框架，它起的作用更多是在黏合众多的框架、简化各种烦琐配置，给开发者一个“开箱即用”的体验。

记得当初写 Spring 时要配置好多 XML（在当时还是相对先进的模式），虽然实现了松耦合，但这些 XML 却又成为项目甩不掉的负担——随着项目越做越大，这些 XML 的可读性和可维护性极差。后来受.Net 平台中 Annotation 的启发，Java 世界中也引入了元数据的修饰符，

Spring 也可以使用这种方式进行配置。到了近些年，随着 Ruby on Rails 的兴起而流行的惯例优先原则（convention over configuration）开始深入人心。那什么是惯例优先原则呢？简单来说就是牺牲一部分的自由度来减少配置的复杂度，打个比方就是如果遵从我定义的一系列规则，那么你要配置的东西就非常简单甚至可以零配置。既然已经做到这个地步，各种“脚手架”项目就纷纷涌现了，目的只有一个：让你更专注于代码的编写，而不是浪费在各种配置上。近两年前端也有类似趋势，各种前端框架的官方 CLI 纷纷登场：create-react-app、angular-cli、vue-cli 等。

那么 Spring Boot 就是 Spring 框架的“脚手架”了，它可以帮你快速搭建、发布一个 Spring 应用。官网列出了 Spring Boot 的几个主要目标。

- 提供一种快速和广泛适用的 Spring 开发体验。
- 开箱即用却又可以适应各种变化。
- 提供一系列开发中常用的“非功能性”的特性（比如嵌入式服务器、安全、度量、自检及外部配置等）。
- 不生成任何代码，不需要 XML 配置。

本书的后端服务主要使用 Spring Boot 进行搭建。

3.1 创建一个 Spring Boot 工程

创建一个 Spring Boot 工程以下几种方式。

（1）通过 Gradle 或 Maven 进行创建。

- 优势：不依赖任何 IDE 或操作系统，和各种持续集成、自动化测试框架结合形成。
- 劣势：学习成本较高，需要对 Gradle 或 Maven 进行团队的培训。

（2）通过 IDE 进行创建。

- 优势：创建过程非常简单。
- 劣势：严重依赖 IDE。

这里我们会分别介绍用以上方式创建 Spring Boot 工程的步骤，笔者更推荐第一种方式。

3.1.1 通过 Gradle 创建

本书后面的章节均采用 Gradle 来做工程组织，这种方式与 Maven 方式差不多，只不过个人感觉 Gradle 比 Maven 的 XML 形式的配置文件可读性要好很多。

1. 安装 Gradle

首先当然要先安装 Gradle，这一步只有第一次当系统中没有 Gradle 时才需要，以后的项目创建并不需要这一步骤。

- Linux: 在 terminal 中输入 sdk install gradle。假如你没有安装 sdkman，请参考第 1 章内容进行安装。
- macOS：在 terminal 中输入 brew install gradle，或者你也可以按 Linux 的安装方式进行。
- Windows：在命令行下执行 scoop install gradle，如果你没有安装 scoop，请参考第 1 章相关内容。

要确认是否安装成功，可以在打开一个命令行窗口，输入 gradle -v，如果你可以看到类似下面的输出，那么就恭喜你了，安装成功！

```
------------------------------------------------------------
Gradle 4.4.0
------------------------------------------------------------

Build time:   2017-10-02 15:36:21 UTC
Revision:     a88ebd6be7840c2e59ae4782eb0f27fbe3405ddf

Groovy:       2.4.12
Ant:          Apache Ant(TM) version 1.9.6 compiled on June 29 2015
JVM:          1.8.0_111 (Oracle Corporation 25.111-b14)
OS:           Mac OS X 10.13.3 x86_64
```

2. 创建 Gradle 工程

创建一个 Gradle 只需要两步。

- 新建一个工程目录 gtm。
- 在此目录下使用 Gradle 进行初始化 gradle init（就和在 git 中使用 git init 的效果类似，这一动作初始化的生成一个工程）。

这个命令帮我们建立一个一个使用 Gradle 进行管理的模板工程，下面的列表中给出了这个命令为我们创建了哪些文件和文件夹，以及它们的作用：

- build.gradle：有过 Android 开发经验的读者可能觉得很亲切，这就是用于管理和配置工程的核心文件。
- gradlew：用于 *nix 环境下的 Gradle Wrapper 文件。
- gradlew.bat：用于 Windows 环境下的 Gradle Wrapper 文件
- setting.gradle：用于管理多项目的 Gradle 工程时使用，单项目时可以不做理会。

- gradle 目录：Gradle Wrapper 的 jar 和属性设置文件所在的文件夹。

3．什么是 Gradle Wrapper

你是否有过这样的经历？在安装/编译一个工程时需要一些先决条件，需要安装一些软件或设置一些参数。如果这一切比较顺利还好，但很多时候我们会发现这样那样的问题，比如版本不对、参数没设置等。Gradle Wrapper 就是这样一个让你不会浪费时间在配置问题上的方案。它会对应一个开发中使用的 Gradle 版本，以确保任何人任何时候得到的结果是一致的。

- ./gradlew <task>: 在 *nix 平台上运行，例如 Linux 或 macOS ×。
- gradlew <task> 在 Windows 平台运行（是通过 gradlew.bat 来执行的）。

（1）Gradle Wrapper 的配置

当我们在 terminal 执行 Gradle Wrapper 生成相关文件时，可以为其指定一些参数，来控制 gradle wrapper 的生成，比如依赖的版本等。很多时候，我们可能需要不同版本的 Gradle，如 Spring Boot 2.0 需要 Gradle 4.0 以上版本，如果要指定使用 4.4 版本，就可以使用下面的命令。

```
gradle wrapper --gradle-version 4.4
```

这个命令其实是更改了 gradle/wrapper/gradle-wrapper.properties 文件中的 distributionUrl，从下面的代码可以看出我们使用的是 gradle-4.4-bin.zip。

```
distributionBase=GRADLE_USER_HOME
distributionPath=wrapper/dists
zipStoreBase=GRADLE_USER_HOME
zipStorePath=wrapper/dists
distributionUrl=https\://services.gradle.org/distributions/gradle-4.4-
bin.zip
```

那么，接下来还有一个问题就是，有时我们可能希望使用的是含 Gradle 源代码的版本，也就是文件名带有 all 的那种发行版本。这种情况则可以使用另一个参数 --distribution-type：

```
gradle wrapper --gradle-version 4.4 --distribution-type all
```

这样的命令就会使得 distributionUrl 指向一个带有源代码的完全版本：

```
distributionBase=GRADLE_USER_HOME
distributionPath=wrapper/dists
zipStoreBase=GRADLE_USER_HOME
zipStorePath=wrapper/dists
distributionUrl=https\://services.gradle.org/distributions/gradle-4.4-
all.zip
```

（2）执行 Gradle Wrapper 时为什么一直卡在那里

这个问题，其实很常见。可以采用一个国内的 Gradle 镜像站点。如果采用镜像站，或者干脆到官网下载 zip 包，架设到公司内网的某台服务器，这种情况下，就需要用到参数 --gradle-distribution-url。当然这种情况就不用写--gradle-version 和 --distribution-type 了，因为 URL 里面都已经含有这些信息了。

```
gradle wrapper --gradle-distribution-url "https://local.dev/gradle-4.4-all.zip"
```

4. Gradle 工程概述

那么下面我们打开默认生成的 build.gradle 文件，将其改造成下面的样子：

```
/*
 * 这个 build.gradle 文件是由 Gradle 的'init'任务生成的。
 *
 * 更多关于在 Gradle 中构建 Java 项目的信息可以查看 Gradle 用户文档中的
 * Java 项目快速启动章节
 */
// 在这个区块中你可以声明你的 build 脚本需要的依赖和解析下载该依赖所使用的仓储位置
buildscript {
  ext {
    springBootVersion = '2.0.0.RELEASE'
  }
  repositories {
    jcenter()
  }
}
// 从 Spring Boot 2.0 起，可以直接在 plugins 区块中声明 SpringBoot 插件
plugins {
  id 'org.springframework.boot' version '2.0.0.RELEASE'
}
/*
 * 在这个区块中你可以声明使用哪些插件
 * apply plugin: 'java' 代表这是一个 Java 项目，需要使用 Java 插件
 * 如果想生成一个 Intellij IDEA 的工程，类似的如果要生成
 * eclipse 工程，就写 apply plugin: 'eclipse'
 * 同样地，我们要学习的是 Spring Boot，所以应用 Spring Boot 插件
 */
apply plugin: 'java'
apply plugin: 'idea'
apply plugin: 'eclipse'
apply plugin: 'io.spring.dependency-management'
```

```
    // 在这个区块中可以声明编译后的 Jar 文件信息
    jar {
      baseName = 'gtm-backend'
      version = '0.0.1-SNAPSHOT'
    }

    // 在这个区块中可以声明在哪里可以找到你的项目依赖
    repositories {
        // 使用'jcenter'作为中心仓储查询解析你的项目依赖
        // 你可以声明任何 Maven/Ivy/file 类型的依赖类库仓储位置
        // 如果遇到下载速度慢，则可以换成阿里镜像
        // maven {url
"http://maven.aliyun.com/nexus/content/repositories/central/"}
        mavenCentral()
    }

    // 在这个区块中可以声明源文件和目标编译后的 Java 版本兼容性
    sourceCompatibility = 1.8
    targetCompatibility = 1.8

    // 在这个区块可以声明你的项目的开发和测试所需的依赖类库
    dependencies {
      // 在 gradle 4.x 中 compile 已经 deprecated，请使用 implementation
      implementation('org.springframework.boot:spring-boot-starter-web')
      testImplementation('org.springframework.boot:spring-boot-starter-test')
    }
```

使用 Gradle 作为工程脚本有以下优势。

- 提供 Gradle 脚本系统中的 Spring Boot 支持。
- 简化执行和发布：它可以把所有 classpath 的类库构建成一个单独的可执行 jar 文件（fat jar），这样可以简化执行和发布等操作，或者按传统方式打包成 war。
- 自动搜索入口文件：它会扫描 public static void main() 函数并且标记这个函数的宿主类为可执行入口。
- 简化依赖：一个典型的 Spring 应用还是需要很多依赖类库的，想要配置正确并不容易，所以这个插件提供了内建的依赖解析器，可以自动匹配和当前 Spring Boot 版本匹配的依赖库版本。

当我们在脚本中应用了 io.spring.dependency-management 插件时，插件会自动导入当前要使用的 Spring Boot 版本的依赖清单，这个体验和 Spring Boot 在使用 Maven 时保持了高度一致。我们不必指定具体声明当依赖的版本号，比如:

```
dependencies {
  implementation('org.springframework.boot:spring-boot-starter-web')
  testImplementation('org.springframework.boot:spring-boot-starter-test')
}
```

（1）引入 Spring Boot 依赖但不应用 Spring Boot 插件

有些时候，我们可能不希望创建一个 Spring Boot 工程，但希望引入 Spring Boot 的依赖。比如我们在进行一个类库工程的开发时，这个类库本身的工程其实并不可以直接启动，而是会被其他工程引用。这种情况下，我们就会希望只是单纯地引入 Spring Boot 的依赖。此时需要在 build.gradle 中的插件区块中使用 apply false。

```
plugins {
  id 'org.springframework.boot' version '{version}' apply false
}
```

应用 dependency-management 插件，并且在 dependencyManagement 区块中导入来料清单（Maven POM）中的依赖。

```
apply plugin: 'io.spring.dependency-management'

dependencyManagement {
  imports {
    mavenBom '
org.springframework.boot.gradle.plugin.SpringBootPlugin.BOM_COORDINATES
  }
}
```

（2）可选依赖插件

由于 Gradle 没有提供和 Maven 等价的 optional 或 provided 形式的依赖类型，这个插件是 Spring 团队开发用于扩充 Gradle 的，以使 Maven 和 Gradle 的项目构建方式更加一致。

```
buildscript {
    ext {
        // 省略其他部分
        propDepsVersion = '0.0.9.RELEASE' //增加 propdeps 的版本号定义
    }
    repositories {
        // 省略其他
        maven { setUrl('http://repo.spring.io/plugins-release') }
// 增加 Spring Gradle 插件的软件仓库
    }
    dependencies {
        classpath("io.spring.gradle:propdeps-plugin:${propDepsVersion}")
// 增加依赖
```

```
        }
    }
    // 省略其他部分
    subprojects {
        // 省略其他部分
        apply plugin: 'java'
        // 省略其他部分
        apply plugin: 'propdeps' // 应用插件
        apply plugin: 'propdeps-idea' // 应用插件
        // 省略其他部分
        dependencies {
            optional("org.springframework.boot:spring-boot-configuration-
processor") // 使用可选依赖
            testImplementation("org.springframework.boot:spring-boot-
starter-test")
        }
        compileJava.dependsOn(processResources)
    }
```

安装这个插件并如上设置后，我们以后就可以应对可选依赖了。

3.1.2 通过 Maven 创建

Maven 和 Gradle 非常类似，只不过 Maven 采用了 XML 格式的配置文件，这主要是由于 Maven 诞生时，XML 还是被广泛认为是可读性很好的一种文件格式。

1. 安装 Maven

和 Gradle 类似，这一步只有第一次当系统中没有 Maven 时才需要，以后的项目创建并不需要这一个步骤。

- Linux: 在 terminal 中输入 sdk install maven。假如你没有安装 sdkman，则请参考第 1 章内容进行安装。
- Mac OS：在 terminal 中输入 brew install maven，或者也可以按 Linux 的方式进行安装。
- Windows：在命令行下执行 scoop install maven，如果没有安装 scoop，则请参考第 1 章相关内容。

要确认是否安装成功，可以打开一个命令行窗口，输入 mvn -v，如果可以看到类似以下的输出内容，那么就恭喜你了，安装成功！

```
    Apache Maven 3.5.2 (138edd61fd100ec658bfa2d307c43b76940a5d7d; 2017-10-
18T15:58:13+08:00)
    Maven home: /usr/local/Cellar/maven/3.5.2/libexec
    Java version: 1.8.0_111, vendor: Oracle Corporation
```

```
Java home:
/Library/Java/JavaVirtualMachines/jdk1.8.0_111.jdk/Contents/Home/jre
Default locale: zh_CN, platform encoding: UTF-8
OS name: "mac os x", version: "10.13.3", arch: "x86_64", family: "mac"
```

2. 创建 Maven 工程

使用 mvn 命令创建一个新的工程，与 Gradle 的区别是，我们是在上一级目录执行下面的命令，也就是说工程目录会在执行后被创建出来。

```
mvn -B archetype:generate \
  -DarchetypeGroupId=org.apache.maven.archetypes \
  -DgroupId=dev.local.gtm \
  -DartifactId=gtm-backend
```

这个命令帮我们建立一个一个使用 Maven 进行管理的模板工程，下面的列表中给出了这个命令为我们创建了哪些文件和文件夹，以及它们的作用。

- pom.xml：用于组织工程结构和项目依赖的 XML 配置文件。
- src 目录：用于存放源文件，如图 3-1 所示。

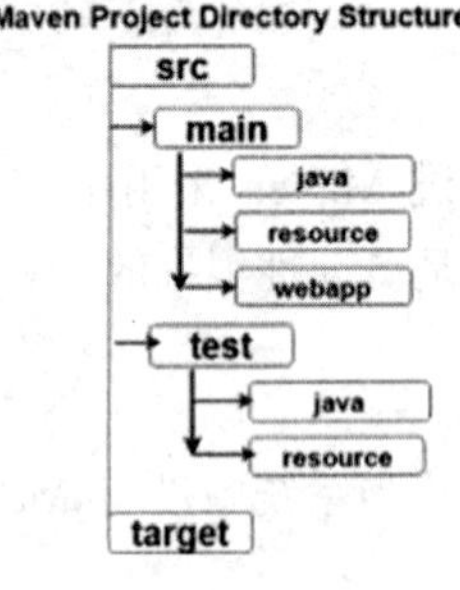

图 3-1

一个典型的 pom.xml 文件如下：

```
<project xmlns="http://maven.apache.org/POM/4.0.0"
xmlns:xsi="http://www.w3.org/2001/XMLSchema-instance"
  xsi:schemaLocation="http://maven.apache.org/POM/4.0.0
http://maven.apache.org/maven-v4_0_0.xsd">
  <modelVersion>4.0.0</modelVersion>
  <groupId>dev.local.gtm</groupId>
  <artifactId>gtm-backend</artifactId>
  <packaging>jar</packaging>
  <version>1.0-SNAPSHOT</version>
  <name>my-app</name>
  <url>http://maven.apache.org</url>
```

```
    <dependencies>
      <dependency>
        <groupId>junit</groupId>
        <artifactId>junit</artifactId>
        <version>3.8.1</version>
        <scope>test</scope>
      </dependency>
    </dependencies>
  </project>
```

加入 Spring Boot 的支持只需要略做改动一下即可：

```
  <project xmlns="http://maven.apache.org/POM/4.0.0"
xmlns:xsi="http://www.w3.org/2001/XMLSchema-instance"
    xsi:schemaLocation="http://maven.apache.org/POM/4.0.0
http://maven.apache.org/maven-v4_0_0.xsd">
    <modelVersion>4.0.0</modelVersion>
    <groupId>dev.local.gtm</groupId>
    <artifactId>gtm-backend</artifactId>
    <packaging>jar</packaging>
    <version>1.0-SNAPSHOT</version>
    <name>my-app</name>
    <url>http://maven.apache.org</url>
    <parent>
      <groupId>org.springframework.boot</groupId>
      <artifactId>spring-boot-starter-parent</artifactId>
      <version>2.0.0.RELEASE</version>
    </parent>
    <dependencies>
      <dependency>
          <groupId>org.springframework.boot</groupId>
          <artifactId>spring-boot-starter-web</artifactId>
      </dependency>
    </dependencies>
    <build>
      <plugins>
        <plugin>
          <groupId>org.springframework.boot</groupId>
          <artifactId>spring-boot-maven-plugin</artifactId>
        </plugin>
      </plugins>
    </build>
  </project>
```

Maven 也有 Wrapper 的概念，如果要生成 Wrapper，则在工程目录下执行下面的命令：

```
mvn -N io.takari:maven:wrapper
```

与 Gradle 类似，会生成 mvnw、mvnw.cmd 及一个 .mvn 目录。

- mvnw：用于 *nix 环境下的 maven wrapper 文件。
- mvnw.cmd：用于 Windows 环境下的 maven wrapper 文件。

使用下面命令验证一下是否成功：

```
./mvnw spring-boot:run
```

如果你看到类似下面的输出，那就说明一切正常。

```
[INFO] Scanning for projects...
[INFO]
[INFO] ------------------------------------------------------------------------
[INFO] Building gtm-backend 1.0-SNAPSHOT
[INFO] ------------------------------------------------------------------------
[INFO]
[INFO] >>> spring-boot-maven-plugin:2.0.0.RELEASE:run (default-cli) > test-compile @ gtm-backend >>>
[INFO]
[INFO] --- maven-resources-plugin:3.0.1:resources (default-resources) @ gtm-backend ---
[INFO] Using 'UTF-8' encoding to copy filtered resources.
[INFO] skip non existing resourceDirectory gtm-backend/src/main/resources
[INFO] skip non existing resourceDirectory gtm-backend/src/main/resources
[INFO]
[INFO] --- maven-compiler-plugin:3.7.0:compile (default-compile) @ gtm-backend ---
[INFO] Nothing to compile - all classes are up to date
[INFO]
[INFO] --- maven-resources-plugin:3.0.1:testResources (default-testResources) @ gtm-backend ---
[INFO] Using 'UTF-8' encoding to copy filtered resources.
[INFO] skip non existing resourceDirectory gtm-backend/src/test/resources
[INFO]
[INFO] --- maven-compiler-plugin:3.7.0:testCompile (default-testCompile) @ gtm-backend ---
[INFO] Nothing to compile - all classes are up to date
[INFO]
```

```
[INFO] <<< spring-boot-maven-plugin:2.0.0.RELEASE:run (default-cli) < test-compile @ gtm-backend <<<
[INFO]
[INFO]
[INFO] --- spring-boot-maven-plugin:2.0.0.RELEASE:run (default-cli) @ gtm-backend ---
Hello World!
[INFO] ------------------------------------------------------------------------
[INFO] BUILD SUCCESS
[INFO] ------------------------------------------------------------------------
[INFO] Total time: 3.733 s
[INFO] Finished at: 2018-04-01T14:49:31+08:00
[INFO] Final Memory: 21M/227M
[INFO] ------------------------------------------------------------------------
```

3.1.3 通过 IDE 创建

在 Intellij IDEA 中创建 Spring Boot 工程，只需要选择左侧的“Spring Initializr”作为工程模板，如图 3-2 所示。

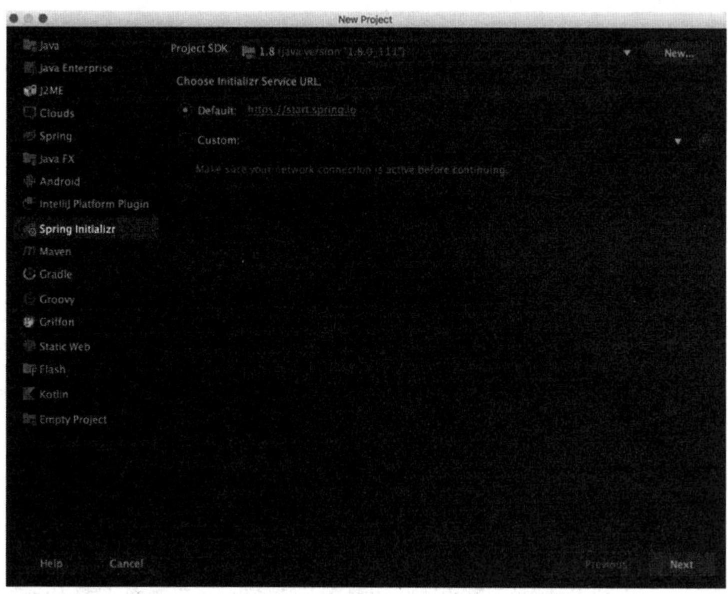

图 3-2

点击“Next”按钮，在接下来的向导页面中填写参数，如图 3-3 所示。

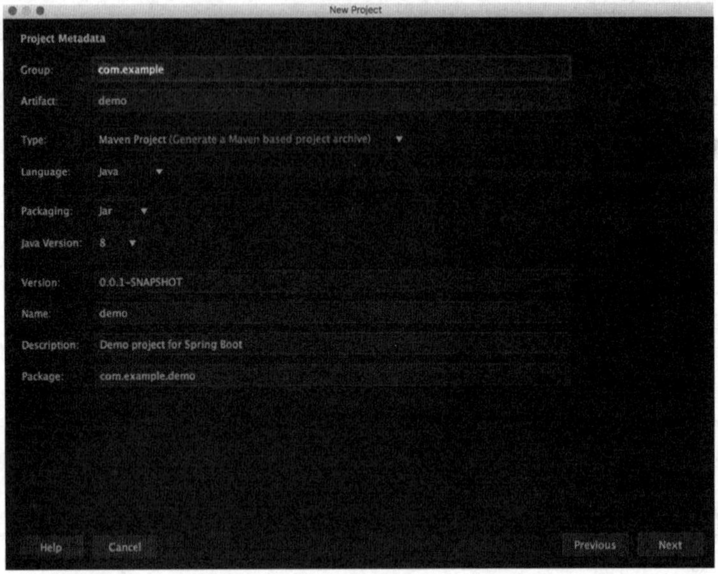

图 3-3

选择项目的依赖，如图 3-4 所示。

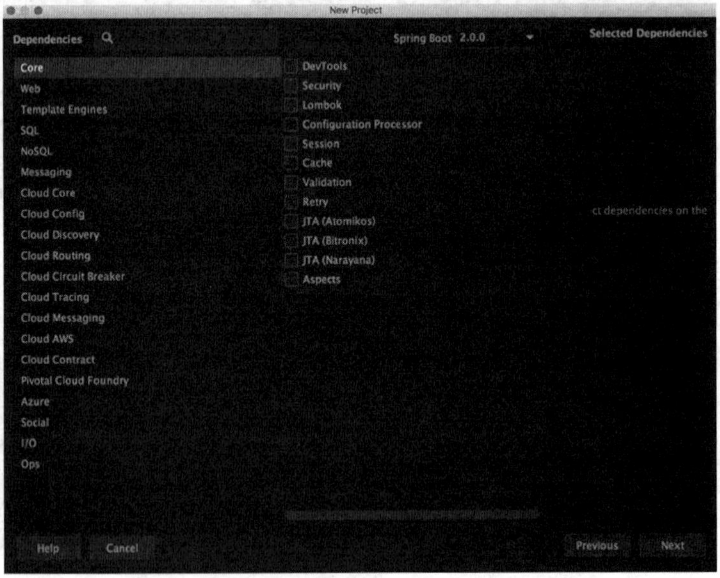

图 3-4

最后填写一下项目位置的有关信息就可以完成了，如图 3-5 所示。

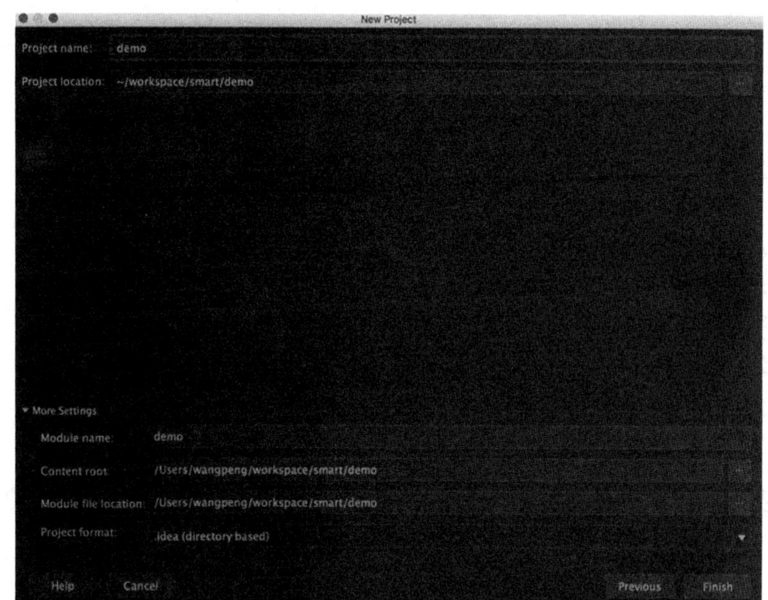

图 3-5

3.1.4 工程项目的组织

在大型软件开发中，一个项目解决所有问题显然不可取，因为存在太多的开发团队共同协作，所以对项目进行拆分，形成多个子项目的形式是普遍存在的。而且一个子项目只专注自己的逻辑也易于维护和拓展，现在随着容器和微服务的理念逐渐获得大家的认可，多个子项目分别发布到容器形成多个微服务也逐渐成为趋势。

我们的项目使用 Gradle 来处理多项目的构建，包括大项目和子项目的依赖管理及容器的建立等。对于 Gradle 项目，会有一个根项目，这个根项目下会建立若干子项目，具体文件结构如下：

```
|--backend（根项目）
|----common（共享子项目）
|------src（子项目源码目录）
|--------main（子项目开发源码目录）
|----------java（子项目开发 Java 类源码目录）
|----------resources（子项目资源类源码目录）
|------build.gradle（子项目 gradle 构建文件）
|----api（API 子项目）
|------src
```

```
|--------main
|----------java
|----------resources
|------build.gradle
|----report（报表子项目）
|------src
|--------main
|----------java
|----------resources
|------build.gradle
|--build.gradle（根项目构建文件）
|--settings.gradle（根项目设置文件）
```

要让 Gradle 支持多项目，首先需要把 settings.gradle 改成：

```
include 'common'
include 'api'
include 'report'
rootProject.name = 'gtm-backend'
```

这样 gtm 就成为根项目，而 common、api 和 report 就是其之下的子项目。接下来，我们看一下根项目的 build.gradle，对于多项目构建来说，根项目的 build.gradle 中应该尽可能地配置各子项目中共同的配置，从而让子项目只配置自己不同的东西。

```
// 一个典型的根项目的构建文件结构
buildscript {
    /*
     * 构建脚本区块可以配置整个项目需要的插件，构建过程中的依赖及依赖类库的版本号等
     */
}

allprojects {
    /*
     * 在这个区块中可以声明对所有项目（含根项目）都适用的配置，比如依赖性的仓储等
     */
}

subprojects {
    /*
     * 在这个区块中可以声明适用于各子项目的配置（不包括根项目）
     */
    version = "0.0.1"
}

/*
```

```
 * 对于子项目的特殊配置
 */
project(':common') {

}

project(':api') {

}

project(':report') {

}
```

其中，buildscript 区块用于配置 Gradle 脚本生成时需要的东西，比如配置整个项目需要的插件，构建过程中的依赖及在其他部分需要引用的依赖类库的版本号等，就像下面这样，我们在 ext 中定义了一些变量来集中配置了所有依赖的版本号，无论是根项目还是子项目都可以使用这些变量指定版本号。这样做的好处是当依赖的版本更新时，我们无须四处更改散落在各处的版本号。此外，在这个区块中我们还提供了项目所需的第三方 Gradle 插件所需的依赖：spring-boot-gradle-plugin 和 dependency-management-plugin，因此，各子项目可以简单地使用诸如 apply plugin: 'io.spring.dependency-management' 等。

```
buildscript {
    ext {
        springBootVersion = '2.0.0.RELEASE'
        gradleDockerVersion = '1.2'
    }
    repositories {
        maven
{ setUrl('http://maven.aliyun.com/nexus/content/groups/public/') }
        maven
{ setUrl('http://maven.aliyun.com/nexus/content/repositories/jcenter') }
    }
    dependencies {
        classpath("org.springframework.boot:spring-boot-gradle-
plugin:${springBootVersion}")
    }
}
```

allprojects 中可以声明对于所有项目（含根项目）都适用的配置，比如依赖性的仓储等。而 subprojects 和 allprojects 的区别在于 subprojecrts 只应用到子项目，而非根项目。所以大部分通用型配置可以通过 subprojects 和 allprojects 来完成。下面列出的样例配置中，我们为所有的项目包括根项目配置了依赖仓储及软件的 group，同时为每个**子项目**配置了 Java 和 IDEA 两个插件、版本号和通用的测试依赖。

```
    allprojects {
        group = 'dev.local.gtm'
        repositories {
            maven
{ setUrl('http://maven.aliyun.com/nexus/content/groups/public/') }
            maven
{ setUrl('http://maven.aliyun.com/nexus/content/repositories/jcenter') }
        }
    }

    subprojects {
        version = "0.0.1"
        tasks.withType(Jar) {
            baseName = "$project.name-$version"
        }
        apply plugin: 'java'
        apply plugin: 'idea'
        apply plugin: 'eclipse'
        apply plugin: 'io.spring.dependency-management'
        sourceCompatibility = 1.8
        targetCompatibility = 1.8
        dependencies {
            testImplementation("org.springframework.boot:spring-boot-
starter-test")
        }
    }
```

除此之外，为了展示 project 的用法，我们这个例子里把每个子项目的依赖放到根 build.gradle 中的 project(':子项目名') 中列出，这样做有好处也有坏处，好处是依赖性的管理统一在根 build.gradle 中完成，对于依赖的情况一目了然。当然坏处是每个项目更改依赖时都会造成根 build.gradle 的更新，这样如果一个项目有非常多的子项目时，则会在协作上出现一些问题。所以请根据具体情况决定把依赖放到根 build.gradle 中的 project(':子项目名') 中还是放到各子项目的 build.gradle 中。

```
    project(':common') {
        dependencies {
            implementation("org.springframework.boot:spring-boot-starter-
data-rest")
            implementation("org.springframework.boot:spring-boot-starter-
data-mongodb")
```

```
            implementation("org.projectlombok:lombok:${lombokVersion}")
        }
    }

    project(':api') {
        dependencies {
            implementation project(':common')
            implementation("org.springframework.boot:spring-boot-devtools")
            implementation("org.springframework.boot:spring-boot-starter-
security")
            implementation("io.jsonwebtoken:jjwt:${jjwtVersion}")
            implementation("org.projectlombok:lombok:${lombokVersion}")
        }
    }

    project(':report') {
        dependencies {
            implementation project(':common')
            // the following 5 are required by jasperreport rendering
            implementation files(["lib/simsun.jar"])
            implementation("org.springframework.boot:spring-boot-devtools")
            implementation("org.springframework.boot:spring-boot-starter-
web")
            implementation("org.springframework:spring-context-
support:${springCtxSupportVersion}")

implementation("net.sf.jasperreports:jasperreports:${jasperVersion}")
            implementation("com.lowagie:itext:${itextVersion}")
            implementation("org.apache.poi:poi:${poiVersion}")
            implementation("org.olap4j:olap4j:${olap4jVersion}")
        }
    }
```

3.2 API 的构建可以如此简单

3.2.1 API 工程结构

我们 API 是整个后端工程组中的一个子项目，也是我们这个工程的第一个子项目，请按照下面的目录结构方式建立 API 项目目录——在 backend 目录下建立一个子目录 API，然后建立对应的目录和文件。

```
    |--backend（后端根项目）
```

```
|----api（API 子项目）
|------src
|--------main
|----------java
|----------resources
|------build.gradle
```

将 API 目录下的 build.gradle 修改成如下所示的内容。

```
apply plugin: 'org.springframework.boot'

dependencies {
    implementation('org.springframework.boot:spring-boot-starter-web')
}
```

同时，请确保父一级的 build.gradle 已经更新成如下所示的内容。

```
buildscript {
    ext {
        springBootVersion = '2.0.0.RELEASE'
    }
    repositories {
        maven
{ setUrl('http://maven.aliyun.com/nexus/content/groups/public/') }
        maven
{ setUrl('http://maven.aliyun.com/nexus/content/repositories/jcenter') }
    }
    dependencies {
        classpath("org.springframework.boot:spring-boot-gradle-
plugin:${springBootVersion}")
    }
}

allprojects {
    group = 'dev.local.gtm'
    repositories {
        maven
{ setUrl('http://maven.aliyun.com/nexus/content/groups/public/') }
        maven
{ setUrl('http://maven.aliyun.com/nexus/content/repositories/jcenter') }
    }
}

subprojects {
    version = "0.0.1"
    tasks.withType(Jar) {
```

```
            baseName = "$project.name-$version"
        }
        apply plugin: 'java'
        apply plugin: 'idea'
        apply plugin: 'eclipse'
        apply plugin: 'io.spring.dependency-management'
        sourceCompatibility = 1.8
        targetCompatibility = 1.8
        dependencies {
            testImplementation("org.springframework.boot:spring-boot-
starter-test")
        }
    }
```

更新 backend 目录下的 settings.gradle 以便包含 API 作为子项目。

```
include 'api'
rootProject.name = 'gtm-backend'
```

3.2.2 领域对象

源代码目录在哪里呢？我们得手动建立一个，这个目录一般情况下是 src/main/java。好的，下面我们要开始第一个 RESTful 的 API 搭建了，首先还是在 src/main/java 下新建一个包（package）。既然是本机的就叫 dev.local.gtm。首先尝试建立一个 Web API，在 dev.local.gtm.api 下建立一个子 package:domain，然后创建一个 Task 的领域对象：

```
package dev.local.gtm.api.domain;

/**
 * Task 是一个领域对象（domain object）
 */
public class Task {
    private String id;
    private String desc;
    private boolean completed;

    public String getId() {
        return id;
    }

    public void setId(String id) {
        this.id = id;
    }

    public String getDesc() {
```

```
        return desc;
    }

    public void setDesc(String desc) {
        this.desc = desc;
    }

    public boolean isCompleted() {
        return completed;
    }

    public void setCompleted(boolean completed) {
        this.completed = completed;
    }
}
```

这个对象很简单，只是描述了 Task 的几个属性：id、desc 和 completed。API 返回或接受的参数就是以这个对象为模型的类或集合。

3.2.3 构造 Controller

我们经常看到的 RESTful API 是这样的：http://local.dev/tasks、http://local.dev/tasks/1。而 Controller 就是要暴露这样的 API 给外部使用。现在我们同样在 Task 下建立一个叫 TaskController 的 Java 文件:

```
package dev.local.gtm.api.controller;

// 省略导入

/**
 * 使用@RestController 来标记这个类是 Controller
 */
@RestController
public class TaskController {
    // 使用@RequstMapping 指定可以访问的 URL 路径
    @RequestMapping("/tasks")
    public List<Task> getAllTasks() {
        List<Task> tasks = new ArrayList<>();
        Task item1 = new Task();
        item1.setId("1");
        item1.setCompleted(false);
        item1.setDesc("go swimming");
        tasks.add(item1);
        Task item2 = new Task();
```

```
        item2.setId("2");
        item2.setCompleted(true);
        item2.setDesc("go for lunch");
        tasks.add(item2);
        return tasks;
    }
}
```

上面这个文件也比较简单，但注意到以下几个事情。

- @RestController 和@RequestMapping 这两个是注解，这种方式原来在 .Net 中很常见，后来 Java 也引进过来。一方面它们可以增加代码的可读性，另一方面也有效减少了代码量。具体机理就不讲了，简单来说就是利用 Java 的反射机制和 IoC 模式的结合把注释的特性或属性注入被注释的对象中。
- 我们看到 List<Task> getAllTasks()方法中简单地返回了一个 List，并未做任何转换成 json 对象的处理，这个是 Spring 会自动利用 Jackson 这个类库的方法将其转换成 JSON 形式。

到这里就基本接近成功了，但是现在缺少一个入口，那么在 dev.local.gtm.api 包下面建立一个 Applicaiton.java 吧。

```
package dev.local.gtm.api;

import org.springframework.boot.SpringApplication;
import org.springframework.boot.autoconfigure.SpringBootApplication;

/**
 * Spring Boot 应用的入口文件
 */
@SpringBootApplication
public class Application {

    public static void main(String[] args) {
        SpringApplication.run(Application.class, args);
    }
}
```

同样，我们只需简单地标注这个类是@SpringBootApplication 就可以了，习惯了 Spring 烦琐的 XML 配置的读者可能需要适应一下。

3.2.4 启动服务

使用下面任意一种方法去启动 Web 服务。

- 命令行中：./gradlew bootRun。
- 在 IDEA 中，选择 Application，右击选择 Run 'Application'，如图 3-6 所示。

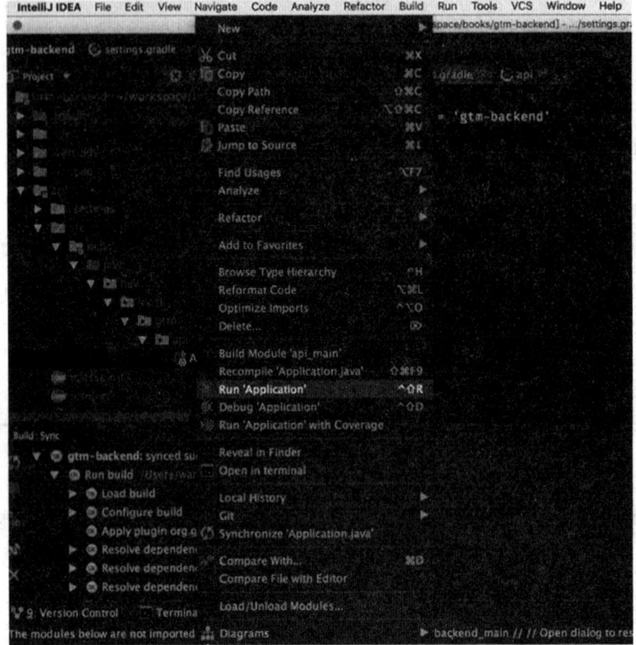

图 3-6

如果使用命令行，则可以输入下面的命令启动 API 服务。

```
./gradlew :api:bootRun
```

注意，上面的命令是在 backend 根目录执行的，而我们的 API 是整个项目的一个子项目，所以首先采用 :api 这种方式指明是哪个子项目，然后使用 :bootRun 指定执行什么任务。所以，运行命令就是如下所示的样子。

```
./gradlew :<子项目名称>:<任务名称>
```

执行命令后，如果你可以看到类似下面的输出，那就说明服务启动成功了。

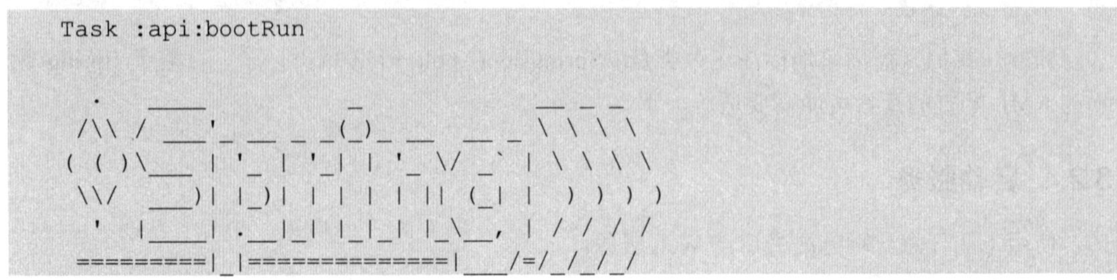

```
Task :api:bootRun

  .   ____          _            __ _ _
 /\\ / ___'_ __ _ _(_)_ __  __ _ \ \ \ \
( ( )\___ | '_ | '_| | '_ \/ _` | \ \ \ \
 \\/  ___)| |_)| | | | | || (_| |  ) ) ) )
  '  |____| .__|_| |_|_| |_\__, | / / / /
 =========|_|==============|___/=/_/_/_/
```

```
    :: Spring Boot ::        (v2.0.0.RELEASE)

    2018-04-02 09:53:09.506  INFO 48808 --- [           main]
dev.local.gtm.api.Application            : Starting Application on
wangpengdeMacBook-Pro.local with PID 48808
    ...
    AnnotationConfigServletWebServerApplicationContext@c8e4bb0: startup date
[Mon Apr 02 09:53:09 CST 2018]; root of context hierarchy
    2018-04-02 09:53:11.640  INFO 48808 --- [           main]
    <=============> 75% EXECUTING [11m 11s]
    > :api:bootRun
    > IDLE
    > IDLE
    > IDLE
```

启动后，打开浏览器访问 http://localhost:8080/tasks ，就可以看到 JSON 形式的返回结果了，如图 3-7 所示。

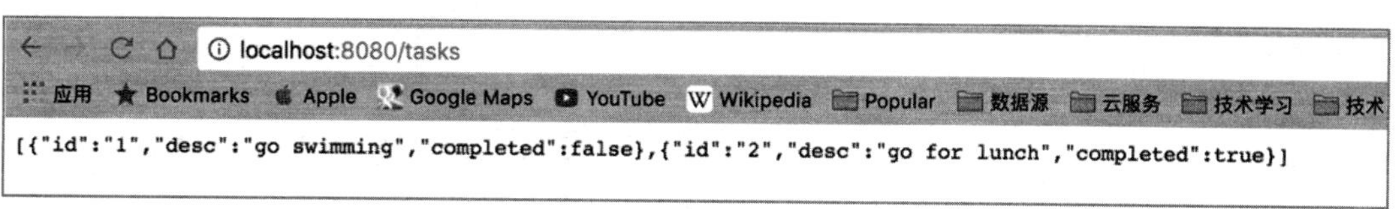

图 3-7

是不是有种感觉，做一个 API 也太容易了？对了，Spring Boot 的一个主要作用就是将我们从烦琐的配置和大量模式化代码中解放出来。

3.2.5 测试 API

上面的测试只是通过浏览器测试一个简单的 GET 请求，但复杂一些的请求就没办法这么测试了。“工欲善其事，必先利其器”，我们得有一个合适的测试工具才好。*nix 用户可能已经比较熟悉 curl，对于新手或者不熟悉 curl 的读者，笔者推荐 Postman，一个有图形化界面的专业 HTTP API 测试工具，支持 macOS/Linux/Windows，大家可以下载对应平台的版本。

Postman 的界面非常友好，基本上看一下就知道大概怎么用了，图 3-8 中标识了大部分测试 API 中最常见的一些参数配置区域和返回的显示区域。

从图 3-8 中也可以很清晰地看到第一个 API 返回的结果，由于 Postman 会很智能地识别返回的格式（这个例子里面是 JSON 形式），显示的效果比浏览器好。

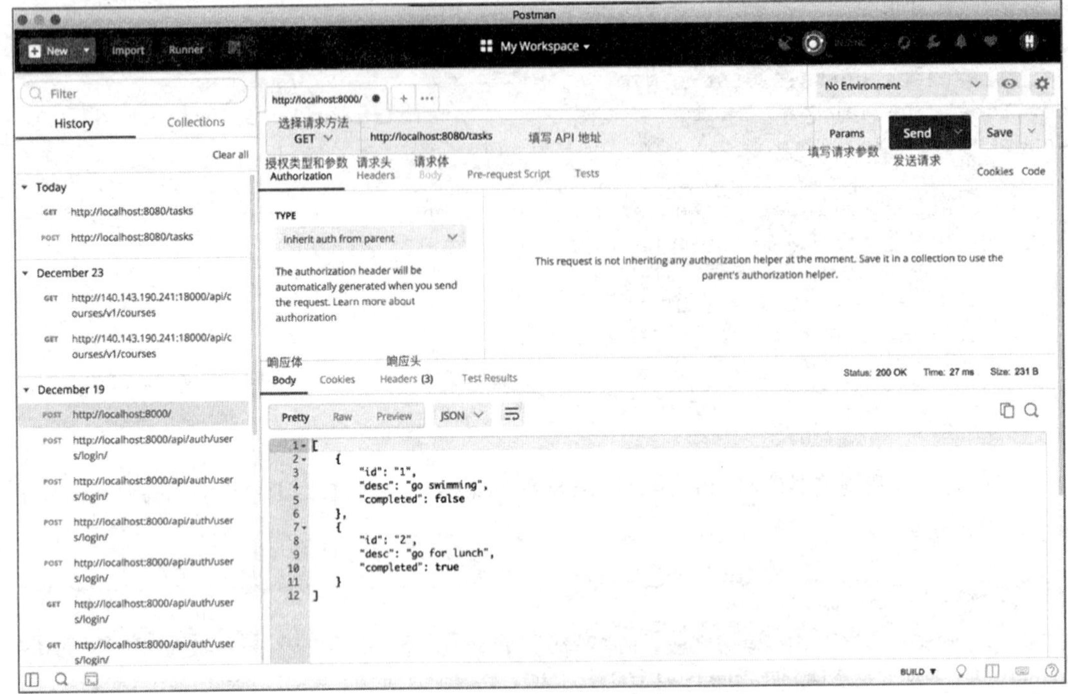

图 3-8

3.3 MongoDB 支撑的 API

一般来说，做后端数据库是必不可少的，我们提供的 API 很多也是需要查询或修改数据库中的某些数据的。数据库这些年的发展也日新月异，但近 10 年来最显著的变化之一应该是 NoSQL 数据库的兴起了。

3.3.1 什么是 NoSQL

NoSQL 是 Not Only SQL 的缩写，而不是许多人印象中的 No SQL，其实更正式的名字应该称作非关系型数据库。关系型数据库大家都很熟知了，比如 MySQL、Oracle、SQL Server 等，而 NoSQL 数据库泛指不能划归到关系型数据库中的其他数据库。这个类别可是很广泛的，也就是说，即使都是 NoSQL 数据库，差异还是很大的。表 3-1 列出了一些典型的 NoSQL 数据库和它们的分类，当然类型的划分只是一个大概的参考，它们之间没有绝对的分界，也有交叉的情况。

表 3-1

类　型	典型数据库	特　点
列存储	Hbase、Cassandra	按列存储数据，特点是方便存储结构化和半结构化数据，方便做数据压缩，对针对某一列或者某几列的查询有非常大的 I/O 优势
文档存储	MongoDB、CouchDB	用类似 JSON 的格式存储，存储的内容是文档型的。这样也就有机会对某些字段建立索引，实现关系型数据库的某些功能
键值存储	MemcacheDB、Redis	键值对形式，通过 Key 快速查询到其 Value 值
图存储	Neo4J	图形关系的最佳存储

1. 为什么要使用 NoSQL

要牢记的一点是，NoSQL 数据库并没有带来新的功能，传统关系型数据库其实从功能角度完全可以支持任何业务场景。NoSQL 兴起的原因是在某些领域，关系型数据库遇到了性能瓶颈。传统关系型数据库一般在遇到性能问题时都会采用分表分库、主从复制、异构复制等。但花费的精力和投入的资源越来越大，这正是 NoSQL 要解决的问题。

（1）NoSQL 的优势

- 大数据量，高性能——NoSQL 数据库都具有非常高的读写性能，尤其在大数据量下，同样表现优秀。这得益于它的无关系性，数据库的结构简单。
- 灵活的数据模型——NoSQL 无须事先为要存储的数据建立字段，随时可以存储自定义的数据格式。而在关系型数据库里，增删字段是一件非常麻烦的事情。如果是非常大数据量的表，增加字段简直就是一个噩梦。
- 高可用——NoSQL 在不太影响性能的情况，就可以方便地实现高可用的架构。

2. SQL 和 NoSQL

需要注意的一点是，数据库在 SQL 和 NoSQL 之间并不是非此即彼的关系，相反，NoSQL 数据库与 SQL 数据库往往并存于企业的数据架构中。所以从使用角度来说，要分析业务场景，哪些适合关系型数据库，哪些适合非关系型数据库。

非关系型数据库和关系型数据库结合使用。举例来说，业务系统如果可以使用关系型数据库，而且没有性能问题就不必使用非关系型数据库，但现在客户需要一些用户行为分析的特性，而这个部分我们可以确定以后需要统计的维度越来越多，而且需要进行海量数据的查询，那么采用非关系型数据库就可以很好地进行这一块业务的设计，而关系型数据库也节省了大量 I/O。

非关系型数据库替代关系型数据库。假设现在要构建一个社交网络，用户可以发布状态信息（信息内容包括文本、视频、音频和图片等），那么可以画出一个如图 3-9 所示的表关系结构。

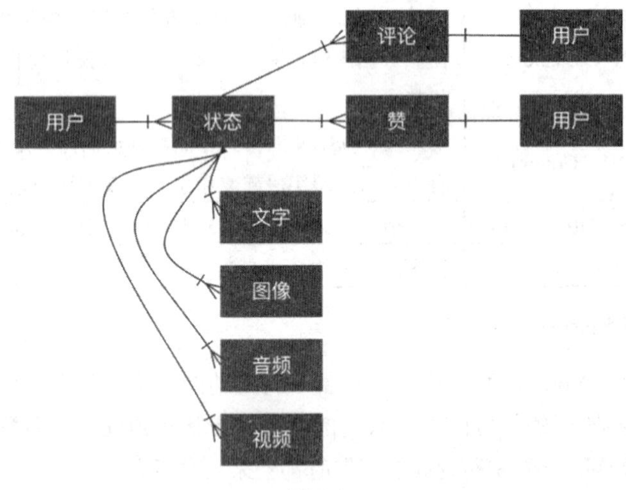

图 3-9

在这种情况下，这样状态的结构怎么在页面中显示，如果我们希望显示状态的文字，以及关联的图片、音频、视频、用户评论、赞和用户的信息，则需要关联 8 个表取得自己想要的数据。如果我们有这样的状态列表，而且是随着用户的滚动动态加载，同时需要监听是否有新内容的产生。这样一个任务我们需要太多这种复杂的查询了。

NoSQL 中的文档型数据库解决这类问题的思路是，抛弃传统的表结构直接存储和传输一个这样的数据，如下所示。

```
{
    "id":"5894a12f-dae1-5ab0-5761-1371ba4f703e",
    "title":"2017 年的 Spring 发展方向",
    "date":"2017-01-21",
    "body":"这篇文章主要探讨如何利用 Spring Boot 集成 NoSQL",
    "createdBy":User,

"images":["http://dev.local/myfirstimage.png","http://dev.local/mysecondimag
e.png"],
    "videos":[
        {"url":"http://dev.local/myfirstvideo.mp4", "title":"The first
video"},
        {"url":"http://dev.local/mysecondvideo.mp4", "title":"The second
video"}
    ],
```

```
        "audios":[
            {"url":"http://dev.local/myfirstaudio.mp3", "title":"The first audio"},
            {"url":"http://dev.local/mysecondaudio.mp3", "title":"The second audio"}
        ]
    }
```

NoSQL 一般情况下是没有严格的 Schema 约束的，这也给开发带来较大的自由度。因为在关系型数据库中，一旦 Schema 确定，以后更改 Schema，维护 Schema 是很麻烦的一件事。但反过来说，Schema 对于维护数据的完整性却是非常必要的。

一般来说，如果你在做一个 Web、物联网等类型的项目，你应该考虑使用 NoSQL。如果你要面对的是一个对数据的完整性、事务处理等有严格要求的环境（比如财务系统），那么应该考虑使用关系型数据库。

3. 如何选择 NoSQL 数据库

这么多类型的 NoSQL 数据库，到底怎么选择呢？影响选择的因素有很多，而选择也可能包含多种数据库，一般要分析具体的业务场景和需求。

- 数据结构特点：包括结构化、半结构化、字段是否可能变更、是否有大文本字段、数据字段是否可能变化等。
- 数据写入的特点：包括插入（insert）和更新（update）比例、是否经常更新数据的某一部分，是否有原子更新要求等。
- 数据查询特点：包括查询的条件、查询的范围。

3.3.2 MongoDB 的集成

MongoDB 是一个分布式的文档型数据库，由 C++ 语言编写，旨在提供可扩展的高性能数据存储解决方案。是非关系型数据库当中功能比较丰富，比较像关系型数据库的，因此也是 NoSQL 中应用较广泛的一种。

1. 安装 MongoDB

想要安装 MongoDB，当然可以去官网下载，手动安装。但本书建议使用 Docker 进行安装。这样做的优点是，你可以轻松管理多个版本的 MongoDB，比如这个项目使用了 MongoDB 2.2，那个项目使用了 MongoDB 3.6，如果采用本地安装，那么可能会比较头疼了，但使用 Docker 就安全没有压力。

```
## 直接抓取最新的 mongoDB 镜像
docker pull mongo
```

```
## 和上面的效果一样，显式声明了版本
docker pull mongo:latest
## 拉取 3.6.3 版本的 mongoDB
docker pull mongo:3.6.3
```

通过 docker images 命令可以查看已经拉取的镜像列表，下面的输出就是笔者在电脑上执行该命令后的结果。

```
REPOSITORY      TAG                       IMAGE ID       CREATED         SIZE
gtm_nginx       latest                    522c49bf937c   42 hours ago    17.9MB
nginx           latest                    7f70b30f2cc6   11 days ago     109MB
elasticsearch   2.4.6                     cb429ea8b6e8   2 weeks ago     574MB
memcached       latest                    e4aa3d5e67a3   4 weeks ago     60.8MB
wordpress       latest                    80a6fca6cc6a   6 weeks ago     407MB
rabbitmq        3.7-management-alpine     21d4ff1bda19   2 months ago    85.5MB
openjdk         8-jre-alpine              b1bd879ca9b3   2 months ago    82MB
nginx           1.13.8-alpine             bb00c21b4edf   2 months ago    16.8MB
node            8-alpine                  7c2983dfbf98   2 months ago    68.1MB
mongo           3.4.10                    e905a87e116d   3 months ago    360MB
openjdk         8u121-jdk-alpine          630b87931295   10 months ago   101MB
```

拉取镜像之后，你就可以根据镜像建立并启动容器了，为了避免麻烦，我们还是映射 MongoDB 默认端口 27017 到本地的 27017。

```
## 根据 mongoDB 3.6.3 镜像版本建立一个叫作 mymongo 的容器
docker run -p 27017:27017 --name mymongo -d mongo:3.6.3
```

如果要测试一下，则可以通过以下命令登录到容器中：

```
docker exec -it mymongo bash
```

此时，你应该可以看到终端的提示符改变了，变成了类似 root@a0b609105000: 这种，这说明你已经登录到容器中。在容器中执行 mongo，可以看到如下输出：

```
MongoDB shell version v3.4.10
connecting to: mongodb://127.0.0.1:27017
MongoDB server version: 3.4.10
Welcome to the MongoDB shell.
For interactive help, type "help".
For more comprehensive documentation, see
 http://docs.mongodb.org/
Questions? Try the support group
 http://groups.google.com/group/mongodb-user
Server has startup warnings:
2018-04-02T10:09:55.572+0000 I STORAGE  [initandlisten]
2018-04-02T10:09:55.572+0000 I STORAGE  [initandlisten] ** WARNING:
Using the XFS filesystem is strongly recommended with the WiredTiger storage
```

```
engine
   2018-04-02T10:09:55.572+0000 I STORAGE  [initandlisten] **          See
http://dochub.mongodb.org/core/prodnotes-filesystem
   2018-04-02T10:09:56.017+0000 I CONTROL  [initandlisten]
   2018-04-02T10:09:56.017+0000 I CONTROL  [initandlisten] ** WARNING:
Access control is not enabled for the database.
   2018-04-02T10:09:56.018+0000 I CONTROL  [initandlisten] **          Read
and write access to data and configuration is unrestricted.
   2018-04-02T10:09:56.018+0000 I CONTROL  [initandlisten]
```

接下来，我们看看有哪些数据库，输入 show databases，不出意外的话，应该可以看到两个默认内置的数据库。

```
admin  0.000GB
local  0.000GB
```

至此，MongoDB 测试成功，输入 exit 退出容器。

2. 配置 Spring Boot 使用 MongoDB

Spring 的应用一般通过外部配置文件，比如 *.properties 或者 *.yml 文件来对应用进行配置。Spring 默认情况下寻找配置文件会依照以下顺序进行。

- 当前目录下的 config 文件夹。
- 当前目录。
- 位于 classpath 中命名为 config 的包。
- classpath 的根路径。一个典型的 Web 应用的资源 classpath 是如下位置，所以一般我们习惯把 *.properties 或者 *.yml 放在 src/main/resources 目录下。

```
/META-INF/resources/
/resources/
/static/
/public/
```

下面，在 src/main/resources 目录中建立一个 application.yml 文件，并在其中指定要连接的 MongoDB 的数据库名称。注意，其实不指定这个数据库名称，Spring Boot 也可以正常启动并连接 MongoDB，只不过这时连接的数据库默认是 test，而这种方式在正式开发中要尽量避免。

.yml 格式非常简单，以缩进来表示层级关系，比如我们要给 spring.data.mongodb. database 这个属性设置为 gtm-api，也就是设置数据库名称为 gtm-api，就可以通过下面的. yml 文件完成。

```
## application.yml
spring:
  data:
    mongodb:
```

```
      database: gtm-api
```

这个数据库现在还是不存在的，但是当第一次写操作时，这个 gtm-api 会自动被建立。

3. 改写 Task

在我们刚刚的项目中，集成 MongoDB 很简单，只需更改一下子项目 API 的 build.gradle，增加两个依赖：

```
implementation("org.springframework.boot:spring-boot-starter-data-rest")
implementation("org.springframework.boot:spring-boot-starter-data-
mongodb")
```

是的，就这么简单，这就是 Spring Boot 给开发者提供的开箱即用的体验。当然这个只是集成 MongoDB，如果要使用，则还得做一点工作，当然也很简单。

首先，需要 Task 这个领域对象做一点小修改，在原有的 Id 属性前加一个 @Id 的注解，以标识这个字段是数据库中的 Id 字段。修改后的 Task 如下：

```java
package dev.local.gtm.api.domain;

import org.springframework.data.annotation.Id;

public class Task {
    @Id
    private String id;
    private String desc;
    private boolean completed;

    public String getId() {
        return id;
    }

    public void setId(String id) {
        this.id = id;
    }

    public String getDesc() {
        return desc;
    }

    public void setDesc(String desc) {
        this.desc = desc;
    }

    public boolean isCompleted() {
```

```
        return completed;
    }

    public void setCompleted(boolean completed) {
        this.completed = completed;
    }
}
```

领域对象修改完，我们得有数据访问，所以建立一个和 domain 平行的包，叫 repository。然后在下面建一个接口叫 TaskRepo：

```
package dev.local.gtm.api.repository;

// 省略导入

@RepositoryRestResource(collectionResourceRel = "tasks", path = "tasks")
public interface TaskRepo extends MongoRepository<Task, String> {
}
```

这个接口非常简单，继承了 MongoRepository ，这样一个简单的继承的能量会大到让你吃惊，常见的增、删、改、查，包括分页、排序等操作，完全不用写一行代码。

注意一下@RepositoryRestResource 这个注解，collectionResourceRel 指定了集合资源的名称，怎么理解呢？好比说，我们现在要得到 tasks 的列表，访问 API 后返回的 JSON 中会有 tasks: [...]，这个 tasks 就是在 collectionResourceRel 中的值。如果改成 collectionResourceRel = "todos"，那么返回的 JSON 中就变成了 todos: [...]。而 path 指的是这个资源的相对路径，path = "tasks" 就会为 respository 生成一个 API 路径：http://localhost:8080/ tasks。

看到这里你应该也知道了，TaskController 已经没用了，因为有能力更强的顶替了它，那么删掉 TaskController.java，然后终止之前启动的应用。如果你是在 terminal 中启动的，就按 CTRL+C 组合键；如果是 IDE，就按停止按钮，最后重新启动应用。

此时，测试一下 API，还是在 Postman 中输入 http://localhost:8080/tasks，可以看到如图 3-10 所示类似的输出。

API 是可以访问的，但返回的 JSON 的样子看起来有点奇怪，这个其实是一个符合 HyperMedia 的 Rest API 的返回结果，是一个 Rest 之上的进一步的封装标准，又叫 HATEOAS，当然这是一个缩写词：Hypermedia As The Engine Of Application State。那么究竟什么叫 HATEOAS 呢？我们一起往下看。

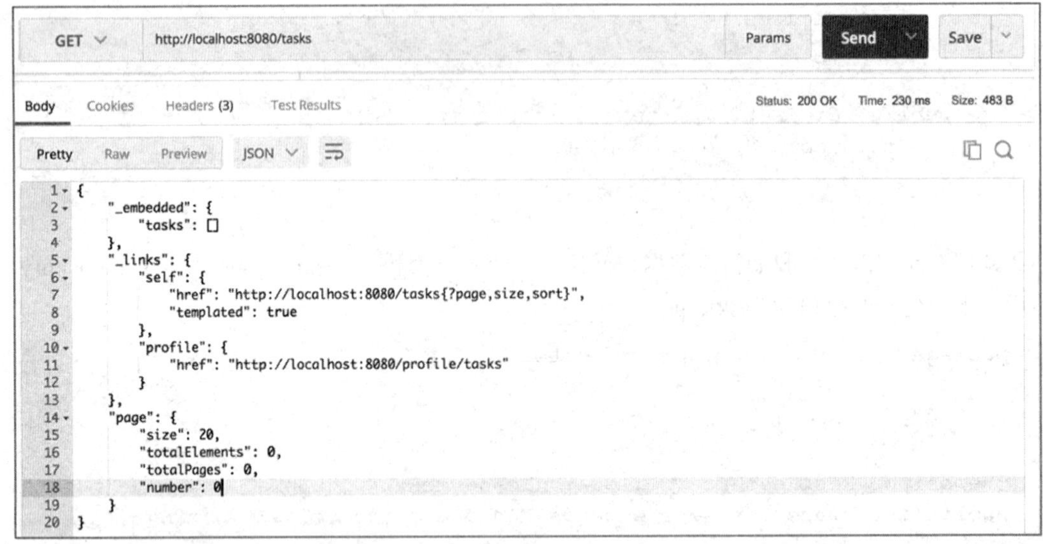

图 3-10

3.3.3 HATEOAS

简单来说，HATEOAS 可以让客户端清晰地知道自己可以做什么，而无须依赖服务器端指示你做什么。原理也很简单，通过返回的结果中不仅包括数据本身，也包括指向相关资源的链接。拿上面的例子来说（虽然这种默认状态生成的内容不是很有代表性）：links 中有一个 profiles，我们看看这个 profile 的链接 http://localhost:8080/profile/tasks 执行的结果是什么：

```
{
    "alps": {
        "version": "1.0",
        "descriptors": [
            {
                "id": "task-representation",
                "href": "http://localhost:8080/profile/tasks",
                "descriptors": [
                    {
                        "name": "desc",
                        "type": "SEMANTIC"
                    },
                    {
                        "name": "completed",
                        "type": "SEMANTIC"
                    }
                ]
```

```
            },
            {
                "id": "create-tasks",
                "name": "tasks",
                "type": "UNSAFE",
                "rt": "#task-representation"
            },
            {
                "id": "get-tasks",
                "name": "tasks",
                "type": "SAFE",
                "rt": "#task-representation",
                "descriptors": [
                    {
                        "name": "page",
                        "doc": {
                            "value": "The page to return.",
                            "format": "TEXT"
                        },
                        "type": "SEMANTIC"
                    },
                    {
                        "name": "size",
                        "doc": {
                            "value": "The size of the page to return.",
                            "format": "TEXT"
                        },
                        "type": "SEMANTIC"
                    },
                    {
                        "name": "sort",
                        "doc": {
                            "value": "The sorting criteria to use to
calculate the content of the page.",
                            "format": "TEXT"
                        },
                        "type": "SEMANTIC"
                    }
                ]
            },
            {
                "id": "update-task",
                "name": "task",
                "type": "IDEMPOTENT",
```

```
                    "rt": "#task-representation"
                },
                {
                    "id": "patch-task",
                    "name": "task",
                    "type": "UNSAFE",
                    "rt": "#task-representation"
                },
                {
                    "id": "delete-task",
                    "name": "task",
                    "type": "IDEMPOTENT",
                    "rt": "#task-representation"
                },
                {
                    "id": "get-task",
                    "name": "task",
                    "type": "SAFE",
                    "rt": "#task-representation"
                }
            ]
        }
    }
```

虽然暂时不能完全地理解这个对象，但大致可以猜出来，这个是 Task API 的元数据描述，告诉我们这个 API 中定义了哪些操作和接受哪些参数等。

其实，Spring 使用了一个叫 ALPS 的专门描述应用语义的数据格式。摘出下面这一小段代码来分析一下，其描述了一个 get 方法，类型是 SAFE，表明这个操作不会对系统状态产生影响（因为只是查询），而且这个操作返回的结果格式定义在 task-representation 中。

```
{
  "id": "get-task",
  "name": "task",
  "type": "SAFE",
  "rt": "#task-representation"
}
```

还是不太理解？没关系，我们再来做一个实验，用 Postman 构建一个 POST 请求，首先如图 3-11 所示，添加一个键值对到“Headers”中，别忘了选择左侧的下拉框中的 POST 方法。

选择 Body，单击“raw”单选项，输入如图 3-12 所示的 JSON，当作请求体。

POST　http://localhost:8080/tasks　Params　Send　Save

Authorization　Headers (1)　Body ●　Pre-request Script　Tests　Cookies　Code

Key	Value	Description
Content-Type	application/json	

图 3-11

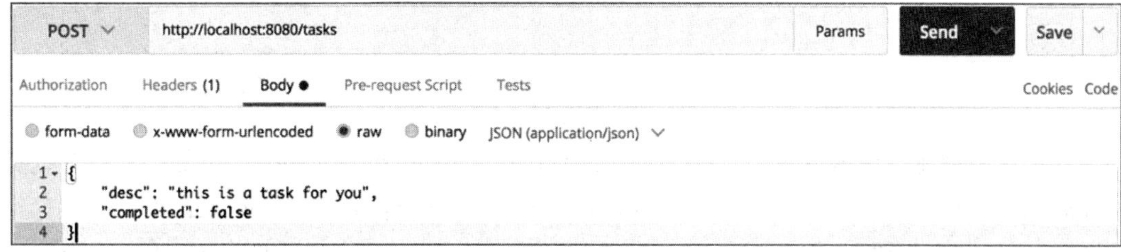

图 3-12

单击“Send”按钮，可以看到如图 3-13 所示输出。

Body　Cookies　Headers (4)　Test Results　Status: 201 Created　Time: 481 ms　Size: 468 B

Pretty　Raw　Preview　JSON

```
{
    "desc": "this is a task for you",
    "completed": false,
    "_links": {
        "self": {
            "href": "http://localhost:8080/tasks/5ac21918446d71e6f417f225"
        },
        "task": {
            "href": "http://localhost:8080/tasks/5ac21918446d71e6f417f225"
        }
    }
}
```

图 3-13

可以看到返回的链接中包括了刚刚新增的 Task 的 API 链接 http://localhost:8080/tasks/5ac21918446d71e6f417f225（5ac21918446d71e6f417f225 就是数据库自动为这个 Task 自动生成的 Id），这样客户端可以方便地知道指向刚刚生成的 Task 的 API 链接。

我们可以试一下，在 Postman 中单击这个返回结果中的链接，Postman 会自动为这个链接新建一个标签，单击“Send”按钮，就可以得到刚刚创建的 Task 了，如图 3-14 所示。

再举一个现实一些的例子，在开发一个“我的”页面时，一般情况下，除取得我的某些信息外，还会有一些可以链接到更具体信息的页面。如果客户端在取得比较概要信息的同时就得到这些详情的链接，那么客户端的开发就比较简单了，而且也更灵活。

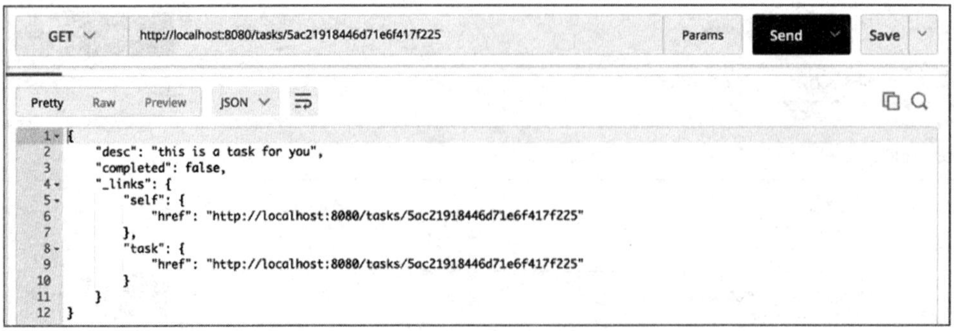

图 3-14

其实这个描述中还告诉我们一些分页的信息，比如每页 20 条记录(size: 20)、总共几页（totalPages：1）、总共多少个元素（totalElements: 1）、当前是第几页（number: 0）。当然你也可以在发送 API 请求时，指定 page、size 或 sort 参数。比如 http://localhost:8080/ tasks?page=0&size=10 就是指定每页 10 条，当前页是第 1 页（从 0 开始）。

1．HAL 浏览器

使用 Postman 查看这种 API 的返回还是有点不直观，那么我们可以在 build.gradle 中添加一个依赖：

```
implementation("org.springframework.data:spring-data-rest-hal-browser")
```

重新启动应用，访问 http://localhost:8080/browser/index.html#/，然后就可以使用这个 HAL 浏览器去实验和观察 API 的请求和返回，如图 3-15 所示。

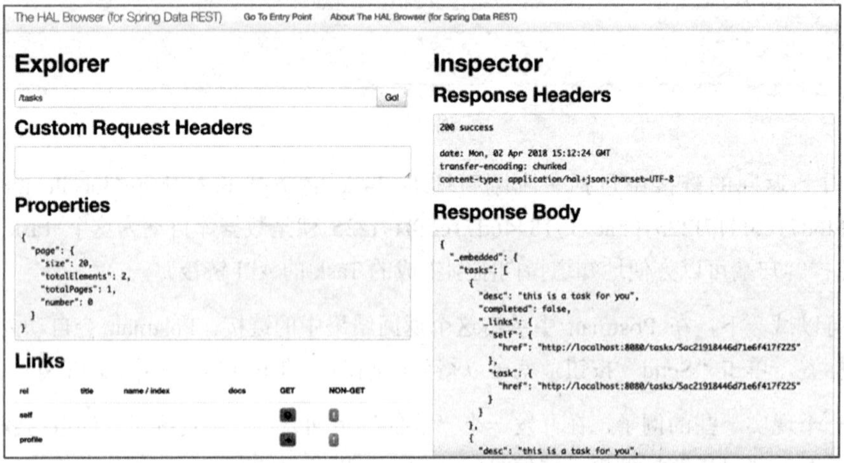

图 3-15

3.3.4 “魔法”的背后

第一个魔法是这么简单就生成了一个有数据库支持的 REST API，看起来比较“魔幻”，但一般这么“魔幻”的事情总感觉不太靠谱，除非我们知道背后的原理是什么。下面再来回顾一下 TaskRepo 的代码：

```
@RepositoryRestResource(collectionResourceRel = "tasks", path = "tasks")
public interface TaskRepo extends MongoRepository<Task, String> {
}
```

Spring 是最早的几个 IoC（控制反转或者叫 DI）框架之一，所以最擅长的就是依赖的注入。这里我们写了一个接口（interface），可以猜到 Spring 一定是有一个接口的实现在运行时注入了进去。如果我们去 spring-data-mongodb 的源码中看一下就知道是怎么回事了，这里只举一个小例子，大家可以去看一下 SimpleMongoRepository.java 文件，由于源码太长，只截取一部分：

```
    public class SimpleMongoRepository<T, ID extends Serializable>
implements MongoRepository<T, ID> {

        private final MongoOperations mongoOperations;
        private final MongoEntityInformation<T, ID> entityInformation;

        /**
            * Creates a new {@link SimpleMongoRepository} for the given
{@link MongoEntityInformation} and {@link MongoTemplate}.
            *
            * @param metadata must not be {@literal null}.
            * @param mongoOperations must not be {@literal null}.
            */
        public SimpleMongoRepository(MongoEntityInformation<T, ID> metadata,
MongoOperations mongoOperations) {

            Assert.notNull(mongoOperations);
            Assert.notNull(metadata);

            this.entityInformation = metadata;
            this.mongoOperations = mongoOperations;
        }

        /*
            * (non-Javadoc)
            * @see
org.springframework.data.repository.CrudRepository#save(java.lang.Object)
            */
        public <S extends T> S save(S entity) {
```

```
            Assert.notNull(entity, "Entity must not be null!");

            if (entityInformation.isNew(entity)) {
                mongoOperations.insert(entity,
entityInformation.getCollectionName());
            } else {
                mongoOperations.save(entity,
entityInformation.getCollectionName());
            }

            return entity;
        }
        ...
        public T findOne(ID id) {
            Assert.notNull(id, "The given id must not be null!");
            return mongoOperations.findById(id,
entityInformation.getJavaType(), entityInformation.getCollectionName());
        }

        private Query getIdQuery(Object id) {
            return new Query(getIdCriteria(id));
        }

        private Criteria getIdCriteria(Object id) {
            return where(entityInformation.getIdAttribute()).is(id);
        }
        ...
    }
```

也就是说，在运行时 Spring 将这个类或者其他具体接口的实现类注入了应用。这个类中有支持各种数据库的操作。一般了解到这步就可以了，有兴趣的读者可以继续深入研究。

虽然不想在具体类上继续研究，但我们还是应该多了解一些关于 MongoRepository 的内容。这个接口继承了 PagingAndSortingRepository（定义了排序和分页）和 QueryByExampleExecutor。而 PagingAndSortingRepository 又继承了 CrudRepository（定义了增、删、改、查等操作）。

第二个魔法就是它直接将 MongoDB 中的集合（本例中的 tasks）映射到了一个 REST URI（tasks）。因此我们连 Controller 都没写就把 API 做出来了，而且还是个 Hypermedia REST。

其实，这个注解 @RepositoryRestResource(collectionResourceRel = "tasks", path = "tasks") 只在你有特殊需求时才需要，比如是否需要将 repository 暴露到 REST 资源？是否需要重新定义这个资源的集合名称和单个名称（比如有时候某些单词的复数形式是特殊的，就需要指定一下 collectionResourceRel 和 path）？本例中如果我们不加 @RepositoryRestResource 这个注解，那同样也可以生成 API，只不过其路径按照默认英语复数定义路径。也就是说如果我们不想按默认路径，那么指定这个注解可以自定义 API 路径，读者可以试试把这个注解去掉，然后重启服务，访问 http://localhost:8080/tasks 看看。

1. REST 的简单介绍

REST 简单来说可以理解成以下几个基本规则。

- 以资源名称构成 API 路径，也就是定义资源时尽量使用名词。
- 以动词定义要对资源执行的操作，也就是使用不同的请求方法（GET/POST/PUT/DELETE 等）访问资源路径就代表着要对资源采用什么操作。一般 GET 表示查询，POST 代表新增，PUT 表示更改，而 DELETE 表示删除。
- 资源的复数形式表示列表，而列表后的资源唯一标识表示取得具体列表中的某一资源。

（1）几个例子

比如 /api/users 代表了 users 这个资源。

- 一般来说，如果以 GET 方法访问 /api/users 表示要得到用户的列表。此处需要注意，列表路径是复数形式，如果没做特殊处理，那么 Spring 会按照英语的语法给出复数形式。
- 以 POST 方法访问 /api/users 表示要新增一个用户，这个要新增的用户信息一般以 JSON 形式作为 Request Body 上传。
- 那么如何得到某一个用户呢？使用 GET 方法访问 /api/users/:id，比如，如果用户 ID 是 1，那么这个用户的 URL 是 /api/users/1。
- 对该用户的更新是通过 PUT 或 PATCH 方法访问 /api/users/:id。同样的修改信息以 JSON 形式作为 Request Body 上传。PUT 和 PATCH 方法的区别在于 PUT 方法要求上传该对象的完整形式，而且更新也是全部进行更新。而 PATCH 方法是可以上传部分改变的信息，比 PUT 方法要节省传输带宽。
- 删除用户即使用 DELETE 方法访问 /api/users/:id。

（2）REST 的成熟度

REST API 的评价标准为 Richardson 成熟度模型（Richardson Maturity Model）。

- Level 0（Swamp of POX）：这一级其实一点都不符合 REST，所以是 Level 0，也就是你使用了 HTTP 或者类似的协议。某一种 HTTP 方法多数情况是 POST 方法。
- Level 1（Resources）：这一级开始定义资源，也就是说对于不同的资源，有不同的 URL 标识，但仍是使用 HTTP 的某一种方法，多数情况是 POST

- Level 2（HTTP Verbs）：这一级除了资源的划分，要用不同的方法实现对不同资源的操作，比如 POST 方法代表新增，PUT 方法代表修改，DELETE 方法代表删除，而 GET 方法表示查询。此外，需要使用协议规定的响应码标识不同的状态，比如在出现错误时返回 200 就是一个错误用法。
- Level 3（Hypermedia Controls）：这一级就是客户端可以自己发现 API，我们上面做出的 HATEOAS 类型的 API 就是符合 Level 3 的。但可惜的是，在现实开发中，目前达到这一级别的团队很少。

3.3.5 让后端也能热更新

热更新，就是在不停止服务的情况下，将代码的改变立即部署到应用程序中。前端工程领域大部分框架都标配了热更新，这样使得开发的体验非常好。其实后端领域 Node.js 和一些 Python、Ruby 框架也是支持热更新的，那么 Java 后端可不可以呢？Spring Boot 作为一个开箱即用的框架当然会考虑到这一点。

1．Spring Boot DevTools

我们只需要在 API 子项目的 build.gradle 中加入一行代码：implementation("org.springframework.boot:spring-boot-devtools")，即可实现热更新。由于 Java 不是一个动态语言，所以这个“热更新”其实是重启服务。但 Spring Boot Devtools 提供的“重启”不是冷启动，而是通过两个类加载器，文件没有改变的作为“基线类加载器”，更新的文件作为“重启类加载器”。只有“重启类加载器”是一次性使用的，用后就丢弃掉，“基线类加载器”并不会每次重新创建，所以这比冷启动要快很多。

```
apply plugin: 'org.springframework.boot'

dependencies {
    implementation("org.springframework.boot:spring-boot-devtools")
    implementation("org.springframework.boot:spring-boot-starter-web")
    implementation("org.springframework.boot:spring-boot-starter-data-
rest")
    implementation("org.springframework.boot:spring-boot-starter-data-
mongodb")
    implementation("org.springframework.data:spring-data-rest-hal-
browser")
}
```

在 IDEA 中选择“Preference→Compiler → Build Project automatically”，如图 3-16 所示。

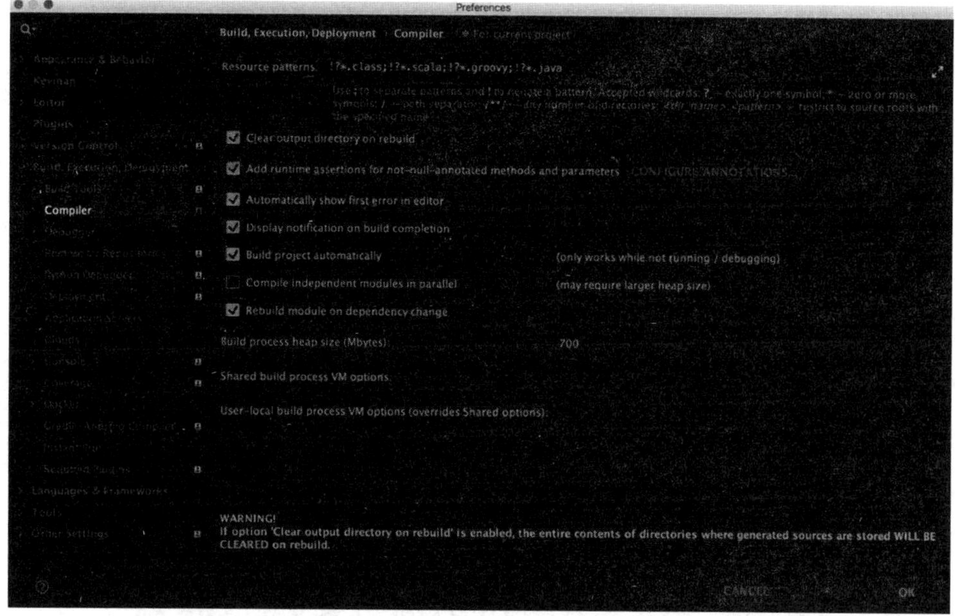

图 3-16

默认情况下，IDEA 对于运行中的项目不会自动编译，我们需要修改一下。按住 cmd + Shift + A（Windows 系统中使用 Ctrl + Shift + A）组合键选择“Registry”，如图 3-17 所示。

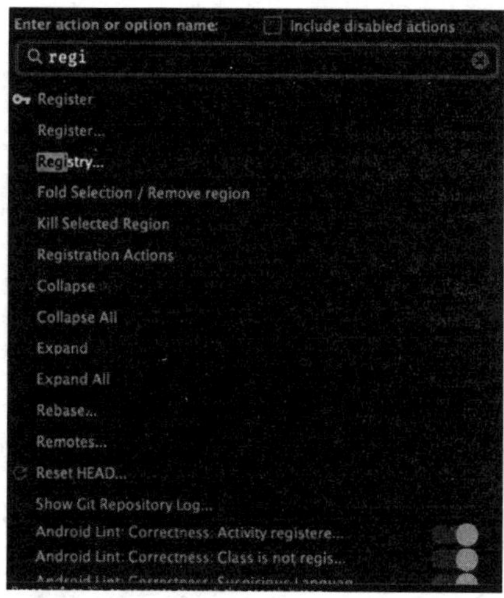

图 3-17

找到 complier.automake.allow.when.app.running 这一项，勾选后面的复选框，如图 3-18 所示。

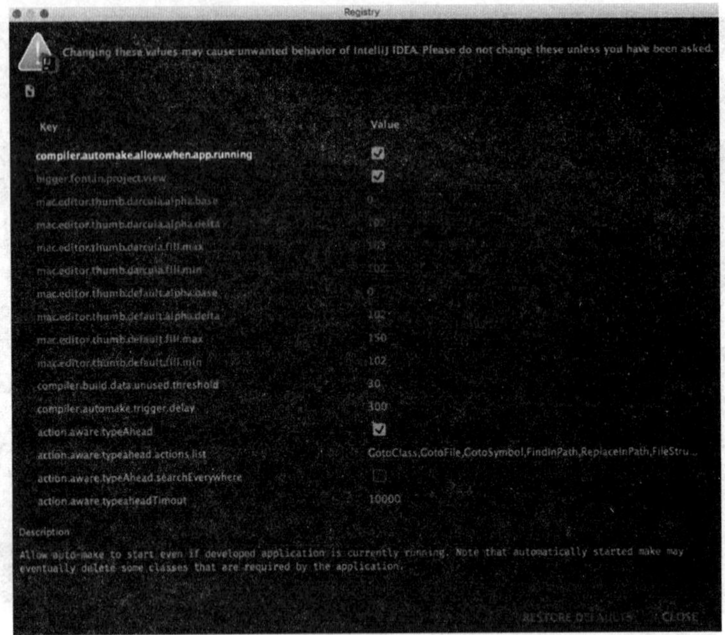

图 3-18

在 IDEA 中，鼠标右击选择“Application→Run Application”。现在尝试更新一些文件，在 Console 中就可以看到服务自动重启了，如图 3-19 所示。

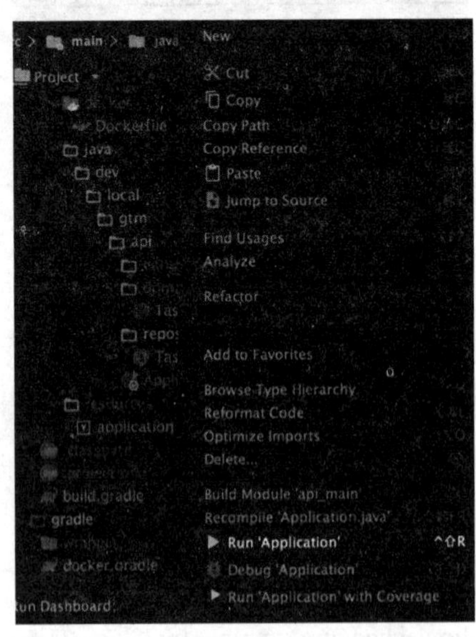

图 3-19

Spring Boot 默认对 classpaht 下的文件更新会触发重启，但对于静态资源，一般情况下不会引起热更新，Spring Boot 默认对以下文件夹的变化不会触发更新：/META-INF/maven, 、/META-INF/resources、/resources、/static、/public 和 /templates。你可以通过调整 spring. devtools. restart.exclude 属性来决定哪些文件被排除在更新触发器之外。同样也可以指定哪些不再 classpath 下的文件会触发更新，只需设置属性 spring.devtools.restart.additional-paths 即可。

对于静态资源，Spring Boot 提供了 LiveReload 的方式进行热更新。但这个在我们的前后端分离项目中没有意义，所以就不在这里阐述了。

3.4 容器化后端

就像在 Angular 中使用镜像一样，我们也可以将后端应用容器化。

3.4.1 手动创建镜像

首先在 backend/api/src/main 下建立 docker 文件夹，然后在 docker 文件夹中创建一个 Dockerfile 文件，如图 3-20 所示。

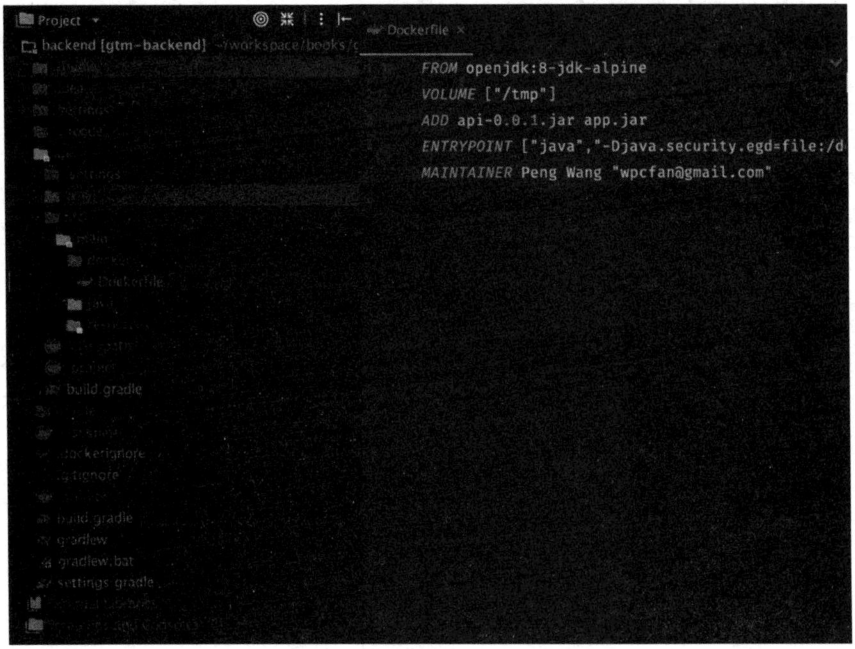
Project
backend [gtm-backend]
Dockerfile
FROM openjdk:8-jdk-alpine
VOLUME ["/tmp"]
ADD api-0.0.1.jar app.jar
ENTRYPOINT ["java","-Djava.security.egd=file:/d
MAINTAINER Peng Wang "wpcfan@gmail.com"
build.gradle
.dockerignore
.gitignore
build.gradle
gradlew
gradlew.bat
settings.gradle

图 3-20

代码如下：

```
// 以 Java 8 jdk 镜像为基础
FROM openjdk:8-jdk-alpine
VOLUME ["/tmp"]
// 将编译后的 jar 文件加入容器
ADD api-0.0.1.jar app.jar
// 入口执行的方法
ENTRYPOINT ["java","-Djava.security.egd=file:/dev/./urandom","-jar","/app.jar"]
// 镜像维护者的信息
MAINTAINER Peng Wang "wpcfan@gmail.com"
```

简单说明一下，既然是一个 Java 工程，我们就以 jdk 的镜像为基础来构建我们自己的镜像，要做的事情也很简单，就是把打包好的 jar 文件加入镜像。ENTRYPOINT 起的作用其实是相当于把数组中的各个元素拼成一个命令执行：

```
java -Djava.security.egd=file:/dev/./urandom -jar /app.jar
```

我们可以测试一下这个 Dockerfile 是否可以成功创建镜像。在 Dockerfile 所在目录执行：

```
docker build -t gtm/api:0.0.1 .
```

如果不出意外的话，那么你会遇到一个错误：

```
Sending build context to Docker daemon  2.048kB
Step 1/5 : FROM openjdk:8-jdk-alpine
 ---> 224765a6bdbe
Step 2/5 : VOLUME ["/tmp"]
 ---> Using cache
 ---> 60efb1149bd3
Step 3/5 : ADD api-0.0.1.jar app.jar
ADD failed: stat /var/lib/docker/tmp/docker-builder039755918/api-0.0.1.jar: no such file or directory
```

这有两个原因，首先我们没有将 jar 进行打包，其次 Dockerfile 中的 ADD 默认找的路径就是 Dockerfile 所在路径，而这个路径下没有这个 jar。找不到 jar，因此镜像创建失败了。知道了原因，我们就进行一下编译，在项目 backend 目录下执行：

```
./gradlew :api:bootJar
cp ./api/build/libs/api-0.0.1.jar ./api/src/main/docker
```

然后回到 Dockerfile 所在目录执行：

```
docker build -t gtm/api:0.0.1 .
```

这次，我们应该可以看到镜像创建成功了，但是 docker 目录里面的 jar 其实没什么用了，删掉它。

```
Sending build context to Docker daemon  21.54MB
Step 1/5 : FROM openjdk:8-jdk-alpine
 ---> 224765a6bdbe
Step 2/5 : VOLUME ["/tmp"]
 ---> Using cache
 ---> 60efb1149bd3
Step 3/5 : ADD api-0.0.1.jar app.jar
 ---> c5cfd21a8410
Step 4/5 : ENTRYPOINT ["java","-
Djava.security.egd=file:/dev/./urandom","-jar","/app.jar"]
 ---> Running in 10a4837b2cf7
Removing intermediate container 10a4837b2cf7
 ---> a7853ee194f0
Step 5/5 : MAINTAINER Peng Wang "wpcfan@gmail.com"
 ---> Running in c3b210b305d2
Removing intermediate container c3b210b305d2
 ---> a5bd57615fd4
Successfully built a5bd57615fd4
Successfully tagged gtm/api:0.0.1
```

这么操作也太麻烦了，嗯，笔者也这么觉得，所以我们要把这个过程进行自动化。

3.4.2 使用 Gradle 自动化 Docker 任务

使用 Gradle 或 Maven 这种脚本的好处之一在于你拥有了无限扩展的可能性，有很多插件可以扩充这个脚本的能力。这里我们要使用的是一个 Gradle Docker 插件。这个插件专门给 Gradle 脚本添加了很多 Docker 的处理能力。这里我们只使用了几个：创建 Dockerfile，复制文件到镜像构建文件夹和创建镜像。

首先，在 backend 下的 Gradle 子目录中新建一个 docker.gradle 文件。

```
buildscript {
    repositories {
        jcenter()
    }

    dependencies {
        classpath 'com.bmuschko:gradle-docker-plugin:3.2.6'
    }
}

apply plugin: com.bmuschko.gradle.docker.DockerRemoteApiPlugin
```

```
    task createDockerfile(type:
com.bmuschko.gradle.docker.tasks.image.Dockerfile, dependsOn: ['bootJar']) {
        description = "自动创建 Dockerfile"
        destFile = project.file('src/main/docker/Dockerfile')
        from 'openjdk:8-jdk-alpine'
        volume '/tmp'
        addFile "${project.name}-${project.version}.jar", "app.jar"
        instruction { 'ENTRYPOINT [' +
                '"java", ' +
                '"-Djava.security.egd=file:/dev/./urandom", ' +
                '"-jar","/app.jar"]'}
        maintainer 'Peng Wang "wpcfan@gmail.com"'
    }

    task copyDockerFiles(type: Copy, dependsOn: 'createDockerfile') {
        description = "复制 Dockerfile 和相应数据到镜像构建文件夹"
        from 'src/main/docker'
        from "${project.buildDir}/libs"
        into { "${project.buildDir}/docker" }
        include "*"
        exclude "**/*.yml"
    }

    task buildDocker(type:
com.bmuschko.gradle.docker.tasks.image.DockerBuildImage, dependsOn:
'copyDockerFiles') {
        description = "打包应用构建镜像"
        group = "twigcodes"
        inputDir = project.file("${project.buildDir}/docker")
        tags = ["${project.name}:latest".toString(),
"${project.name}:${project.version}".toString()]
    }
```

这个文件是专门处理 Docker 相关任务的，所以我们单独建立一个文件，以后其他类似情况也会为专门的任务建立 Gradle 脚本文件，这样有助于我们管理一个日趋庞大的项目。

由于这个插件做的工作大部分是将之前的手动完成的动作自动化，读者可以参照官方文档和我们的代码自行体会一下，这里就不展开讲述了。

有了这个文件，我们还需要在主文件中处理，所以在 backend/build.gradle 中的 subprojects 区块添加一行代码：apply from: '../gradle/docker.gradle'，这相当于为每个子项目添加了 Docker 任务。

```
    // 略去其他部分
```

```
subprojects {
    // 略去
    apply plugin: 'java'
    apply plugin: 'idea'
    apply from: '../gradle/docker.gradle'
    apply plugin: 'eclipse'
    apply plugin: 'io.spring.dependency-management'
    // 略去
}
```

现在，我们的任务可以完全自动化了，可以自动生成 Dockerfile：

```
./gradlew :api:createDockerfile
```

可以自动化生成镜像：

```
./gradlew :api:buildDocker
```

由于 buildDocker 依赖 copyDockerFiles，而 copyDockerFiles 依赖 createDockerfile，所以其实只需要执行 buildDocker，就会先生成 Dockerfile，然后创建镜像。

镜像已经建立了，接下来，我们创建一个容器，来测试一下：

```
docker run --rm --name api api:0.0.1
```

我们会发现出现了一些异常，如果使用 Postman 进行 API 测试也会返回错误。

```
    2018-04-03 15:03:53.912  INFO 1 --- [localhost:27017]
org.mongodb.driver.cluster: Exception in monitor thread while connecting to
server localhost:27017

    com.mongodb.MongoSocketOpenException: Exception opening socket
        at com.mongodb.connection.SocketStream.open(SocketStream.java:62)
~[mongodb-driver-core-3.6.3.jar!/:na]
        at com.mongodb.connection.InternalStreamConnection.
open(InternalStreamConnection.java:126) ~[mongodb-driver-core-3.6.3.jar!/:na]
        at  com.mongodb.connection.DefaultServerMonitor$ServerMonitorRunnable.
run(DefaultServerMonitor.java:114) ~[mongodb-driver-core-3.6.3.jar!/:na]
        at java.lang.Thread.run(Thread.java:748) [na:1.8.0_151]
    Caused by: java.net.ConnectException: Connection refused (Connection
refused)
        at java.net.PlainSocketImpl.socketConnect(Native Method)
~[na:1.8.0_151]
        at java.net.AbstractPlainSocketImpl.doConnect
(AbstractPlainSocketImpl.java:350) ~[na:1.8.0_151]
        at java.net.AbstractPlainSocketImpl.connectToAddress
(AbstractPlainSocketImpl.java:206) ~[na:1.8.0_151]
        at java.net.AbstractPlainSocketImpl.connect(AbstractPlainSocketImpl.
```

```
java:188) ~[na:1.8.0_151]
        at java.net.SocksSocketImpl.connect(SocksSocketImpl.java:392)
~[na:1.8.0_151]
        at java.net.Socket.connect(Socket.java:589) ~[na:1.8.0_151]
        at com.mongodb.connection.SocketStreamHelper.initialize
(SocketStreamHelper.java:59) ~[mongodb-driver-core-3.6.3.jar!/:na]
        at com.mongodb.connection.SocketStream.open(SocketStream.java:57)
~[mongodb-driver-core-3.6.3.jar!/:na]
        ... 3 common frames omitted
```

这是由于我们将 API 容器化之后，它无法访问到 MongoDB 了，这就有点奇怪了，我们的 MongoDB 一直运行着，一直都在 27017 端口，为会么容器化之后就访问不到了？可以这样理解，容器化相当于把服务关进一个小屋，服务可以看到这个小屋中的各种号码牌（把端口想像成号码牌），但是它看不到屋外的号码牌，MongoDB 在屋外立了一个 27017 的号码牌，但是在小黑屋里看不到啊。

这个问题怎么解决呢？数据库的连接一般都是通过某种协议访问的，而 MongoDB 也不例外，它可以通过 TCP 连接：mongodb://[IP/HOSTNAME]: [端口号]/[数据库名]。那么我们可以更改 docker 任务，传递这个数据库连接信息到 Spring Boot。

```
    task createDockerfile(type:
com.bmuschko.gradle.docker.tasks.image.Dockerfile) {
        description = "自动创建 Dockerfile"
        destFile = project.file('src/main/docker/Dockerfile')
        from 'openjdk:8-jdk-alpine'
        volume '/tmp'
        addFile "${project.name}-${project.version}.jar", "app.jar"
        instruction { 'ENTRYPOINT ["java", "-
Dspring.data.mongodb.uri=mongodb://mymongo/taskmgr", "-
Djava.security.egd=file:/dev/./urandom", "-jar","/app.jar"]'}
        maintainer 'Peng Wang "wpcfan@gmail.com"'
    }
```

增加的这行代码-Dspring.data.mongodb.uri=mongodb://mongo/taskmgr 是会设置 spring.data.mongodb.uri 为 mongodb://mymongo/taskmgr 的。注意，这里采用了主机名的形式，这是因为我们会在一台宿主机器上运行，并没有多个 IP。看到这里，你可能会问那主机名不也是有问题吗，怎么让小屋里的知道外部的主机名呢？Docker 提供了一个 link 参数，在启动容器时使用这个参数让容器可以和其他容器互相通信。

```
    docker run --rm --name api api:0.0.1 --link mymongo
```

如果服务多起来，这样处理还是很麻烦的，这就需要使用对于多个服务的容器脚本 docker-compose。

3.4.3 使用 docker-compose 组合服务

使用 docker-compose 的几个明显的好处如下所示。

- 避免每次写带有很多参数的命令，脚本化之后效率会提高。
- 需要启动多个服务，而且彼此有依赖时候，这个构建过程会非常复杂，脚本可以帮助我们清晰地定义关系，简化流程。

使用 docker-compose 需要在 backend 目录下建立一个 docker-compose.yml。在 services 区块下定义各项服务及服务的各项参数。

```
version: '3.2'
services:
  mongo:
    image: mongo:3.4.10
    ports:
      - "27017:27017"
    volumes:
      - api_db:/data/db
  api-server:
    image: api
    ports:
      - "8080:8080"
    links:
      - mongo
volumes:
  api_db:
```

这个.yml 文件中的内容都比较浅显，我就不做详细解释了，具体 docker-compose 的用法可以参照官网文档进行学习。

需要注意的一个地方是 volumes: 的作用，这个是做容器的文件夹和宿主的文件夹映射，主要作用有共享数据和保存数据两个方面。就数据库而言，一般建议将数据库的数据文件夹映射到宿主机器，否则容器销毁时，数据就全没了。

此外，docker-compose 中的各个服务会单独创建容器，不会使用系统中已存在的容器。所以我们需要改写一下 createDockerfile，将 mongodb://mymongo/taskmgr 改成 mongodb://mongo/taskmgr，也就是把 docker-compose.yml 中的 MongoDB 的服务名改成 mongo。

```
task createDockerfile(type:
com.bmuschko.gradle.docker.tasks.image.Dockerfile) {
    description = "自动创建 Dockerfile"
    destFile = project.file('src/main/docker/Dockerfile')
    from 'openjdk:8-jdk-alpine'
```

```
        volume '/tmp'
        addFile "${project.name}-${project.version}.jar", "app.jar"
        instruction { 'ENTRYPOINT ["java", "-
Dspring.data.mongodb.uri=mongodb://mongo/taskmgr", "-
Djava.security.egd=file:/dev/./urandom", "-jar","/app.jar"]'}
        maintainer 'Peng Wang "wpcfan@gmail.com"'
    }
```

docker-compose 可以启动一组服务，在 docker-compose.yml 所在的目录执行下面命令，可以启动所有在 yml 中定义的服务。

```
docker-compose up -d
```

结束这一组服务，也很简单：

```
docker-compose down
```

3.4.4 IDEA 中的 Gradle 支持

在 IDEA（Java 编程语言开发的集成环境）中，我们可以通过“View→Tool Window→Gradle”调出 Gradle 工具窗口，如图 3-21 所示。在工具窗口双击某个具体任务执行，如图 3-22 所示。

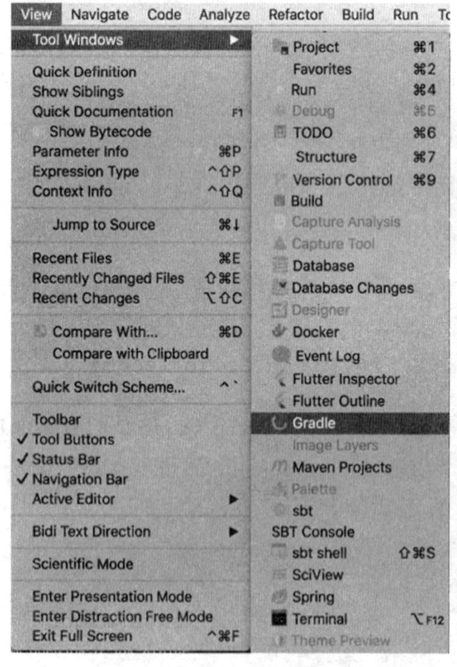

图 3-21

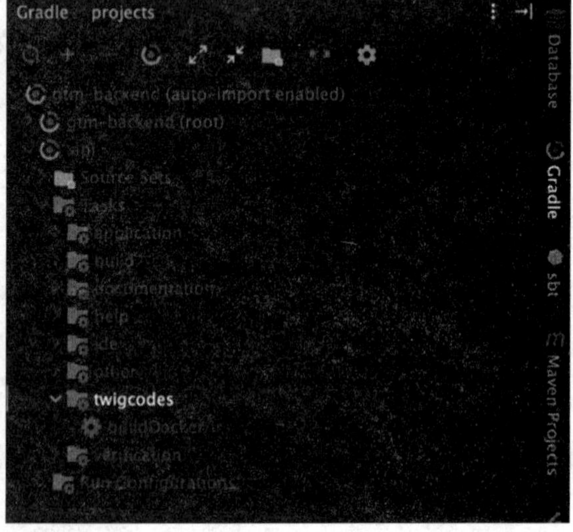

图 3-22

3.4.5 在容器中调试

形成容器之后，部署是比较方便了，那可能有读者要问，调试怎么办呢？其实这个也很简单，容器其实就是一个类似虚拟机的存在，那么主流 IDE（集成开发环境）对于远程调试都有良好的支持，我们可以把容器看作一个远程主机，所以调试当然也不成问题。

我们唯一要做的是在创建 Dockerfile 时，增加 ENTRYPOINT 的一个参数，-agentlib:jdwp=transport=dt_socket,address=5005,server=y,suspend=n，这其实是为了远程调试而设置数据传输的协议和监听端口等。现在 createDockerfile 如下所示。

```
task createDockerfile(type:
com.bmuschko.gradle.docker.tasks.image.Dockerfile) {
      description = "自动创建 Dockerfile"
      destFile = project.file('src/main/docker/Dockerfile')
      from 'openjdk:8-jdk-alpine'
      volume '/tmp'
      addFile "${project.name}-${project.version}.jar", "app.jar"
      instruction { 'ENTRYPOINT ["java", "-
agentlib:jdwp=transport=dt_socket,address=5005,server=y,suspend=n", "-
Dspring.data.mongodb.uri=mongodb://mongo/taskmgr", "-
Djava.security.egd=file:/dev/./urandom", "-jar","/app.jar"]'}
      maintainer 'Peng Wang "wpcfan@gmail.com"'
}
```

设置好这些，重新生成镜像：

```
./gradlew :api:buildDocker
```

由于在容器内多了一个监听远程调试的端口，我们也需要在 docker-compose.yml 中多加一个 5005:5005 的端口映射：

```
version: '3.2'
services:
  mongo:
    # 略过
  api-server:
    image: api
    ports:
      - "8080:8080"
      - "5005:5005"
    links:
      - mongo
## 略过
```

我们可以在 IDEA 中，选择 RUN→Edit Configurations，新建一个 Remote 类型的

Run/Debug Configuration，如图 3-23 所示。

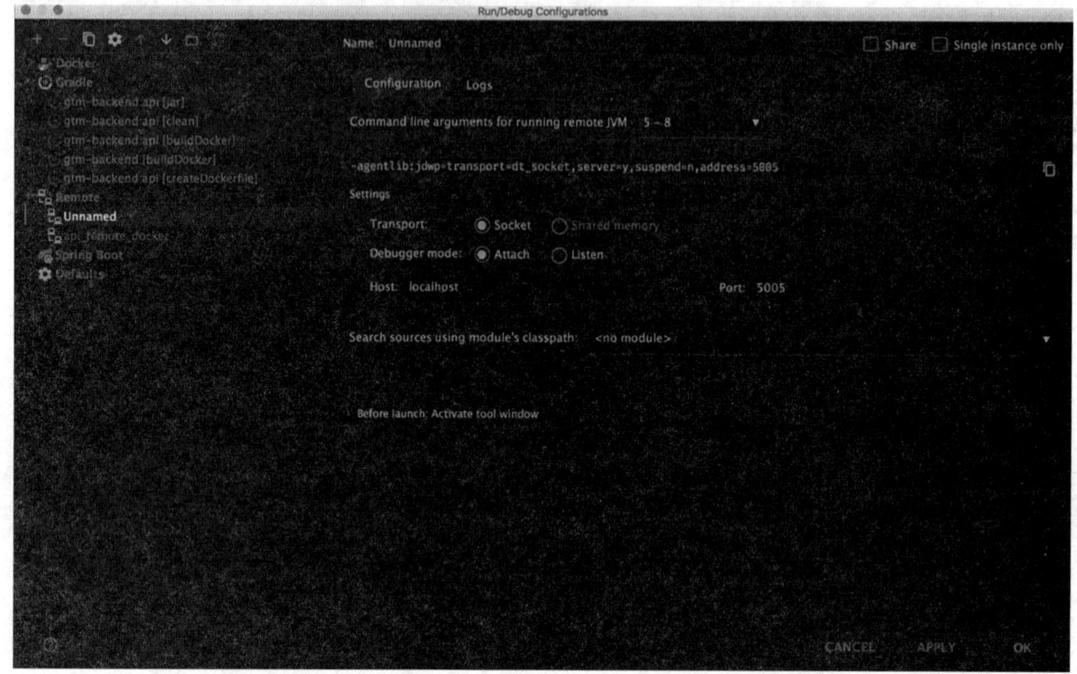

图 3-23

此时，如果你还没有启动 docker-compose，那么请执行 docker-compose up -d。然后启动 IDE，选中刚刚建立的 Remote 配置，点击调试。如果你看到下面这句话，那么我们的配置就全部完成了（见图 3-24）。

```
Connected to the target VM, address: 'localhost:5005', transport:
'socket'
```

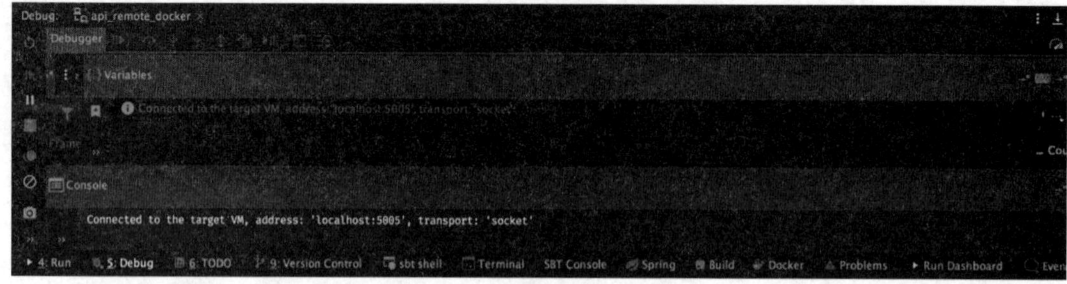

图 3-24

第 4 章 登录鉴权功能的构建

其实，我们完全可以使用普通的 CSS 和 HTML 去构建一个 Angular 应用，而使用类似 Angular Material 这种 UI 框架的好处在于，可以利用比较成型的 UI 组件快速开发，而不是花费精力重复制造轮子。所以在 4.1 节我们会继续学习几个 Angular Material 的组件，其中包括表单控件 FormField 和 MatInput；布局控件 GridList 和 Stepper；以及自定义表单控件的开发，我们会开发两个表单控件，图片选择器 ImagePicker 和验证手机号控件 VerifyMobile。

本章要学习的不仅仅是 Angular Material 的一些知识，还要学习 Angular 的模块化、服务层的开发、响应式编程的概念，以及响应式编程框架 RxJS，最后还会应用以上技巧打造一个较复杂的交互应用——忘记密码。

4.1 模块化和组件化

不论是企业级的应用还是面向消费市场的应用，一般都会有登录、注册等功能。我们要构建的是一个团队效率工具，所以这些功能也是不可缺少的。

4.1.1 登录的领域模型构建

开始之前需要注意，我们会把领域对象都建立在 src/app/domain 目录下，一般情况下每个领域对象建一个文件，有时几个对象关系特别紧密，我们也会把它们统一放在一个文件中。一个比较基础的用户对象如下所示。

```
export interface User {
  // 唯一标识
  id: string;
```

```
  // 用户名
  login: string;
  // 姓名
  name: string;
  // 密码
  password?: string
  // 手机号
  mobile: string;
  // 头像
  avatar: string;
  // 电子邮件
  email: string | null;
}
```

有的读者可能会注意到，密码这个字段是可选的。这是由于密码属于比较敏感的字段，一般在商用系统中不会暴露出来。所以从服务器读取用户信息时一般不会返回此字段。

这样，一个对象在注册时或者登录成功后返回时是比较适合的，但对于登录本身，其实只需用户名和密码即可，那么再新建一个 Auth 对象。为什么不复用 User 呢，这里就涉及编程中的一个重要原则：单一责任原则。任何对象或者方法都应该只负责一件事。鉴权这件事其实不是 User 对象应该负责的，因为鉴权时还不知道你是否是系统用户，需要判断用户名和密码之后才能确定你是系统用户，而这个判断中只不过恰好需要 User 中的用户名和密码两个字段而已。

```
export interface Auth {
  login: string;
  password: string;
}
```

那么有读者又有疑问了，为什么 Auth 对象是 login 的属性，而不是直接写 username 呢？这个是由于我们之后可能会有不止一种的登录方式，比如用户可不可以使用手机号登录呢？手机号登录现在基本是国内登录的标配了。那么 E-mail 登录或社交账号登录呢？为了以后不用频繁修改名字，我们就叫这个登录名为 login。

除了这两个对象还有什么对象？有的，但是这不是登录必需的，而是笔者自己添加的。因为干巴巴的一个登录，显得好无趣啊，页面也很空，所以笔者想将其作为一个效率工具，给每个人每天加加油。

于是，就有了这个“佳句”Quote 对象：我们会有一个图片链接、一个中文佳句（或者英文的翻译）和一个英文佳句（或对中文的翻译）。

```
export interface Quote {
  cn: string;
```

```
    en: string;
    imgUrl: string;
  }
```

4.1.2 前端页面设计

在开始构建前端之前，为了让读者有初步的印象，我们先看看成品页面。

登录页面由两个卡片构成，左边是佳句，右边是登录表单。登录表单下部有两个路由链接，可以跳转到注册和忘记密码页面，如图 4-1 所示。

图 4-1

注册页面比登录页面要复杂，复杂主要体现在表单验证比较复杂，包括必填字段验证、最小长度验证、最大长度验证、正则表达式验证，以及异步验证用户名、手机和电子邮件是否唯一等。这个页面还涉及一个自定义表单控件，就是选取头像的这个控件，如图 4-2 所示。

忘记密码页面采用了 Angular Material 的 Stepper 组件开发，Stepper 类似一个向导页面，一步一步地指引用户完成某个特定的业务。这个组件里面会使用两个表单，也就是每个步骤都是一个表单。里面还有一个较复杂的自定义控件，就是发送短信验证码的控件，如图 4-3 所示。

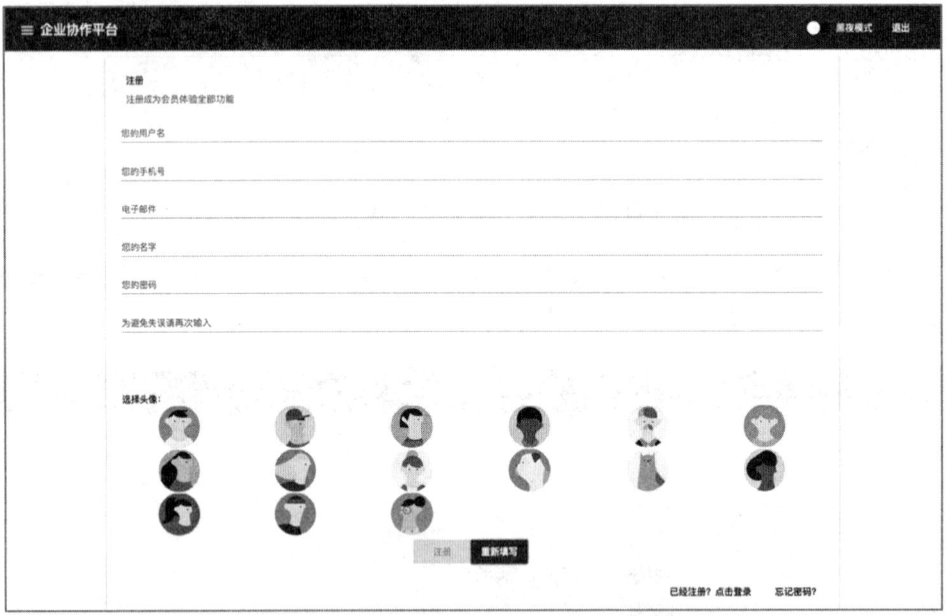
企业协作平台
黑夜模式
退出
注册
注册成为会员体验全部功能
您的用户名
您的手机号
电子邮件
您的名字
您的密码
为避免失误请再次输入
选择头像：
注册
重新填写
已经注册？点击登录
忘记密码？

图 4-2

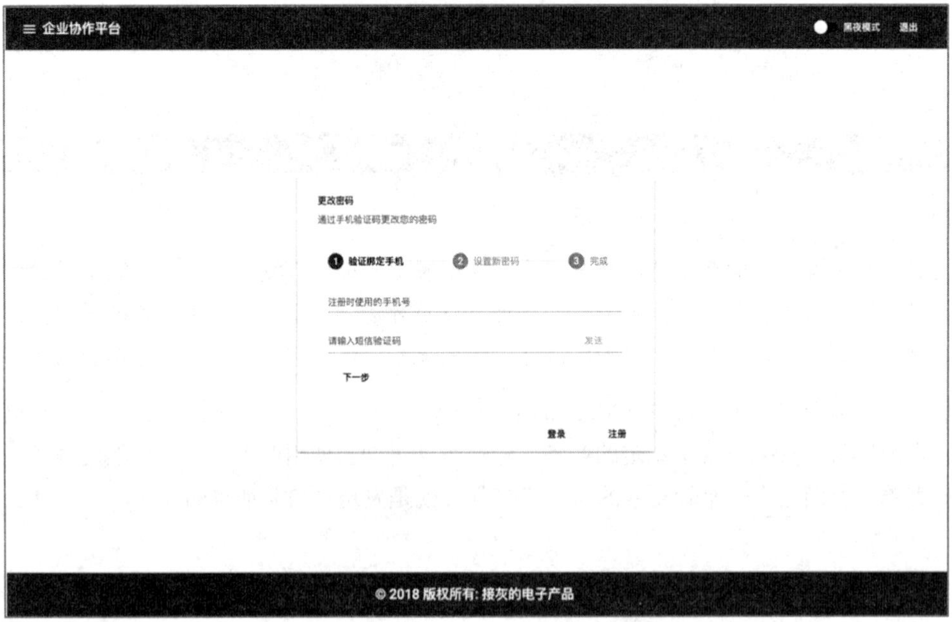
企业协作平台
黑夜模式
退出
更改密码
通过手机验证码更改您的密码
1 验证绑定手机
2 设置新密码
3 完成
注册时使用的手机号
请输入短信验证码
发送
下一步
登录
注册
© 2018 版权所有：接灰的电子产品

图 4-3

1．使用 CLI 生成模块

在 frontend 目录下输入：

```
ng g m auth --routing
```

Angular CLI 就会为我们生成 3 个文件，下面的就是执行命令后的输出，你会发现多了一个 auth 目录，这个目录里面就是这 3 个文件。

```
CREATE src/app/auth/auth-routing.module.ts (247 bytes)
CREATE src/app/auth/auth.module.spec.ts (259 bytes)
CREATE src/app/auth/auth.module.ts (271 bytes)
```

其中，auth.module.ts 是 Auth 模块文件，auth.module.spec.ts 是模块测试文件，而 auth-routing.module.ts 是这个路由的子模块文件。对于 auth.module.ts，我们把默认生成的 imports 数组中的 CommonModule 去掉，替换成 SharedModule。

```
import { NgModule } from '@angular/core';
import { AuthRoutingModule } from './auth-routing.module';
import { SharedModule } from '../shared/shared.module';

@NgModule({
  imports: [SharedModule, AuthRoutingModule]
})
export class AuthModule {}
```

接下来，我们在 auth 目录下创建 3 个子目录。

- components 目录——“笨组件”存放在这个目录，这个目录中组件，请记得设置 changeDetection: ChangeDetectionStrategy.OnPush。
- containers 目录——“聪明组件”存放在这个目录。
- services 目录——服务和路由守卫存放在此目录。

2．创建“笨组件”

本章涉及的“笨组件”都位于 src/app/auth/components 目录中，如果没有这个目录，请手动创建它。这也会是我们后期前端工程中的一个规范——任何一个模块的 components 目录中的都是这个模块的“笨组件”。

（1）登录表单组件

首先，我们还是创建一个 login-form 的“笨组件”，注意下面命令中的 --module auth/auth.module.ts 选项可以在 AuthModule 中声明新创建组件。

```
ng g c auth/components/login-form --module auth/auth.module.ts
```

这个命令会在 auth/components 目录下新创建一个叫 login-form 的目录，此目录下会生成 4 个文件：

- login-form.component.ts ——组件文件
- login-form.component.html——组件模板文件
- login-form.component.scss——组件样式文件
- login-form.component.spec.ts——组件单元测试文件

那么，我们首先更改其模板，模板是一个卡片布局：卡片的头部有标题和副标题；内容部分是一个简单的响应式表单，有"登录名"和"密码"两个输入框；而卡片的尾部操作部分是两个链接，可以导航到注册页面或者忘记密码页面。

```
<mat-card>
  <mat-card-header>
    <mat-card-title> {{ title }} </mat-card-title>
    <mat-card-subtitle> {{ subtitle }} </mat-card-subtitle>
  </mat-card-header>
  <mat-card-content>
    <form fxLayout="column" fxLayoutAlign="stretch center"
[formGroup]="form" (ngSubmit)="submit(form, $event)">
      <mat-form-field class="full-width">
        <input matInput type="text" placeholder="用户名"
formControlName="login">
        <mat-error>用户名是必填项哦</mat-error>
      </mat-form-field>
      <mat-form-field class="full-width">
        <input matInput type="password" placeholder="您的密码"
formControlName="password">
        <mat-error>密码不正确哦</mat-error>
      </mat-form-field>
      <div fxLayout="row" fxLayoutAlign="end stretch" class="full-width">
        <button mat-raised-button color="primary" type="submit"
[disabled]="!form.valid">登录</button>
      </div>
    </form>
  </mat-card-content>
  <mat-card-actions>
    <div fxLayout="row" fxLayoutAlign="end stretch">
      <a mat-button routerLink="/auth/register"> {{ regBtnText }} </a>
      <a mat-button routerLink="/auth/forgot"> {{ forgotBtnText }} </a>
    </div>
  </mat-card-actions>
</mat-card>
```

对应的组件文件中定义了页面显示的几个文本的输入型属性，以及一个事件 submitEvent，用于在表单提交时将表单数据发送给调用者。

```
// 省略导入

@Component({
  selector: 'app-login-form',
  templateUrl: './login-form.component.html',
  styleUrls: ['./login-form.component.scss'],
  changeDetection: ChangeDetectionStrategy.OnPush
})
export class LoginFormComponent implements OnInit {
  form: FormGroup;
  @Input() title = '登录';
  @Input() subtitle = '使用您的用户名密码登录';
  @Input() regBtnText = '还没有注册？';
  @Input() forgotBtnText = '忘记密码？';
  @Output() submitEvent = new EventEmitter<Auth>();

  constructor(private fb: FormBuilder) {}

  ngOnInit() {
    this.form = this.fb.group({
      login: [
        '',
        Validators.compose([
          Validators.required,
          Validators.pattern(usernamePattern)
        ])
      ],
      password: [
        '',
        Validators.compose([
          Validators.required,
          Validators.minLength(8),
          Validators.maxLength(50)
        ])
      ]
    });
  }

  submit({ value, valid }: FormGroup, ev: Event) {
    if (!valid) {
      return;
```

```
    }
    this.submitEvent.emit(value);
  }
}
```

（2）佳句组件

我们前面提到过在登录页面还需要有一个佳句，这个佳句也是以卡片形式提供的，和登录的卡片并列放置。我们使用 Angular CLI 创建一个 Quote 组件：

```
➜ ng g c auth/components/quote --module auth/auth.module.ts
CREATE src/app/auth/components/quote/quote.component.scss (0 bytes)
CREATE src/app/auth/components/quote/quote.component.html (24 bytes)
CREATE src/app/auth/components/quote/quote.component.spec.ts (621 bytes)
CREATE src/app/auth/components/quote/quote.component.ts (261 bytes)
UPDATE src/app/auth/auth.module.ts (1012 bytes)
```

更改 Quote 模板为一个卡片形式：中文为卡片头部的副标题，英文为卡片内容，而图片使用 matCardImage 指令拉伸到卡片宽度。

```
<mat-card>
  <mat-card-header>
    <mat-card-title> {{ title }} </mat-card-title>
    <mat-card-subtitle> {{ chineseQuote}} </mat-card-subtitle>
  </mat-card-header>
  <img matCardImage [src]="quoteImg">
  <mat-card-content>
    <p> {{ englishQuote }}</p>
  </mat-card-content>
</mat-card>
```

而它的对应组件文件比较简单:

```
import { Component, Input } from '@angular/core';

@Component({
  selector: 'app-quote',
  templateUrl: './quote.component.html',
  styleUrls: ['./quote.component.scss']
})
export class QuoteComponent {
  @Input() title = '佳句';
  @Input() chineseQuote = '';
  @Input() englishQuote = '';
  @Input() quoteImg = '';
}
```

（3）注册表单组件

相对来说，注册表单属于比较复杂的一个，下面的文件就是这个组件的模板。

```
<mat-card>
  <mat-card-header>
    <mat-card-title> {{ title }} </mat-card-title>
    <mat-card-subtitle> {{ subtitle }} </mat-card-subtitle>
  </mat-card-header>
  <mat-card-content>
    <form [formGroup]="form" (ngSubmit)="submit(form, $event)">
      <mat-form-field class="full-width">
        <input matInput placeholder="您的用户名" formControlName="login">
        <mat-error> {{ usernameErrors }} </mat-error>
      </mat-form-field>
      <mat-form-field class="full-width">
        <input matInput placeholder="您的手机号" formControlName="mobile">
        <mat-error> {{ mobileErrors }}</mat-error>
      </mat-form-field>
      <mat-form-field class="full-width">
        <input matInput placeholder="电子邮件" formControlName="email">
        <mat-error> {{ emailErrors }}</mat-error>
      </mat-form-field>
      <mat-form-field class="full-width">
        <input matInput type="text" placeholder="您的名字"
formControlName="name">
        <mat-error> {{ nameErrors }}</mat-error>
      </mat-form-field>
      <ng-container formGroupName="passwords">
        <mat-form-field class="full-width">
          <input matInput type="password" placeholder="您的密码"
formControlName="password">
          <mat-error> {{ passwordErrors }} </mat-error>
        </mat-form-field>
        <mat-form-field class="full-width">
          <input matInput type="password" placeholder="为避免失误请再次输入"
formControlName="repeat">
          <mat-error> {{ repeatErrors }}</mat-error>
        </mat-form-field>
      </ng-container>
      <div fxLayout="row" fxLayoutAlign="center stretch" class="full-
width">
        <button mat-raised-button color="primary" type="submit"
[disabled]="!form.valid">注册</button>
        <button mat-raised-button color="primary" type="reset">重新填写
```

```
</button>
          </div>
        </form>
      </mat-card-content>
      <mat-card-actions>
        <div fxLayout="row" fxLayoutAlign="end stretch">
          <a mat-button routerLink="/auth/login"> {{ loginBtnText }} </a>
          <a mat-button routerLink="/auth/forgot"> {{ forgotBtnText }} </a>
        </div>
      </mat-card-actions>
    </mat-card>
```

如果仔细看一下，就会发现我们漏掉了一个选择头像的功能。这是由于 Angular Material 内建组并没有提供。如果要实现这个功能，则需要使用 GridList 和一些 HTML 标签以及对应逻辑完成。这样的代码如果直接写在注册表单中，显然违背了“单一责任”原则。而且从代码的可读性来说也是非常不好的。

如果我们实现一个普通组件是否可以呢？当然会比前一种方案好很多，但是我们观察到表单控件，比如 input 是可以通过指定其 formControlName，然后在组建文件中初始化这个控件的，这个控件的验证状态和值变化等也可以通过 control.stateChanges 或 control.valueChanges 得到。那么我们有没有可能做成一个具备同样特性的控件呢？当然是可以了，下面就一起学习如何做一个表单控件。

首先用 Angular CLI 创建一个组件，但这个组件放在 auth 模块中显然是有点问题的。因为这个组件的复用可能性比较高，所以应该放在 shared 模块中，这样所有模块都可以使用这个组件了。当然这个还是个“笨组件”，如果 shared 目录下还没有 components 子目录，则请先创建。这里将这个组件命名为 ImagePicker。

```
ng g c shared/components/image-picker
```

注意，这次我们没有指定组件，这是因为 SharedModule 中由于导入导出的东西太多，做了一些特殊处理。我们不希望 Angular CLI 自动声明组件，所以只好手动添加。在 shared.module.ts 中更新 COMPONENTS 数组，添加 ImagePicker。

```
// 省略其他部分
const COMPONENTS = [ImagePicker];

@NgModule({
  declarations: COMPONENTS,
  imports: [...MODULES],
  exports: [...MODULES, ...COMPONENTS]
})
export class SharedModule {}
```

然后，看一下要实现的效果，这个表单控件分为上下两部分，下面是一个备选图片区域，上面是一个选中图片的显示区域。鼠标划过备选图片区域的图片时，图片会叠加一个遮罩，用来提示用户鼠标指向当前的图片。而由于备选图片有时可能会很多，所以这个备选区域应该是一个可滚动的区域，如图 4-4 所示。

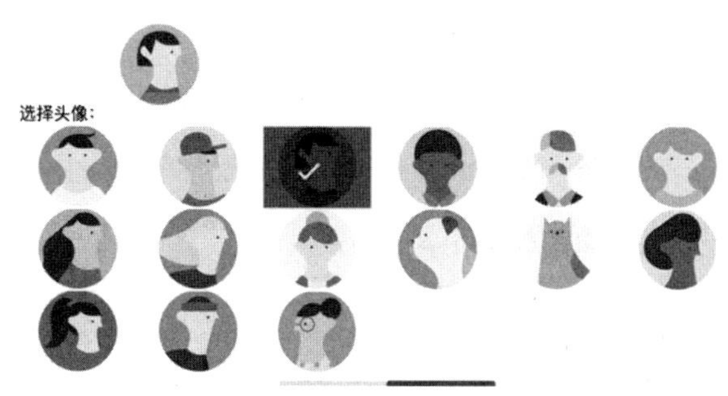

图 4-4

知道了需求，我们看看设计一下组件的模板，这里我们使用了一个新的 Angular Material 组件 MatGridList。这个组件和我们在 HTML 中见过的 table 有点像，都是以行列形式展现数据的。而且数据可以通过指定 colspan 和 rowspan 来占据多列或多行。

```
<mat-grid-list cols="3" rowHeight="2:1">
  <mat-grid-tile>1</mat-grid-tile>
  <mat-grid-tile colspan="2">2</mat-grid-tile>
  <mat-grid-tile rowspan="2">3</mat-grid-tile>
  <mat-grid-tile rowspan="2" colspan="2">4</mat-grid-tile>
</mat-grid-list>
```

效果如图 4-5 所示。

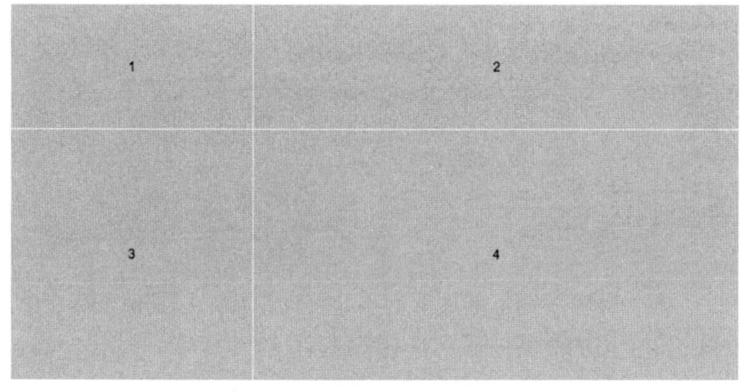

图 4-5

可以看到，有两个组件构成了 GridList：<mat-grid-list> 和 mat-grid-tile。<mat-grid-list> 必须指定 cols 属性用来确定网格中的列数。

每行的高度可以通过 rowHeight 属性进行设置，行高有以下几种形式。

- 固定高度：可以是 px、em 或者 rem 几种单位。
- 比例：这里的比例指的是“列宽：行高”。
- 自适应：设置为 fit 会依据 GridList 的高度或其容器高度自动为其分配行数。

此外，<mat-grid-list> 还有一个 gutterSize 属性用来设置单元格直接的间隔。

知道这些就可以开始设计组件模板了。

```
<div>
  <span>{{ title }}</span>
  <mat-icon class="avatar" [svgIcon]="selected" *ngIf="useSvgIcon else
imgSelected"></mat-icon>
  <ng-template #imgSelected>
    <img class="cover" [src]="selected">
  </ng-template>
</div>
<div class="scroll-container">
  <mat-grid-list [cols]="cols" [rowHeight]="rowHeight">
    <mat-grid-tile *ngFor="let item of items; let i = index;">
      <div class="image-container" (click)="selectImage(i)">
        <mat-icon class="avatar" [svgIcon]="item" *ngIf="useSvgIcon else
imgItem"></mat-icon>
        <ng-template #imgItem>
          <img [src]="item" [ngStyle]="{'width': itemWidth }">
        </ng-template>
        <div class="after">
          <span class="zoom">
            <mat-icon>checked</mat-icon>
          </span>
        </div>
      </div>
    </mat-grid-tile>
  </mat-grid-list>
</div>
```

值得注意的是，我们兼容了图标的选取而不是仅仅图片，这是由于两者的选取逻辑完全一致，仅仅是文件格式和使用的渲染控件上有所区别。

```
<mat-icon class="avatar" [svgIcon]="selected" *ngIf="useSvgIcon else
imgSelected">
```

```
  </mat-icon>
  <ng-template #imgSelected>
    <img class="cover" [src]="selected">
  </ng-template>
```

上面这段代码在 Angular 中是非常常见的范式——如果怎样就渲染什么，否则就渲染什么。而在 *ngIf="useSvgIcon else imgSelected" 中使用的 else 指定的是一个模板引用名 imgSelected，就是下面 ng-template 的引用名。

重点来了，我们看看一个表单控件应该怎么写。Angular 中提供了一个 ControlValueAccessor 接口，实现这个接口并注册为 NG_VALUE_ACCESSOR 就可以做出一个表单控件了。

在具体介绍这个接口之前，我们先回顾一些 Angular 中 FormControl 的概念。FormControl 是一个用于追踪表单控件的值和验证状态的实体类，无论你使用模板驱动型表单还是响应式表单，这个 FormControl 都会被创建。只不过在响应式表单中要指定 formControlName 指令去绑定这个 FormControl 到一个 HTML 原生控件中；而在模板驱动型表单中，这个绑定过程是隐含的，通过 ngModel 的绑定。因为 ngModel 的这个指令内部初始化了 FormControl，下面的 ngModel 源码片段会告诉我们原因。

```
@Directive({
  selector: '[ngModel]...',
  ...
})
export class NgModel ... {
  _control = new FormControl();    <---------------- 这里
```

所以，FormControl 是 Angular 中追踪和改变 HTML 原生控件的方式，而 ControlValueAccessor 接口就是 Angular Form API 和 HHTML 原生控件直接互相交互以及状态同步的桥梁。

接下来到要实现的表单控件，如果现在只是实现一个普通控件，那么其组件代码如下，非常简单，不做解释了。

```
// 省略导入

@Component({
  selector: 'app-image-picker',
  templateUrl: './image-picker.component.html',
  styleUrls: ['./image-picker.component.scss']
})
export class ImagePicker {
  selected: string | null = null;
  @Input() title = '选择封面：';
  @Input() items: string[] = [];
```

```
  @Input() cols = 8;
  @Input() rowHeight = '64px';
  @Input() itemWidth = '80px';
  @Input() useSvgIcon = false;
  @Output('itemChange') itemChange = new EventEmitter<string>();
  // 列表元素选择发生改变触发
  selectImage(i: number) {
    this.selected = this.items[i];
    this.itemChange.emit(this.items[i]);
  }
}
```

我们现在要做的就是将这个普通控件转化成表单控件，改造后的代码如下：

```
// 省略导入

@Component({
  selector: 'app-image-picker',
  templateUrl: './image-picker.component.html',
  styleUrls: ['./image-picker.component.scss'],
  providers: [
    {
      provide: NG_VALUE_ACCESSOR,
      useExisting: forwardRef(() => ImagePicker),
      multi: true
    },
    {
      provide: NG_VALIDATORS,
      useExisting: forwardRef(() => ImagePicker),
      multi: true
    }
  ]
})
export class ImagePicker implements ControlValueAccessor {
  selected: string | null = null;
  @Input() title = '选择封面：';
  @Input() items: string[] = [];
  @Input() cols = 8;
  @Input() rowHeight = '64px';
  @Input() itemWidth = '80px';
  @Input() useSvgIcon = false;
  @Output('itemChange') itemChange = new EventEmitter<string>();

  // 这里是做一个空函数体，真正使用的方法在 registerOnChange 中
  // 由框架注册，然后我们使用它把变化发回表单
```

```
  // 注意，尽管和 EventEmitter 很像，但发送回的对象不同
  private propagateChange = (_: any) => {};

  // 写入控件值
  public writeValue(obj: any) {
    this.selected = obj ? String(obj) : null;
  }

  // 当表单控件值改变时，函数 fn 会被调用
  // 这也是我们把变化发送回表单的机制
  public registerOnChange(fn: any) {
    this.propagateChange = fn;
  }

  // 验证表单，验证结果正确则返回 null，否则返回一个验证结果对象
  public validate(c: FormControl) {
    return this.selected
      ? null
      : {
          imageListSelect: {
            valid: false
          }
        };
  }

  // 这里没有使用，用于注册 touched 状态
  public registerOnTouched() {}

  // 列表元素选择发生改变触发
  selectImage(i: number) {
    this.selected = this.items[i];
    // 更新表单
    this.propagateChange(this.items[i]);
    this.itemChange.emit(this.items[i]);
  }
}
```

要实现 ControlValueAccessor，需要实现 4 个方法。

- registerOnChange(fn)——这个函数中传入的是值变化的回调函数 fn，我们需要在本地保存它的引用（this.propagateChange = fn），并在控件值变化时调用这个本地引用，将值传递出去 this.propagateChange(this.items[i])。
- writeValue(obj) ——外部写入控件值时会调用此方法，所以我们把外部传入的值写入控件 this.selected = obj ? String(obj) : null;

- registerOnTouched(fn) ——传入是否点击控件的回调函数，如果需要的话，也像 registerOnChange(fn) 的处理那样做一个本地保存。这个函数的目的是设置一个当控件接受到点击事件时可以被调用的回调函数。这个例子中，我们对于控件的点击事件并不关心，所以设置为空。
- setDisabledState(isDisabled: boolean) ——这个方法是可选的，顾名思义它是用来设置 disabled 状态的。

你可能观察到了，我们在装饰器中的 providers 中使用了一些陌生的关键字，比如 forwardRef 和 multi。

首先来看 forwardRef：官方文档的定义是“允许引用一个尚未定义的引用”，这么说似乎还是懵懵懂懂的，让我们来举个例子：

```
import { Component } from '@angular/core';

@Component({
  selector: 'app-root',
  template: '<h1> {{ hello }}, World! </h1>'
})
class AppComponent {
  hello: string;

  constructor(helloService: HelloService) {
    this.hello = helloService.getHello();
  }
}

class HelloService {
  getHello () {
    return 'Hello';
  }
}
```

如果你试图运行上面这段代码，则会发现这段代码根本无法跑起来，在 console 中会看到如下错误信息：

```
Uncaught Error: Can't resolve all parameters for AppComponent: (?).
```

这是由于在 AppComponent 的构造函数中，我们注入了 HelloService，但是 HelloService 在此时还未定义。如果把 HelloService 移动到 AppComponent 的上方就没有错误了。难道在 Angular 中只能把类定义写在引用之前码？当然不是了，可以用 forwardRef 来拯救，像下面这样改造后就没问题了。

```
import {Component, Inject, forwardRef} from '@angular/core';

@Component({
  selector: 'app-root',
  template: '<h1> {{ hello }}, World! </h1>'
})
class AppComponent {
  hello: string;

  constructor(@Inject(forwardRef(() => HelloService)) helloService) {
    this.hello = helloService.getHello();
  }
}

class HelloService {
  getHello () {
    return 'Hello';
  }
}
```

再说回到我们的例子，这里是自己引用自己，所以当然得用 forwardRef。那么接下来这个 multi 又是什么？

先简单地复习一下依赖注入，一般情况下，同样的 token 只能对应一个依赖，如果使用同样的 token，那么最后注册成功的那个胜出。

```
class Example { }
class AnotherExample { }

let injector = Injector.create([
  { provide: Example },
  { provide: Example, useClass: AnotherExample }
]);

let example = injector.get(Example);
// 一般情况下，同样的 token，最后注册成功的那个胜出
// 这里是 AnotherExample 胜出
```

有些情况下，我们需要一个 token 对应多个依赖，比如验证器，我们会为同一个控件指定多个验证器，既有内建的也有外部自定义的。这种情况下，就需要一个 token 对应多个依赖。所以设置 multi 为 true 就是告诉系统，这里使用了一个 token（这里是 NG_VALIDATORS）对应多个依赖，我们声明的是其中之一。当你注册依赖时，输入如下代码。

```
providers: [
  provide: NG_VALIDATORS,
```

```
    useValue: (formControl) => {
      // validation happens here
    },
    multi: true
  ]
```

这个自定义表单控件就做完了，在注册表单组件中使用看看，现在看上去是不是很和谐，和其他的表单组件一模一样。把如下代码复制到 src/app/auth/components/register-form/register-form. component.html 文件中去。

```
  <app-image-picker [useSvgIcon]="true" [cols]="6" [title]="'选择头像：'"
[items]="avatars" formControlName="avatar">
  </app-image-picker>
```

接下来，看看组件的写法，还是通过注入 FormBuilder 的方式来快速构建表单。

```
  // 省略导入

  @Component({
    selector: 'register-form',
    templateUrl: './register-form.component.html',
    styleUrls: ['./register-form.component.scss']
  })
  export class RegisterFormComponent implements OnInit {
    @Input() avatars: string[] = [];
    @Input() title = '注册';
    @Input() subtitle = '注册成为会员体验全部功能';
    @Input() loginBtnText = '已经注册？点击登录';
    @Input() forgotBtnText = '忘记密码？';
    @Input() usernameValidator: AsyncValidatorFn;
    @Input() emailValidator: AsyncValidatorFn;
    @Input() mobileValidator: AsyncValidatorFn;
    @Output() submitEvent = new EventEmitter();
    form: FormGroup;
    private readonly avatarName = 'avatars';
    constructor(private fb: FormBuilder) {}

    ngOnInit(): void {
      this.avatars = _.range(1, 16)
        .map(i => `${this.avatarName}:svg-${i}`)
        .reduce((r: string[], x: string) => [...r, x], []);
      this.form = this.fb.group({
        login: [
          '',
          [
```

```
        Validators.required,
        Validators.minLength(3),
        Validators.maxLength(50),
        Validators.pattern(usernamePattern)
      ],
      this.usernameValidator
    ],
    mobile: [
      '',
      [Validators.required, Validators.pattern(mobilePattern)],
      this.mobileValidator
    ],
    email: [
      '',
      [Validators.required, Validators.pattern(emailPattern)],
      this.emailValidator
    ],
    name: [
      '',
      [
        Validators.required,
        Validators.minLength(2),
        Validators.maxLength(50),
        Validators.pattern(humanNamePattern)
      ]
    ],
    avatar: [],
    passwords: this.fb.group(
      {
        password: [
          '',
          [
            Validators.required,
            Validators.minLength(8),
            Validators.maxLength(20)
          ]
        ],
        repeat: ['', Validators.required]
      },
      { validator: this.matchPassword }
    )
  });
}
```

```
matchPassword(c: AbstractControl) {
  const password = c.get('password').value;
  const repeat = c.get('repeat').value;
  if (password !== repeat) {
    c.get('repeat').setErrors({ notMatchPassword: true });
    return { notMatchPassword: true };
  } else {
    return null;
  }
}

submit({ valid, value }: FormGroup, ev: Event) {
  if (!valid) {
    return;
  }
  this.submitEvent.emit(value);
}

get nameErrors() {
  const name = this.form.get('name');
  if (!name) {
    return '';
  }
  return humanNameErrorMsg(name);
}

get usernameErrors() {
  const username = this.form.get('username');
  if (!username) {
    return '';
  }
  return usernameErrorMsg(username);
}

get emailErrors() {
  const email = this.form.get('email');
  if (!email) {
    return '';
  }
  return emailErrorMsg(email);
}

get mobileErrors() {
  const mobile = this.form.get('mobile');
```

```
    if (!mobile) {
      return '';
    }
    return mobileErrorMsg(mobile);
  }

  get passwordErrors() {
    const password = this.form.get('passwords').get('password');
    if (!password) {
      return '';
    }
    return password.hasError('required')
      ? '密码为必填项'
      : password.hasError('minlength')
        ? `不能少于 ${password.errors['minlength'].requiredLength} 个字符`
        : password.hasError('maxlength')
          ? `不能超过 ${password.errors['maxlength'].requiredLength} 个字符`
          : '';
  }

  get repeatErrors() {
    const repeat = this.form.get('passwords').get('repeat');
    if (!repeat) {
      return '';
    }
    return repeat.hasError('required')
      ? '密码为必填项'
      : repeat.hasError('notMatchPassword') ? `密码不匹配` : '';
  }
}
```

这里面需要注意的有以下几个方面。

- 使用 lodash 快速创建一个 range 数组，用来构建图片的数据。使用 lodash 需要安装 yarn add lodash 以及其类型 yarn add @types/lodash –dev。
- 对正则表达式类型的验证器，我们没有把正则表达式具体写在这个组件中，而是放在了 src/app/utils/regex.ts 中，这样其他需要此类正则表达式的类只需要引用这些工具函数，也方便我们后期对正则表达式的更改和优化。
- 对于错误的显示处理，我们把对 name、username、mobile 和 email 的错误显示都抽取出去做了专门的函数，位于 src/app/utils/validate-errors.ts，这是由于这几个错误显示不只在这个地方会用到。

- 对于涉及两个控件的验证器，我们在这个例子中的 matchPassword 采用的是对 FormGroup 构造验证器，然后在验证器中使用 FormGroup 的 GET 方法得到对应的 FormControl 进行数据验证。

此外，我们为 username、mobile 和 email 等需要验证数据唯一性的控件预留了异步验证器。为什么需要异步验证呢？因为数据的唯一性需要到后端数据库进行查找比对，而连接后端是一个异步过程，这时我们就需要使用异步验证。再有这个组件是一个“笨组件”，所以我们把这几个异步验证器设置为输入型属性（对的，函数也可以作为属性），留给父组件使用这个组件时设置。

```
@Input() usernameValidator: AsyncValidatorFn;
@Input() emailValidator: AsyncValidatorFn;
@Input() mobileValidator: AsyncValidatorFn;
```

我们还遗留了“忘记密码”组件，这个组件比较复杂，放到 4.2 节介绍。

4.2 响应式编程初探

这里就不谈响应式编程的官方概念了，有兴趣的读者可以去维基百科查看。官方概念太理论化了，笔者来谈谈个人的一点粗浅认识，响应式编程关注随着时间变化产生的事件或数据，这些事件和数据可以看成随着时间形成的数据流或事件流。估计读者觉得这个解释也看不懂，没关系，请记住这个概念，在随后的学习中你会发现这个概念是很容易理解的。

当然，在本书中的响应式编程主要是用的 Rx，前端用的是 RxJS，后端是 RxJava。说到 Rx，这本是微软为自家 .Net 平台打造的，Rx 的本义是 Reactive Extension 的缩写，即响应式扩展。距离微软推出 .Net 版本的 Rx 已经 8 年多了，终于逐渐获得了主流社区的认可。从单一的 .Net 支持到现在的支持 18 种编程语言和 3 个平台框架。那为什么社区现在对 Rx 的热情这么高呢？Rx 解决了哪些问题？接下来就一起来学习。

4.2.1 不同的视角

我们先用 4.1 节没做的“忘记密码”功能当作演示页面。由于要尽快地开始 Rx 的练习，就先创建“聪明组件”：

```
ng g c auth/containers/auth -m auth/auth.module.ts
```

更改模板文件，暂时只放一个文本输入框，起一个引用名叫 myInput：

```
<input #myInput type="text">
```

在组件文件中，我们改造成下面的样子：

```
// 省略导入

@Component({
  selector: 'forgot-password',
  templateUrl: './forgot-password.component.html',
  styleUrls: ['./forgot-password.component.scss']
})
export class ForgotPasswordComponent implements OnInit {
  @ViewChild('myInput', { read: ElementRef })
  input: ElementRef;

  ngOnInit(): void {
    const input$ = fromEvent(this.input.nativeElement, 'keyup');
    input$.subscribe(ev => console.log(ev));
  }
}
```

对于上述代码，我们简单做个说明，成员变量 input 是通过 @ViewChild('myInput', { read: ElementRef }) 注解得到了模板中 myInput 的引用。

在 ngOnInit() 中，使用了 fromEvent，这个是 RxJS 提供的用于转换事件为 Observable 的函数。Observable 赋值给一个变量 input$，在 RxJS 领域，一般在 Observable 类型的变量后面加上 $ 标识，这是一个“流变量”（由英文 Stream 得来，Observable 就是一个 Stream，所以用 $ 标识），不是必须的，但是属于约定俗成。

然后我们订阅 Observable，并在“Console”中输出订阅的值。

打开 Chrome 的开发者工具，在输入框中依次输入 12345，同时观察“Console”选项卡中的输出内容，如图 4-6 所示。

可以看到，伴随着每次键盘的抬起事件，“Console”中也输出了一个 KeyboardEvent，由此可见，我们订阅的是这个文本输入框的 keyup 键盘事件——fromEvent(this.input.nativeElement, 'keyup')。

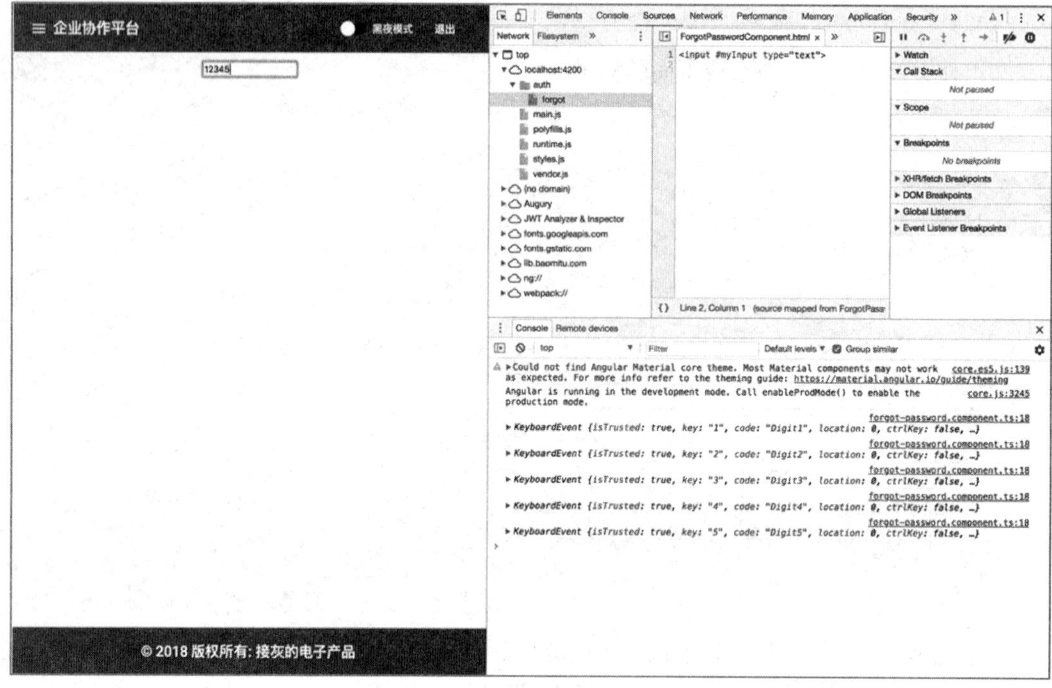

图 4-6

如果我们把时间考虑在内，那上面的这些输出可以看作一个事件流，把这个事件流沿时间轴的变化画成下面的示意图，其中 ke 代表 KeyboardEvent。在不同时间点按了键盘上的数字，从而引发对应的键盘事件，因此我们把它叫作一个事件流。

```
        1        2                3       4              5
------ke-----ke-----------ke----ke---------ke----
```

接下来，对代码稍做更改，体验一下 RxJS 中强大的操作符。上面我们取到的是事件对象，但是这个事件对象当中有太多东西是我们不需要的，我们只关心键盘按下的字符。所以我们采用一个操作符 pluck，这个操作符取出对应事件的属性，在这里就是 ev.target.value。tap 也是一个操作符，就是 RxJS 5.×中的 do，在这里我们在值转换之前，使用这个操作法打印事件对象，这个操作符之所以放在 pluck 之前，是由于执行完 pluck 之后，流就由事件流转换成了字符流（见图 4-7）。

```
const input$ = fromEvent(this.input.nativeElement, 'keyup').pipe(
      tap(ev => console.log(ev)),
      pluck('target', 'value')
    );
input$.subscribe(val => console.log(val));
```

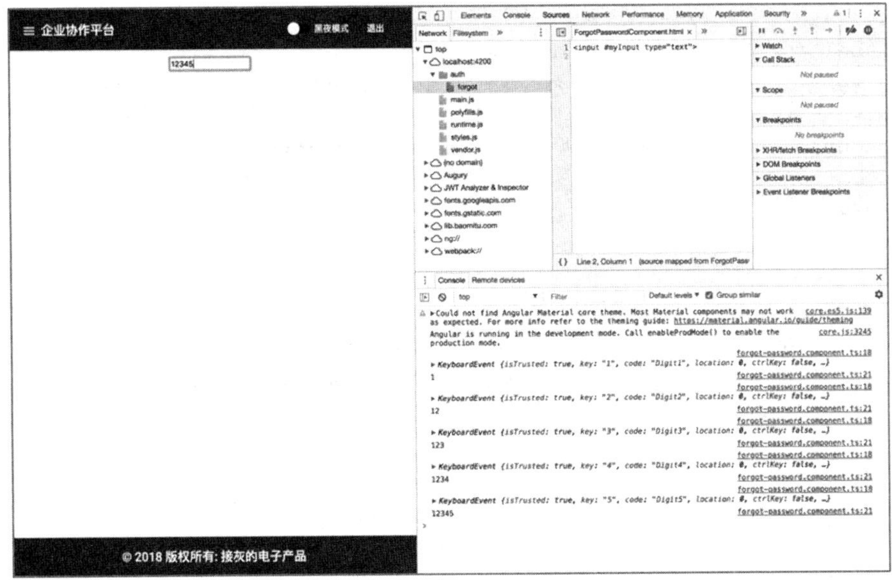

图 4-7

RxJS 最擅长的就是对流进行各种操作变换，下面的示意图可以让我们了解一个事件流到一个字符流的转换过程。

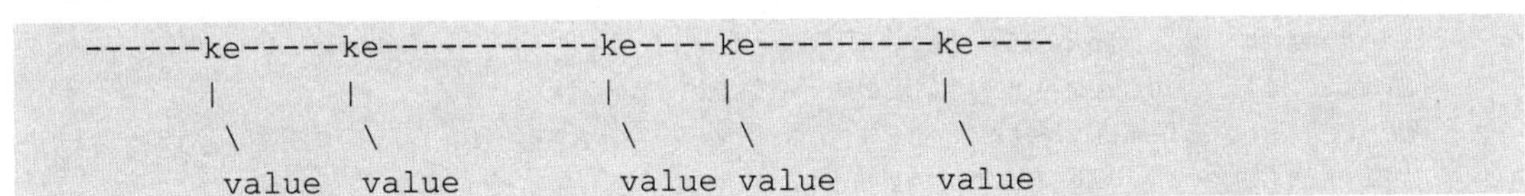

```
------ke-----ke-----------ke----ke---------ke----
      |      |            |     |          |
       \      \            \     \          \
      value  value       value value      value
```

4.2.2 实现一个计数器

为了让读者有进一步的体会，接下来做一个简单的计数器，有 +1 和 -1 两个按钮，点击“+1”按钮计数器就加 1，点击“-1”按钮计数器就减一（见图 4-8）。这个需求很简单，我们看看如何使用 RxJS 来实现。

图 4-8

HTML 模板非常简单，一个圆形的显示数字的区域和两个小按钮。我们为了监听按钮事件，给两个按钮的引用名称分别命名为 increment 和 decrement。

```
<div class="counter-container">
  <label for="qty">计数器</label>
  <div id="qty" class="numberCircle">{{click$ | async}}</div>
  <div>
    <button class="counter" #decrement>-1</button>
    <button class="counter" #increment>+1</button>
  </div>
</div>
```

接下来看一下组件，对应 HTML 中的两个引用，我们创建了两个成员变量 increment 和 decrement。并且定义了一个显示计数器数字的流变量 click$，这也意味着这个流是一个数据为数字的流。这个 async 是 Angular 为异步变量（比如 Observable 或 Promise 类型的变量）定制的管道。

```
// 省略导入

@Component({
  selector: 'app-forgot-password',
  templateUrl: './forgot-password.component.html',
  styleUrls: ['./forgot-password.component.scss']
})
export class ForgotPasswordComponent implements OnInit {
  @ViewChild('increment', { read: ElementRef })
  increment: ElementRef;
  @ViewChild('decrement', { read: ElementRef })
  decrement: ElementRef;
  click$: Observable<number>;

  constructor() {}

  ngOnInit(): void {
    this.click$ = this.getCounterObservable();
  }

  private getCounterObservable(): Observable<number> {
    const increment$ = fromEvent(this.increment.nativeElement,
'click').pipe(
      tap(_ => console.log('increment')),
      mapTo(1)
    );
    const decrement$ = fromEvent(this.decrement.nativeElement,
'click').pipe(
      tap(_ => console.log('decrement')),
      mapTo(-1)
```

```
    );
    return merge(increment$, decrement$).pipe(
      scan((acc: number, curr: number) => acc + curr),
      startWith(0)
    );
  }
}
```

我们在 ngOnInit 时给 click$ 赋值，请注意，为了更好地隔离，抽取出一个专门的函数来做流处理——getCounterObservable。

下面我们看一下如何处理这个流，首先来分析一下原始场景：页面上有两个按钮，点击“+1”按钮会产生一个点击事件，点击“-1”按钮同样也会产生事件。也就是说页面中会产生两个独立事件流，如下所示：

```
+1 按钮：---i----i-------i------i-------i-----i----
-1 按钮：------d-----d-------------d-------d-------
```

这两个事件导致的结果是有区别的，前者会导致数据 +1，而后者会导致数据 -1。如果我们想体现这个结果，那最方便的做法就是在“+1”按钮点击事件产生时，将其转换成数字 1，而在“-1”按钮点击事件产生时，将其转换成 -1，应用 mapTo 操作符，就可以达到这个效果。

```
+1 按钮：---i----i-------i------i-------i-----i----
             \    \       \      \       \     \
              1     1       1      1       1     1
-1 按钮：------d-----d-------------d-------d-------
                \     \             \       \
                -1     -1            -1      -1
```

现在我们面临一个问题，最终在页面上只有一个数字流，但现在却有两个，这怎么办？之前我们说过，Rx 最擅长的就是对流进行各种操作，所以这个问题对于 Rx 来说再简单不过，这两个流需要一个合并操作符。Rx 提供了很多合并操作符，具体用什么操作符需要根据合并的规则，这个例子里面，这两个数据流其实做一个简单合并就可以，简单合并就是保持各自流中的顺序和时间点，可以想象成一个流向另一个做投影，从而产生了一个新的流。

```
+1 按钮：---i----i-------i------i-------i-----i----
-1 按钮：------d-----d-------------d-------d-------
合并流：---i--d-i---d---i------i--d----i--d---i---
            \  \  \  \    \      \   \   \   \   \
            1  -1  1  -1   1      1  -1   1  -1   1
```

接下来又出现了一个问题，我们得到了一个数字流，这个数字流由 1 和 -1 构成，但我们需要的是一个累加值，这个累加值才是应该显示在计算器上的数字。所以就又引入了一个新的操作符 scan，它允许你使用一个累加器函数去操作这个流中的数据，并返回即时的结果。也就是说，原始流中的每一个数据都会产生对应的累加器返回的结果。我们使用的累加器函数非常简单，(acc: number, curr: number) => acc + curr 就是返回“累加器 (acc) + 当前数据 (curr)”的和，而等到下一个数据出现时，累加器的值就是上一次计算后的结果。那么第一个数据产生时，没有累加器的值怎么办呢？没关系，默认值是 0，正好符合我们的需求。

```
+1 按钮：---i----i-------i------i-------i-----i----
-1 按钮：------d-----d-------------d-------d-------
合并流：---i--d-i---d---i------i--d----i--d---i---
           \  \  \  \    \      \   \   \   \   \
           1  -1  1  -1   1      1   -1  1  -1   1
累加器：---1---0--1---0---1-----2----1--2---1---2
```

有一个问题是，如果页面加载后，没有点击任何按钮，这个计数器是没有值的，因为没有事件产生，所以要定初始值的 0，这个就是操作符 startWith 的意义。

当然，前面没有给出样式，这里给出一个样式文件供大家参考，但这个样式和功能是没有什么关系的。

```
:host {
  display: flex;
  flex-direction: column;
  justify-content: center;
  flex: 1 1 auto;
}
.numberCircle {
  border-radius: 50%;
  flex: 1;
  width: 36px;
  height: 36px;
  padding: 8px;
  background: #fff;
  border: 2px solid #666;
  color: #666;
  text-align: center;
  font: 32px Arial, sans-serif;
}
.count-down {
  color: #333;
  text-align: center;
  background-color: #ffd54f;
}
h1 {
  font-weight: normal;
}
```

```
  li {
    display: inline-block;
    font-size: 1.5em;
    list-style-type: none;
    padding: 1em;
    text-transform: uppercase;
  }
  li span {
    display: block;
    font-size: 4.5rem;
  }
  .counter-container {
    /* basics */
    background-color: #444;
    color: #c4be92;
    text-align: center;
    display: flex;
    flex-direction: column;
    align-content: space-around;
    align-items: center;

    /* rounded corners */
    -webkit-border-radius: 12px;
    border-radius: 12px;
    -moz-background-clip: padding;
    -webkit-background-clip: padding-box;
    background-clip: padding-box;
    padding: 0.8em 0.8em 1em;
    width: 8em;
    margin: 0 auto;
    -webkit-box-shadow: 0px 0px 12px 0px #000;
    box-shadow: 0px 0px 12px 0px #000;
  }
  button.counter {
    /* basics */
    color: #444;
    background-color: #b5b198;
    /* rounded corners */
    -webkit-border-radius: 6px;
    border-radius: 6px;
    -moz-background-clip: padding;
    -webkit-background-clip: padding-box;
    background-clip: padding-box;
    font-weight: bold;
```

```
  width: 30px;
}
button.counter:hover,
button.counter:active,
button.counter:focus {
  background-color: #cbc7ae;
}
```

1. 和传统方式的比较

你可能会说，这也没什么啊，感觉并不比传统的事件处理高明，好像还更烦琐一些。那好我们看看传统方式，请看下面这段程序代码。

首先，定义一个成员变量 counterNum，然后给出 +1 对应的处理函数 processIncrement 和 -1 对应的处理函数 processDecrement。这两个函数中分别对成员变量 counterNum 进行 +1 或 -1 的操作。

```
import { Component, OnInit } from '@angular/core';

@Component({
  selector: 'app-forgot-password',
  templateUrl: './forgot-password.component.html',
  styleUrls: ['./forgot-password.component.scss']
})
export class ForgotPasswordComponent implements OnInit {
  counterNum = 0;

  constructor() {}

  ngOnInit(): void {
  }

  processIncrement() {
    ++this.counterNum;
  }

  processDecrement() {
    --this.counterNum;
  }
}
```

然后，在模板中进行事件绑定。

```
<div class="counter-container">
  <label for="qty">计数器</label>
  <div id="qty" class="numberCircle">{{ counterNum }}</div>
```

```
    <div>
      <button class="counter" (click)="processDecrement()">-1</button>
      <button class="counter" (click)="processIncrement()">+1</button>
    </div>
  </div>
```

很简单，不过，让我们增加一个小需求试试看。现在我们希望鼠标一直按着 +1 或 -1 时，数字是会自动连续增加或减少，而且频率是 300ms 更新一个数字。在鼠标抬起时，这个数字就要停止更新。

如果使用传统模式，我们看看需要怎么做。首先需要再引入一个布尔型的成员变量 mouseUp，用这个变量标识是否鼠标已抬起。然后在 processIncrement 和 processDecrement 中加入一个 setInterval 计时器对 counterNum 进行加或减，而且需要判断 mouseUp 为真时取消计时器。在这两个方法中需要将 mouseUp 置为 false，还需要增加一个鼠标抬起的事件处理函数 processMouseUp，在其中设置 mouseUp 为 true。

```
// 省略其他部分
export class ForgotPasswordComponent implements OnInit {
  counterNum = 0;
  mouseUp = false;

  constructor() {}

  ngOnInit(): void {
  }

  processIncrement() {
    this.mouseUp = false;
    const that = this;
    const timer = setInterval(function() {
      ++that.counterNum;
      if (that.mouseUp) {
        clearInterval(timer);
      }
    }, 300);
  }

  processDecrement() {
    this.mouseUp = false;
    const that = this;
    const timer = setInterval(function() {
      --that.counterNum;
      if (that.mouseUp) {
```

```
          clearInterval(timer);
        }
      }, 300);
    }

    processMouseUp() {
      this.mouseUp = true;
    }
  }
```

还需要绑定两个事件 mousedown 和 mouseup。

```
<div class="counter-container">
  <label for="qty">计数器</label>
  <div id="qty" class="numberCircle">{{counterNum}}</div>
  <div>
    <button class="counter" (mousedown)="processDecrement()"
(mouseup)="processMouseUp()">-1</button>
    <button class="counter" (mousedown)="processIncrement()"
(mouseup)="processMouseUp()">+1</button>
  </div>
</div>
```

是不是越来越复杂了，而且引入的标志位多了起来，处理的函数也越来越多，如果又有新的需求，就会导致逻辑越来越难懂，可维护性也越来越差。

下面我们看看使用 Rx 怎么解决这个问题。

2. 简洁而强大的 Rx 解决方案

和最初的 Rx 方案对比，其他文件都不用变化，只是调整一下 src/app/auth/ containers/forgot-password/forgot-password.component.ts 的 getCounterObservable 方法：

```
private getCounterObservable(): Observable<number> {
  const mouseUp$ = fromEvent(document, 'mouseup');
  const increment$ = fromEvent(
    this.increment.nativeElement,
    'mousedown'
  ).pipe(
    switchMap(_ =>
      interval(300).pipe(startWith(1), takeUntil(mouseUp$), mapTo(1))
    )
  );
  const decrement$ = fromEvent(
    this.decrement.nativeElement,
    'mousedown'
```

```
  ).pipe(
    switchMap(_ =>
      interval(300).pipe(startWith(1), takeUntil(mouseUp$), mapTo(-1))
    )
  );
  return merge(increment$, decrement$).pipe(
    scan((acc, curr) => acc + curr),
    startWith(0)
  );
}
```

思考方式和原来类似，只不过现在页面中有 3 个事件流，两个按钮的鼠标按下产生的事件流（从前面的 click 事件改成了 mousedown 事件）以及鼠标抬起的事件流 mouseup。最初我们是直接把两个按钮的事件转换成了数字，现在由于需求的变化，不能直接转换。

先来分析一下，当鼠标按下时，我们希望产生一个计时器，而这个在 Rx 内建提供了 interval，这个计时器每 300ms 产生一个顺序整数，但是第一个数是 300ms 后才发射的，所以使用 startWith(1) 让它一开始就发射一个数据。这么做的原因是希望点击按钮（而不是长按）时可以立即变更该数字。

再有就是这个计时器应该在 mouseup 事件产生时停止，所以使用 takeUntil 操作符，这个操作符的意义在于当其参数的流有数据产生时，外部的流就完成了，也就停止了。比如 interval(300).pipe(startWith(1), takeUntil(mouseUp$))，当 mouseUp$ 有数据时，也就是事件触发时，interval(300) 就停止了。

然后我们还是把流的数据通过 map 操作等变成 1 或 -1，但为什么外层要用 switchMap 呢？这是因为内部的 interval.pipe(...) 得到的是一个 Observable<number> 而不是 number，这就是数据流中又有数据流了，所以使用一个高阶的操作符 switchMap 将它“拍扁”，重新变成 number 型。这一块会在后面讲到 Rx 高阶操作符时详细阐述。

总之，使用 Rx 可以让逻辑更加简洁、清晰地表达出来，而且不需要引入一些标志位。

4.2.3 为什么要使用 Rx

笔者认为 Rx 有两大优点。

- 在 Rx 世界里，一切都是事件流，所以这“逼迫”开发者将时间维度纳入设计的考量，从而需要分析清楚从开始到结束的每个过程，确保过程可控制。
- 提供的各种强大的操作符可以将逻辑非常轻松地组合，减少分散在代码各个角落的逻辑，使得逻辑的可读性更强。

1. RxJS 6.× 和之前版本的区别

本书中的代码是基于 RxJS 6.× 的，和 5.× 有较大区别，从 6.×开始起，不再提倡原来的类似 some$.map(x => x/2).filter(x => x > 1) 这样的链式写法。原因是这种方式容易污染 Observable 原型，而且在 Angular 中无法通过摇树（tree-shaking）机制去删除无用代码。所以现在推荐使用 pipeable 操作符。我们看到前面例子中的 pipe 就是接受一系列的 pipeable 操作符，在这个 pipe 方法中，放入一系列操作符，基本的思路和之前是一致的，只不过写法上比链式写法要稍微烦琐一些。

4.2.4 Observable 的性质

Rx 的本质是观察者模式，所以需要知道 Observable 就是一个可观察对象，与之对应的有 Observer（观察者）和 Subscriber（订阅者）。

Observer 是一个 Observable 所产生的推送消息的消费者。

```
interface Observer<T> {
  closed?: boolean;
  next: (value: T) => void;
  error: (err: any) => void;
  complete: () => void;
}
```

从上面的定义可以看出 Observer 是一个接口，如果我们有一个具体的 observerA 实现了该接口，有一个可观察对象 ob$，那么 ob$.subscribe(observerA) 就是让 observerA 订阅了 ob$。当 ob$ 有新的数据或事件产生时，它会调用 observerA 的 next(value) 方法（这个 next 就是在 Observer 接口中定义的），从而让 observerA 不断地接收到 ob$ 的变化。

我们在上面的例子中没有采用 subscribe，这是由于我们使用了 Angular 提供的 async 管道。其实我们也可以采用 subscribe。把 click$: Observable<number> 改成 click: number;，当然模板的绑定也改成 {{click}}，然后把 ngOnInit 改写成下面的样子。

```
ngOnInit(): void {
  this.getCounterObservable().subscribe({
    next: val => {
      this.click = val;
    }
  });
}
```

运行程序，你会发现效果是一样的。现在来分析一下，this.getCounterObservable() 得到的是一个 Observable，那么使用了 subscribe 方法，这个方法接收的是一个 Observer 类型的参数。

我们也构造了一个对象，尽管只有一个 next 属性。

```
{
  next: val => {
    this.click = val;
  }
}
```

这么写还是有点麻烦，好在我们还有一些语法糖可用，上述形式可以简写为：

```
this.getCounterObservable().subscribe(val => {
      this.click = val;
    });
```

而且在只有一行代码时可以进一步简写：

```
this.getCounterObservable().subscribe(val => this.click = val);
```

这样看起来好多了。那么其他几个参数呢？我们可以给出 3 个参数，也就是说 subscribe 方法提供了一些语法糖的形式，可以不用完整地传入 Observer 对象，而是以参数形式提供 subscribe(nextFn, errorFn, completeFn)，这 3 个参数都是函数形式，而且不用全部提供。

```
this.getCounterObservable().subscribe(
  val => this.click = val,
  err => console.error(err),
  () => console.log('completed'));
```

这 3 个参数分别是 Observable 接口中定义的 next、error 和 complete，都是函数型参数。这个也可以解释任何一个 Observable 都有 3 个状态：等待推送下一个值就是 next，出错了就是 error，流结束了就是 complete。无论是正常结束还是出错结束都会调用 complete。比如我们有一个 from([1,2,3])，这个流有 3 个数据，这 3 个数发送完了，就没有下一步了，就调用 complete，这个就是正常结束。再比如 from([1,2,3]).pipe(map(val => val/val-2)) 在第 2 个数据时由于被除数为 0，会调用 error，然后调用 complete。

直接使用 subscribe 方法会有内存泄漏问题，因为这个订阅一直在那里，怎么取消订阅呢？这就要请出 Subscription 了。subscribe 方法返回的就是一个订阅对象 Subscription，那一般情况下使用一个成员变量存储，然后在组件销毁时 ngOnDestroy 使用 unsubscribe 方法取消订阅。这样才能保证内存不会泄漏，但这样做总会由于比较麻烦导致有时忘记这么做，所以在 Angular 中推荐尽可能地使用 async 管道。

```
import { Observable, fromEvent, merge, interval, Subscription } from
'rxjs';
// 省略其他部分
sub: Subscription
```

```
ngOnInit() {
  this.sub = this.getCounterObservable().subscribe(
    val => this.click = val,
    err => console.error(err),
    () => console.log('completed'));
}

ngOnDestroy {
  if (this.sub) {
    this.sub.unsubscribe();
  }
}
```

4.2.5 RxJS 的调试

在使用 RxJS 时，传统的排除错误的方法（常被称为 debug）往往显得“心有余而力不足”，因为传统的 debug 工具都是基于命令式的编程模型。但对于 RxJS 这种响应式编程模式来说，设置断点跟踪虽然也可以，但由于 RxJS 一个异步模型，这就使得断点跟踪这种同步的 debug 手段不是很顺手，而且有些问题是在这种跟踪模型下根本暴露不出来。

于是，log 大法就出现了，RxJS 提供的 tap 操作符非常适合做 log 输出，因为它并不改变流。就像上面写到的那样，我们可以在 tap 中将希望的内容 log 出来。但这种 log 方式有个问题，就是在开发时可以这样做，如果发布后，就得去掉这些 log。当然也可以自己用环境变量去判断，而不用删掉这些代码。但如果有一个统一的机制能够方便地处理就更好了。

那么有没有更好的 debug 的手段呢？当然有，我们体会到的痛点，社区早就有人体会到了，那么这里介绍一个开源的 RxJS 日志和调试类库 rxjs-spy。

1. 安装和初始化

在 Angular 中使用 rxjs-spy 是很简单的，首先需要安装依赖。

```
yarn add rxjs-spy --dev
```

然后在 src/main.ts 中导入并初始化 rxjs-spy。

```
// 省略导入

const spy = create();

if (environment.production) {
  enableProdMode();
}
```

```
platformBrowserDynamic()
  .bootstrapModule(AppModule)
  .catch(err => console.log(err));
```

2. 给 RxJS 流做标签

在下面的登录鉴权逻辑中，就可以在取得 token 后利用 rxjs-spy 的 tag 方法将其标记为 [AuthService][login][id_token]，这里采用的是以 [类名][方法名][流产生的值名称] 的形式做标记，实际工作中你可以根据自己的规则命名，但最好是团队采用统一的命名方式。

```
// 省略其他 imports
import { Observable } from 'rxjs';
import { pluck, map } from 'rxjs/operators';
import { tag } from 'rxjs-spy/operators';

@Injectable({
  providedIn: 'root'
})
export class AuthService {
  // 省略其他方法
  login(auth: Auth): Observable<string> {
    return this.http
      .post<{ id_token: string }>(`${environment.apiBaseUrl}auth/login`,
JSON.stringify(auth), { headers: this.headers })
      .pipe(
        map((res) => <string>res.id_token),
        tag('[AuthService][login][id_token]')
      );
  }
}
```

做完这个标记之后，我们也不必在生产环境将其删除，可以一直保留，因为我们查看 log 的方式是在 Chrome 的 Console 中去调用 rxSpy.show()，如图 4-9 所示，在 Console 中输入 rxSpy.show() 就会输出对应的 log。如果只想看某个 tag 的 log 可以传入该 tag 值：rxSpy.show('[AuthService][login][id_token]')。

除 show 之外，rxspy-log 还提供了一系列可以用于 Console 中的方法，比如 Stats()是得到调用次数的统计（见图 4-10）。

还有好用的 debug、step 和 undo 等用于 debug 调试的方法。如果在浏览器打开网址[1]，然后开启开发者工具的 Console，就可以体验一下这几个方法，比如首先我们在 Console 中输入

1 https:// cartant.github.io/rxjs-spy/

rxSpy.log("interval");，可以看到这个 interval 的流的值被不断地输出（见图 4-11）。

图 4-9

```
⋮ Console   Remote devices   What's New                                  ×
top ▾   Filter                        Default levels ▾  Group similar
> rxSpy.stats()
▾ Stats                                                  spy-core.js:242
    Subscribes = 325                                     spy-core.js:243
    Root subscribes = 32                                 spy-core.js:245
    Leaf subscribes = 107                                spy-core.js:248
    Flattened subscribes = 92                            spy-core.js:251
    Unsubscribes = 160                                   spy-core.js:253
    Nexts = 648                                          spy-core.js:254
    Errors = 15                                          spy-core.js:255
    Completes = 101                                      spy-core.js:256
    Max. depth = 8                                       spy-core.js:258
    Avg. depth = 2.1                                     spy-core.js:259
```

图 4-10

```
> rxSpy.log("interval");
< undefined
  Tag = interval; notification = next; value = 64       log-plugin.ts:61
  Tag = interval; notification = next; value = 65       log-plugin.ts:61
  Tag = interval; notification = next; value = 66       log-plugin.ts:61
  Tag = interval; notification = next; value = 67       log-plugin.ts:61
  Tag = interval; notification = next; value = 68       log-plugin.ts:61
  Tag = interval; notification = next; value = 69       log-plugin.ts:61
  Tag = interval; notification = next; value = 70       log-plugin.ts:61
  Tag = interval; notification = next; value = 71       log-plugin.ts:61
  Tag = interval; notification = next; value = 72       log-plugin.ts:61
```

图 4-11

而此时，如果我们输入 rxSpy.pause("interval"); 该流会暂停，可以看到 pause 会返回一个 Deck 实例，这个 Deck 是干什么用的？它是用来控制在暂停后实现类似 debug 的 step、skip 等操作的。一般是每个流暂停后都会产生一个 Deck（见图 4-12）。

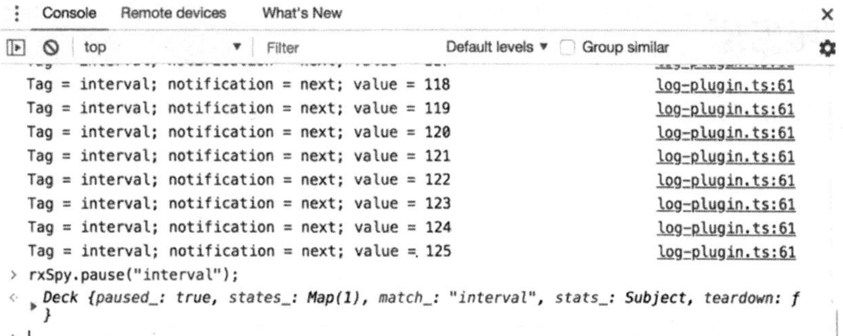

图 4-12

所以还有一个 Deck 方法列出目前暂停的流所产生的 Deck（见图 4-13）。

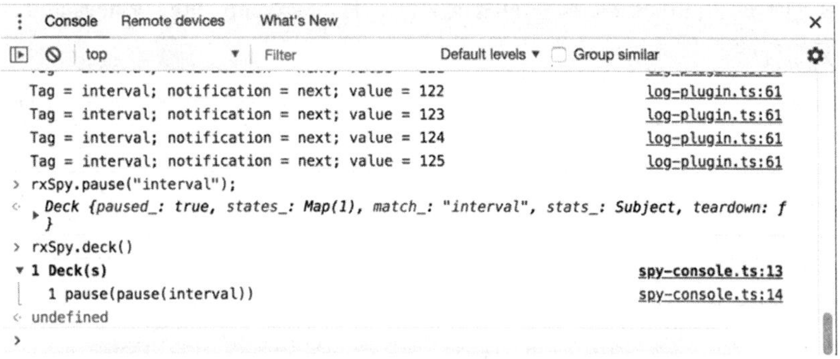

```
Console    Remote devices    What's New
top    Filter    Default levels ▼    Group similar
Tag = interval; notification = next; value = 122    log-plugin.ts:61
Tag = interval; notification = next; value = 123    log-plugin.ts:61
Tag = interval; notification = next; value = 124    log-plugin.ts:61
Tag = interval; notification = next; value = 125    log-plugin.ts:61
> rxSpy.pause("interval");
< Deck {paused_: true, states_: Map(1), match_: "interval", stats_: Subject, teardown: f
  }
> rxSpy.deck()
▼ 1 Deck(s)    spy-console.ts:13
    1 pause(pause(interval))    spy-console.ts:14
< undefined
>
```

图 4-13

那么，我们将第一个赋值给一个变量 var deck = rxSpy.deck(1);，然后就可以通过这个 Deck 操作流，实现 debug 的 step 、 skip 操作了（见图 4-14）。

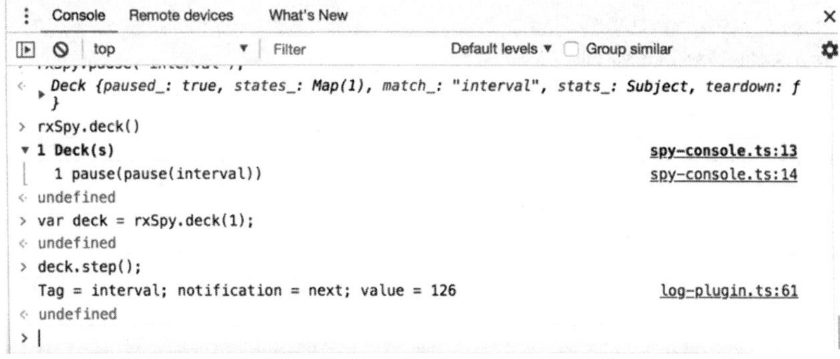

图 4-14

类似地，如果输入 deck.resume(); ，那么流就从暂停状态恢复到正常状态了。

4.3 前端服务层

4.3.1 构建“伪”服务

为什么叫作“伪”服务？这是由于我们还没有和真正的后端 API 对接，暂时写作服务，这些服务里面大多是自己生成一些数据返回，所以称之为“伪”服务。

使用 Angular CLI 生成服务的基础代码，既然是 auth 中的服务，我们就为服务在 auth 下新建一个 services 文件夹。然后输入下面的命令，注意 -m 就是 --module 的缩写形式，指定这个参数将在对应的模块文件中的 providers 数组中自动加上这个生成的服务：

```
ng g s auth/services/auth -m auth/auth.module.ts
```

这会为我们生成 AuthService 和它的测试文件。打开 src/app/auth/services/auth.service.ts，将其改造成下面的样子。

```
// 省略导入

@Injectable()
export class AuthService {
  constructor() {}
  /**
   * 用于用户的登录鉴权
   * @param auth 用户的登录信息，一般是登录名（目前是用户名，以后会允许手机号）和密码
   */
  login(auth: Auth): Observable<User> {
    const user: User = {
      username: 'test',
      mobile: '13012341234',
      email: 'zhangsan@local.dev',
      name: 'Zhang San',
      avatar: 'assets/img/avatar/001.svg',
      roles: []
    };
    return of(user);
  }
  /**
   * 用于用户的注册
   * @param user 用户注册信息
   */
  register(user: User): Observable<User> {
    const user_add: User = { ...user, id: '123abc' };
    return of(user_add);
  }
```

```
  /**
   * 请求发送短信验证码到待验证手机，成功返回空对象 {}
   * @param mobile 待验证的手机号
   */
  requestSmsCode(mobile: string) {
    return of({});
  }
  /**
   * 验证手机号和短信验证码是否匹配
   * @param mobile 待验证手机号
   * @param code 收到的短信验证码
   */
  verifySmsCode(mobile: string, code: string) {
    return of(mobile === '13012341234' && code === '123456');
  }
  /**
   * 检查用户名是否唯一
   * @param username 用户名
   */
  checkUniqueUsername(username: string) {
    return of(username === 'lisi');
  }
  /**
   * 检查电子邮件是否唯一
   * @param email 电子邮件
   */
  checkUniqueEmail(email: string) {
    return of(email === 'lisi@local.dev');
  }
  /**
   * 检查手机号是否唯一
   * @param mobile 手机号
   */
  checkUniqueMobile(mobile: string) {
    return of(mobile === '13112341234');
  }
}
```

上述代码中使用了 RxJS 中的 Observable 作为返回值，读者可以先不用试图深入理解，把它看成一个异步类型（类似 Promise）即可，后面对于 RxJS 我们会进行较详细的阐述。

类似地，我们为“佳句”生成对应服务：

```
ng g s auth/services/quote -m auth/auth.module.ts
```

同样更改其生成的文件 src/app/auth/services/quote.service.ts 为下面的样子，这个服务中，我们先手动构造一个“佳句”数组，然后在 getQuotes 中返回。

```
// 省略导入

@Injectable()
export class QuoteService {
  constructor() {}

  getQuotes(): Observable<Quote[]> {
    const quotes: Quote[] = [
      {
        cn:
          '我突然就觉得自己像个华丽的木偶,演尽了所有的悲欢离合,可是背上总是有无数闪亮的银色丝线,操纵我哪怕一举手一投足。',
        en:
          'I suddenly feel myself like a doll,acting all kinds of joys and sorrows.There are lots of shining silvery thread on my back,controlling all my action.',
        imgUrl: '/assets/img/quotes/0.jpg'
      },
      {
        cn:
          '被击垮通常只是暂时的，但如果你放弃的话，就会使它成为永恒。(Marilyn vos Savant) ',
        en:
          'Being defeated is often a temporary condition. Giving up is what makes it permanent.',
        imgUrl: '/assets/img/quotes/1.jpg'
      },
      {
        cn: '不要只因一次挫败，就放弃你原来决心想达到的梦想。(莎士比亚) ',
        en:
          'Do not, for one repulse, forgo the purpose that you resolved to effect.',
        imgUrl: '/assets/img/quotes/2.jpg'
      },
      {
        cn: '想有发现就要实验，这项实验需要时间。《神盾局特工》 ',
        en:
          'Discovery requires experimentation, and this experiment will take time.',
        imgUrl: '/assets/img/quotes/3.jpg'
      },
```

```
          {
            cn:
              '这世界并不会在意你的自尊，这世界希望你在自我感觉良好之前先要有所成就。',
            en:
              "The world won't care about your self-esteem. The world will
expect you to accomplish something before you feel good about yourself.",
            imgUrl: 'E:\02 责编\全栈技能修炼\201701354-原稿
\Archive(1)/assets/img/quotes/4.jpg'
          },
          {
            cn: '当你最终放开了过去时，更好的事就会到来。',
            en:
              'When you finally let go of the past, something better comes
along.',
            imgUrl: '/assets/img/quotes/5.jpg'
          },
          {
            cn:
              '我们学着放开过去伤害我们的人和事，学着只向前看。因为生活本来就是一往直前的。
',
            en:
              'We learn to let go of things and people that hurt us in the
past and just move on. For life is all about moving on.',
            imgUrl: '/assets/img/quotes/6.jpg'
          },
          {
            cn:
              '绝不要因为怕辛苦而拒绝一个想法、梦想或是目标，成功的路上难免伴随辛苦。
（Bob Proctor）',
            en:
              'Never reject an idea, dream or goal because it will be hard
work. Success rarely comes without it.',
            imgUrl: '/assets/img/quotes/7.jpg'
          },
          {
            cn:
              '我们在人生中会做出许多选择，带着这些选择继续生活，才是人生中最难的一课。
《妙笔生花》',
            en:
              'We all make our choices in life. The hard thing to do is live
with them.',
            imgUrl: '/assets/img/quotes/8.jpg'
          },
          {
```

```
          cn:
            '我总是对新的一天充满喜悦，这是一次新的尝试、一个新的开始，翘首以待，黎明之
后或是惊喜。（约翰·博因顿·普里斯特利）',
          en:
            'I have always been delighted at the prospect of a new day, a
fresh try, one more start, with perhaps a bit of magic waiting somewhere
behind the morning.',
          imgUrl: '/assets/img/quotes/9.jpg'
        }
      ];
      return of(quotes);
    }
  }
```

有了“笨组件”，有了服务，我们就可以构建“聪明组件”了。

4.3.2 构建“聪明组件”

有了“笨组件”，“聪明组件”就非常简单了。在 auth 目录下新建 containers 子目录。

1. 登录页面

首先使用下面命令生成基础文件：

```
ng g c auth/containers/login --module auth/auth.module.ts
```

然后更改 src/app/auth/containers/login/login.component.html，使用前面创建的“佳句”和“登录表单”两个“笨组件”，并设置其对应的属性和处理其产生的事件。这里你又会看到一些奇怪的符号，不要紧，我们稍后会讲。

```
<div fxLayout="row" fxLayout.xs="column" fxLayoutAlign="center space-
around">
  <app-quote [chineseQuote]="(quote$ | async)?.cn"
[englishQuote]="(quote$ | async)?.en" [quoteImg]="(quote$ | async)?.imgUrl">
  </app-quote>
  <app-login-form (submitEvent)="processLogin($event)"></app-login-form>
</div>
更改 src/app/auth/containers/login/login.component.ts 为
// 省略导入

@Component({
  selector: 'login',
  templateUrl: './login.component.html',
  styleUrls: ['./login.component.scss']
})
export class LoginComponent {
```

```
  quote$: Observable<Quote>;
  constructor(
    private quoteService: QuoteService,
    private authService: AuthService
  ) {
    this.quote$ = quoteService
      .getQuotes()
      .pipe(map(quotes => quotes[Math.floor(Math.random() * 10)]));
  }
  processLogin(auth: Auth) {
    this.authService
      .login(auth)
      .pipe(take(1))
      .subscribe(u => console.log(u));
  }
}
```

在这个“聪明组件”的构造函数中，我们注入了两个服务，分别是负责“佳句”的 QuoteService 和负责鉴权的 AuthService。

我们还定义了一个成员变量 quote$: Observable<Quote>;，在 QuoteService 中的 getQuotes 方法中得到的是一个数组 Quote[]，但是在组件中显然需要的是一个数组中的元素，所以成员变量的定义是一个 Quote 类型，而不是 Quote[]。

但是，既然服务中返回的是一个数组，怎么变成了一个元素呢？这就是 map(quotes => quotes[Math.floor(Math.random() * 10)]) 这句代码起的作用，它将一个数组变换成一个元素，具体就是得到数组 Quotes[] 后，返回一个随机位置的元素 Math.floor(Math.random() * 10)。这样我们每次进入这个页面都会显示不同的“佳句”。

对于 Observable 类型的变量有两种方式取得其订阅值。

第一种方式是类似 quote$ 的处理，我们并不直接订阅，而是在模板中使用 quote$ | async。使用 async 管道的好处在于不用显式地订阅和销毁，Angular 会帮你完成这一切。

第二种方式就是显式订阅，就像我们在 processLogin 方法中处理的 subscribe();那样。但是这种显式订阅时一定要注意内存泄漏问题，能显式销毁尽量显式销毁。这个例子中我们没有做销毁动作的原因是使用了 take(1)，这意味着收到一个数据后就完成了这个流的订阅，订阅也就自动销毁了。这一块还是看得云里雾里的话也没关系，我们在讲 RxJS 时会再讨论这个知识点。

2. 注册页面

使用下面命令生成基础文件：

```
ng g c auth/containers/register --module auth/auth.module.ts
```

更改其模板以使用我们的注册表单“笨组件”，并设置几个异步验证器属性 [usernameValidator]="usernameValidator"、[emailValidator]="emailValidator"、[mobileValidator]="mobileValidator" 和处理其表单提交事件 (submitEvent)="processRegister($event)"。

```
<div fxLayout="row" fxLayout.xs="column" fxLayoutAlign="center space-around">
  <register-form (submitEvent)="processRegister($event)"
[usernameValidator]="usernameValidator" [emailValidator]="emailValidator"
[mobileValidator]="mobileValidator">
  </register-form>
</div>
```

在其组件文件中，我们在其构造函数中构建了几个异步验证器，由于在 Angular 中属性的绑定是在 ngOnChanges 时做，而构造函数早于任何一个生命周期钩子，所以我们在构造时进行这个操作，可以让后面的数据绑定正确地进行。

```
// 省略导入

@Component({
  selector: 'register',
  templateUrl: './register.component.html',
  styleUrls: ['./register.component.scss']
})
export class RegisterComponent {
  usernameValidator: AsyncValidatorFn;
  emailValidator: AsyncValidatorFn;
  mobileValidator: AsyncValidatorFn;
  constructor(private service: AuthService, private router: Router) {
    this.usernameValidator =
RegisterValidator.validateUniqueUsername(service);
    this.mobileValidator =
RegisterValidator.validateUniqueMobile(service);
    this.emailValidator = RegisterValidator.validateUniqueEmail(service);
  }

  processRegister(user: User) {
    this.service
      .register(user)
      .pipe(take(1))
```

```
        .subscribe(u => {
          console.log(u);
        });
    }
}
```

我们把使用到的异步验证器抽取到另一个单独文件 src/app/auth/validators/register.validator.ts 中，新建一个 RegisterValidator 类和几个静态方法返回对应的异步验证器。

```
// 省略导入

export class RegisterValidator {
  static validateUniqueUsername(service: AuthService) {
    return (control: AbstractControl) => {
      const val = control.value;
      if (!val) {
        return of(null);
      }
      return service.usernameExisted(val).pipe(
        map(res => {
          return res.existed ? { usernameNotUnique: true } : null;
        })
      );
    };
  }
  static validateUniqueEmail(service: AuthService) {
    return (control: AbstractControl) => {
      const val = control.value;
      if (!val) {
        return of(null);
      }
      return service.emailExisted(val).pipe(
        map(res => {
          return res.existed ? { emailNotUnique: true } : null;
        })
      );
    };
  }
  static validateUniqueMobile(service: AuthService) {
    return (control: AbstractControl) => {
      const val = control.value;
      if (!val) {
        return of(null);
      }
      return service.mobileExisted(val).pipe(
```

```
          map(res => {
            return res.existed ? { mobileNotUnique: true } : null;
          })
        );
      };
    }
  }
```

4.3.3 路由处理

如我们之前所说，每个模块有自己的路由，auth 模块在生成时已经生成了自己的路由模块——src/app/auth/auth-routing.module.ts。

下面需要把这个路由模块中路由数组改造一下，添加我们的路由。

```
// 省略导入

const routes: Routes = [
  {
    path: 'auth',
    redirectTo: 'auth/login',
    pathMatch: 'full'
  },
  {
    path: 'auth/login',
    component: LoginComponent
  },
  {
    path: 'auth/register',
    component: RegisterComponent
  },
  {
    path: 'auth/forgot',
    component: ForgotPasswordComponent
  }
];

@NgModule({
  imports: [RouterModule.forChild(routes)],
  exports: [RouterModule]
})
export class AuthRoutingModule {}
```

当然，别忘了同样更新根路由——src/app/core/app-routing.module.ts，将默认路径指向 /auth。

```
// 省略导入

const routes: Routes = [
  {
    path: '',
    redirectTo: '/auth',
    pathMatch: 'full'
  },
  {
    path: '**',
    component: PageNotFoundComponent
  }
];

@NgModule({
  imports: [RouterModule.forRoot(routes)],
  exports: [RouterModule]
})
export class AppRoutingModule {}
```

最后，由于我们还没有实现懒加载，所以还需要把 auth 模块导入到 CoreModule 中。

```
// 省略
import { AuthModule } from '../auth/auth.module';

@NgModule({
  declarations: [
    // 省略
  ],
  imports: [
    SharedModule,
    HttpClientModule,
    AuthModule, //<------------这里
    AppRoutingModule,
    BrowserAnimationsModule
  ]
})
export class CoreModule {
  // 省略
}
```

4.4 完成忘记密码前端设计

忘记密码是一个相对比较复杂的控件，因为里面涉及两个大的步骤（见图 4-15）。第一个步骤是验证手机的过程，第二个步骤是设置新密码的过程，只有在第一个步骤成功的前提下才能完成第二个步骤。和前面的注册、登录等不太一样，这里不是一个表单就处理完成了，所以需要一个向导组件，还好 Angular Material 提供了内建的支持：Stepper。

图 4-15

除了这个向导，验证手机也是一个相对复杂的控件，这个控件虽然看起来很简单，但是内部的逻辑并不是很容易明白。

4.4.1 使用 RxJS 打造短信验证码控件

使用 Angular CLI 生成短信验证码组件的基础文件，由于这个控件可能会在多处使用到，所以把它创建在 shared 目录中。

```
ng g c shared/components/verify-mobile
```

打开 src/app/shared/components/verify-mobile/verify-mobile.component.html，将其改造成如下所示。

```
<div [formGroup]="form">
  <mat-form-field class="full-width">
    <input matInput placeholder="注册时使用的手机号"
formControlName="regMobile">
    <mat-error> {{ mobileErrors }} </mat-error>
  </mat-form-field>
  <mat-form-field class="full-width">
```

```
      <input matInput type="text" [placeholder]="codePlaceholder"
formControlName="smsCode">
      <button mat-button matSuffix #veriBtn [disabled]="(btnLabel$ |
async).indexOf('发送') === -1 || form.get('regMobile').errors !== null">
        {{ btnLabel$ | async }} </button>
      <mat-error> 验证码不正确 </mat-error>
    </mat-form-field>
  </div>
```

这个组件的 UI 表现形式还是比较简单的，只有两个输入框，上面的是手机号的输入框，下面是验证码的输入框，但是下面这个输入框的右侧会有一个按钮。这个按钮的状态相对复杂。

- 文字状态：一开始时文字是“发送”，点击后，文字变成了一个 60 s 倒计时，倒计时结束后，按钮文字变成“再次发送”
- 禁用/可用状态：在手机号不合法的情况下，按钮应是禁用状态，手机号合法时可用；在点击“发送”后，60 s 倒计时期间，按钮禁用，倒计时结束又变成可用。

此外由于这是一个“笨组件”，所以它并不负责真正的验证码的发送或校检。但是如果交给外部使用它的父组件处理，则要给出一些输出型属性（事件）让外部可以得到相关的信息。

- 手机号：输入的手机号在验证合法后会产生一个事件，将手机号发送给事件处理者。
- “发送”按钮的点击事件应该暴露给外部，以便父组件调用验证码 API 请求发送短信。

再有，这个组件如果不是在注册页面使用的时候，比如是在更改个人信息页面的“重置密码”功能中使用时，我们是不希望用户可以更改手机号的，因为这个手机号就是用户注册的手机号，如果可以更改，那任何人都可以更改密码了。所以我们给组件设置了 @Input() mobile: string | null = null; 属性，如果设置了非 null 的手机号，那么手机号这个输入框就禁用了。

这个表单控件的值是一个对象{mobile: string; code: string;}，外部取得值之后可以自行验证手机号和短信验证码是否匹配。

```
// 省略导入

@Component({
  selector: 'app-verify-mobile',
  templateUrl: './verify-mobile.component.html',
  styleUrls: ['./verify-mobile.component.scss'],
  providers: [
    {
      provide: NG_VALUE_ACCESSOR,
      useExisting: forwardRef(() => VerifyMobileComponent),
      multi: true
    },
```

```
    {
      provide: NG_VALIDATORS,
      useExisting: forwardRef(() => VerifyMobileComponent),
      multi: true
    }
  ],
  changeDetection: ChangeDetectionStrategy.OnPush
})
export class VerifyMobileComponent
  implements ControlValueAccessor, OnInit, OnDestroy {
  @Input() mobilePlaceholder = '绑定手机号';
  @Input() codePlaceholder = '请输入短信验证码';
  @Input() countdown = 60;
  @Input() mobile: string | null = null;
  @Output() requestCode = new EventEmitter<string>();
  @Output() mobileInputEvent = new EventEmitter<string>();
  @ViewChild('veriBtn', { read: ElementRef })
  veriBtn: ElementRef;
  btnLabel$: Observable<string>;
  form: FormGroup;
  private subs: Subscription[] = [];
  private propagateChange = (_: any) => {};

  constructor(private fb: FormBuilder) {}

  ngOnInit() {
    this.form = this.fb.group({
      regMobile: [
        { value: this.mobile, disabled: this.mobile },
        Validators.compose([
          Validators.required,
          Validators.pattern(mobilePattern)
        ])
      ],
      smsCode: [
        '',
        Validators.compose([Validators.required,
Validators.pattern(/^\d{6}$/)])
      ]
    });
    if (!this.mobile) {
      const mobile = this.form.get('regMobile');
      this.subs.push(
        mobile.valueChanges
```

```
            .pipe(filter(_ => mobile.errors === null))
            .subscribe(val => this.mobileInputEvent.emit(val))
        );
      }
      const smsCode = this.form.get('smsCode');
      const countDown$ = interval(1000).pipe(
        map(i => this.countdown - i),
        takeWhile(v => v >= 0),
        startWith(this.countdown)
      );

      this.btnLabel$ = fromEvent(this.veriBtn.nativeElement,
'click').pipe(
        tap(_ => this.requestCode.emit()),
        switchMap(_ => countDown$),
        map(i => (i > 0 ? `还剩 ${i} 秒` : `再次发送`)),
        startWith('发送')
      );

      if (smsCode) {
        const code$ = smsCode.valueChanges;
        this.subs.push(
          code$.pipe(debounceTime(400)).subscribe(v =>
            this.propagateChange({
              mobile: this.mobile
                ? this.mobile
                : this.form.get('regMobile').value,
              code: v
            })
          )
        );
      }
    }
    ngOnDestroy(): void {
      this.subs.forEach(sub => {
        if (sub) {
          sub.unsubscribe();
        }
      });
      this.subs = [];
    }
    // 验证表单，验证结果正确则返回 null，否则返回一个验证结果对象
    validate(c: FormControl): { [key: string]: any } | null {
      const val = c.value;
```

```
      if (!val) {
        return null;
      }
      const smsCode = this.form.get('smsCode');
      if (smsCode) {
        return smsCode.valid
          ? null
          : {
              smsCodeInvalid: true
            };
      }
      return {
        smsCodeInvalid: true
      };
    }
    writeValue(obj: any): void {
      this.mobile = obj;
    }
    registerOnChange(fn: any): void {
      this.propagateChange = fn;
    }
    registerOnTouched(fn: any): void {}

    get mobileErrors() {
      const mobile = this.form.get('regMobile');
      if (!mobile) {
        return '';
      }
      return mobileErrorMsg(mobile);
    }
}
```

这个组件是一个表单组件，所以相关的内容我们就不赘述了，前面有较为详细的分析。这里主要来看一下如何使用 RxJS 解决逻辑问题。

这个组件的页面上有以下的事件/数据流。

- mobile.valueChanges ——手机号输入产生的值的数据流。

这个流先应用 filter 操作符过滤掉不合法的手机号，然后在订阅中发送手机号 this.mobileInputEvent.emit(val)。

- countDown$——倒计时的数据流。

由 interval 每秒自动产生一个顺序增长序列（0, 1, 2, 3, ...）

使用 map 操作符进行变换：map(i => this.countdown - i)，就是用 60 减去 interval 得到的增长序列，这样就变换成一个顺序减少的序列。

倒计时应该到 0 就结束了，所以使用 takeWhile(v => v >= 0) 让其到 0 后自动完成。

interval 是等到时间间隔后才发送值的，所以给出初始值 startWith(this.countdown)。

- btnLabel$——按钮标签文字的数据流。

这个流最开始是由按钮点击事件产生的 fromEvent(this.veriBtn.nativeElement, 'click')

点击之后应该立即发送事件给外部 tap(_ => this.requestCode.emit())。

切换到倒计时流并执行高阶到低阶的转换 switchMap(_ => countDown$)，此时流转换成了倒计时数字的数据流。

通过 map 将数字流转换成按钮要显示的文字 map(i => (i > 0 ? `还剩 ${i} 秒` : `再次发送`))

别忘了给这个流一个初始值，否则刚进入页面时，没有点击事件，这个流就没有数据，这样按钮文字就没有了。因此我们使用 startWith('发送')。

- smsCode.valueChanges——短信验证码输入产生的值的数据流。

这个流使用了 debounceTime(400) 做节流处理，也就是 400ms 内发生的事件都丢弃了，因为用户输入时可能发生错误，我们不想频繁地给父组件发送事件。

在其订阅方法中，将手机号和验证码构成一个对象发射出去。

4.4.2 忘记密码向导“笨组件”

短信验证码表单控件完成后，就可以把它用在“忘记密码向导”中了，首先创建这个组件。

```
ng g c auth/components/forgot-password-form -m auth/auth.module.ts
```

然后开始设计这个向导，这里面最重要的就是 Stepper 组件。Angular Material 提供 mat-horizontal-stepper 和 mat-vertical-stepper 两个向导的父容器，分别设置水平排列的步骤或垂直排列的步骤。在父容器内，使用 mat-step 定义一个步骤。

父容器中可以指定 linear 属性，如果为 true，就必须一步一步执行，不可以点击跳到其他步骤，如果为 false 则允许跳到其他步骤。

mat-step 中的 stepControl 属性是要指定一个 AbastractControl，也就是 FormControl 或 FormGroup，这个 AbastractControl 的验证状态决定了这个步骤是否完成。使用多个 Form 表单配合 Stepper 是一个常见的做法，但也可以单个表单对应多个步骤，甚至可以不用表单。这里就不展开讲了，有兴趣的读者可以去官网了解详情。

```
  <mat-horizontal-stepper [linear]="true">
    <mat-step [stepControl]="mobileForm">
      <form [formGroup]="mobileForm" (ngSubmit)="submit(mobileForm,
$event)">
        <ng-template matStepLabel> 验证绑定手机 </ng-template>
        <app-verify-mobile formControlName="oldCode" [mobile]="mobile"
[mobilePlaceholder]="'绑定手机'" (requestCode)="makeRequest($event)"
          (mobileInputEvent)="mobileInput($event)">
        </app-verify-mobile>
        <div>
          <button mat-button matStepperNext> 下一步 </button>
        </div>
      </form>
    </mat-step>
    <mat-step [stepControl]="newPasswordForm">
      <form [formGroup]="newPasswordForm"
(ngSubmit)="submit(newPasswordForm, $event)">
        <ng-template matStepLabel>设置新密码</ng-template>
        <mat-form-field class="full-width">
          <input matInput placeholder="请输入新密码" type="password"
formControlName="password">
        </mat-form-field>
        <mat-form-field class="full-width">
          <input matInput placeholder="请再次输入" type="password"
formControlName="repeat">
          <mat-error> 两次密码输入不一致 </mat-error>
        </mat-form-field>
        <div>
          <button mat-button matStepperNext> 下一步 </button>
        </div>
      </form>
    </mat-step>
    <mat-step>
      <ng-template matStepLabel> 完成 </ng-template>
      密码修改成功。
      <div>
        <button mat-button matStepperPrevious>返回</button>
      </div>
    </mat-step>
  </mat-horizontal-stepper>
```

接下来看组件文件，由于有两个表单，所以在 ngOnInit 中需要分别进行初始化。为了验证手机和短信验证码是否匹配，我们需要一个异步验证器，这个交给“聪明组件”处理，所以这里设置了一个 @Input() codeValidator: ValidatorFn;。

还需要指出的是，第二个表单同样需要两个密码进行比较，但这里我们并没有采用前面注册表单组件中的做法，而是给出了另一种方式。这里没有在 FormGroup 级别去做验证器，而是给 repeat 做了个 FormControl 的验证器。但比较的思路是基本一致的，这里不过是给大家展示另一种方式：control.parent.get(otherCtrlName)。请注意，在真实的开发过程中，如果遇到一种验证方式是多处需要的时候，最好将验证器剥离出来，或者抽象该组件为公共组件。

```
// 省略导入

@Component({
  selector: 'forgot-password-form',
  templateUrl: './forgot-password-form.component.html',
  styleUrls: ['./forgot-password-form.component.scss']
})
export class ForgotPasswordFormComponent implements OnInit {
  @Input() mobile: string | null = null;
  @Input() codeValidator: ValidatorFn;
  @Output() submitPassword = new EventEmitter<string>();
  @Output() requestCode = new EventEmitter<string>();
  @Output() mobileInputEvent = new EventEmitter<string>();
  mobileForm: FormGroup;
  newPasswordForm: FormGroup;

  constructor(private fb: FormBuilder) {}

  ngOnInit() {
    this.mobileForm = this.fb.group({
      oldCode: [this.mobile, Validators.required, this.codeValidator]
    });

    this.newPasswordForm = this.fb.group({
      password: [
        '',
        Validators.compose([Validators.required,
Validators.minLength(8)])
      ],
      repeat: ['', [Validators.required, this.matchPassword('password')]]
    });
  }

  submit(form: FormGroup, ev: Event) {
    if (!form.valid || !form.value) {
      return;
    }
```

```
      if (form.value.password) {
        this.submitPassword.emit(form.value.password);
      }
    }

    makeRequest(mobile: string) {
      this.requestCode.emit(mobile);
    }

    matchPassword(otherCtrlName: string) {
      let thisControl: FormControl;
      let otherControl: FormControl;
      return (control: FormControl) => {
        if (!control.parent) {
          return null;
        }
        // Initializing the validator.
        if (!thisControl) {
          thisControl = control;
          otherControl = control.parent.get(otherCtrlName) as FormControl;
          if (!otherControl) {
            throw new Error('matchPassword(): 未发现表单中有要比较的控件');
          }
          otherControl.valueChanges.subscribe(() => {
            thisControl.updateValueAndValidity();
          });
        }
        if (!otherControl) {
          return null;
        }
        if (otherControl.value !== thisControl.value) {
          return {
            matchOther: true
          };
        }
        return null;
      };
    }
    mobileInput(mobile: string) {
      this.mobileInputEvent.emit(mobile);
    }
}
```

4.4.3 忘记密码的“聪明组件”

接下来，就是完成忘记密码的“聪明组件”，这就简单多了。

模板部分还是利用卡片组织，把忘记密码的“笨组件”forgot-password-form 嵌入 mat-card-content 之中。卡片尾部提供链接可以返回登录或注册页面。

```
<mat-card>
  <mat-card-header>
    <mat-card-title>
      <span> 更改密码 </span>
    </mat-card-title>
    <mat-card-subtitle>
      通过手机验证码更改您的密码
    </mat-card-subtitle>
  </mat-card-header>
  <mat-card-content>
    <forgot-password-form [codeValidator]="codeValidator"
(codeRequestEvent)="processCodeRequest($event)"
(mobileInputEvent)="processMobile($event)"
      (passwordEvent)="processPassword($event)">
    </forgot-password-form>
  </mat-card-content>
  <mat-card-actions>
    <div fxLayout="row" fxLayoutAlign="end stretch">
      <a mat-button routerLink="/auth/login"> {{ loginBtnText }} </a>
      <a mat-button routerLink="/auth/register"> {{ registerBtnText }}
</a>
    </div>
  </mat-card-actions>
</mat-card>
```

组件文件中，在构造函数中注入 AuthService，同时在构造函数中把“笨组件”需要的异步验证器构建好。

```
// 省略导入

@Component({
  selector: 'forgot-password',
  templateUrl: './forgot-password.component.html',
  styleUrls: ['./forgot-password.component.scss']
})
export class ForgotPasswordComponent implements OnInit {
  loginBtnText = '登录';
  registerBtnText = '注册';
```

```
  codeValidator: AsyncValidatorFn;

  constructor(private service: AuthService) {
    this.codeValidator = SmsValidator.validateSmsCode(service);
  }

  ngOnInit() {}

  processCodeRequest(mobile: string) {
    this.service
      .requestSmsCode(mobile)
      .pipe(take(1))
      .subscribe(val => console.log(val));
  }

  processMobile(mobile: string) {
    console.log(mobile);
  }

  processPassword(password: string) {
    console.log(password);
  }
}
```

注意，上面的异步验证器是抽象成一个独立的类 SmsValidator，文件路径是 src/app/auth/validators/sms.validator.ts。

```
// 省略导入

export class SmsValidator {
  static validateSmsCode(service: AuthService) {
    return (control: AbstractControl) => {
      const val = control.value;
      if (!val.mobile || !val.code) {
        throw new Error('SmsValidator: 没有找到手机号或验证码');
      }
      return service.verifySmsCode(val.mobile, val.code).pipe(
        map(res => {
          return res ? null : { codeInvalid: true };
        })
      );
    };
  }
}
```

第 5 章 构建后端 API

第 3 章只是对 Spring Boot 中的强大功能进行了一个大概介绍，本章我们会就登录鉴权 API 进行详细的讨论。

5.1 HyperMedia API 与传统 API

5.1.1 领域对象

和前端类似，在后端也需要定义领域对象。而且从某种角度说，后端的领域对象往往也影响前端的建模。既然是登录鉴权，那么肯定要建立一个 `User` 对象，这和前端的模型很类似。

```
public class User {
    private String username;
    private String password;
    private String mobile;
    private String name;
    private String email;
    private Gender gender;
    private String avatar;
}
```

上面这个类定义完成了吗？没有，在 Java 中有一个很烦的地方就是要写好多 `getter` 和 `setter`，还有构造函数、`hashCode`，`equals` 和 `toString` 等方法。当然，其实现在 IDE（集成开发环境）都提供了一些快捷方式生成。但是还是有些麻烦，最重要的是，这些方法在查找问题、分析代码和评审代码时会分散注意力。有没有什么方法可以让我们从这些模式化代码中解脱出来呢？有，那就是 Java 社区大名鼎鼎的 `Lombok`，可以访问其官网了解更多特性的详情。

1．使用 Lombok 简化模式化代码的编写

（1）配置 Lombok

在 api/build.gradle 中，加入 Lombok 的依赖，Lombok 也提供了 Spring Boot 的支持，所以不用写版本号。

```
apply plugin: 'org.springframework.boot'

dependencies {
    implementation("org.springframework.boot:spring-boot-devtools")
    implementation("org.projectlombok:lombok")
    implementation("org.springframework.boot:spring-boot-starter-web")
    implementation("org.springframework.boot:spring-boot-starter-data-rest")
    implementation("org.springframework.boot:spring-boot-starter-data-mongodb")
    implementation("org.springframework.data:spring-data-rest-hal-browser")
}
```

在 IDEA（用于 Java 语言的集成环境）中的“Preference > Build, Execution, Deployment > Complier > Annotation Processors”，选中“Enable annotation processors”，然后点击“OK”按钮，如图 5-1 所示。

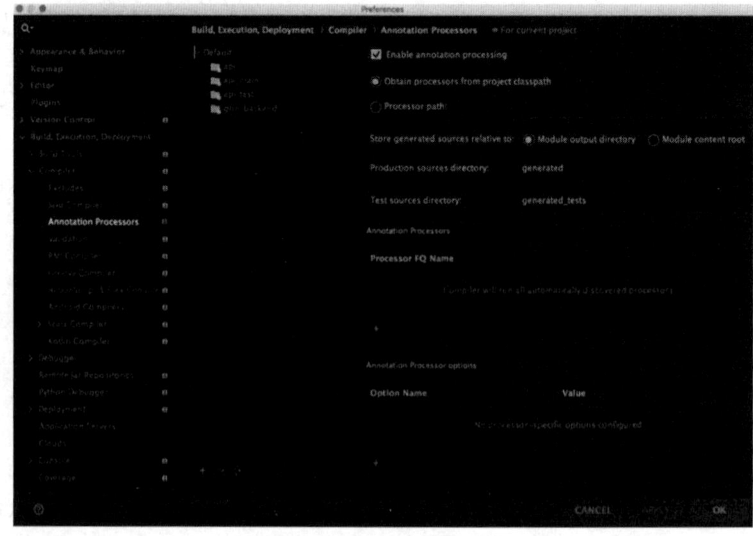

图 5-1

这样就完成了 Lombok 的设置，下面简单地了解一下 Lombok 可以带来哪些“魔法”。

（2）“大杀器”@Data

Lombok 提供了很多注解，但其中最常被使用的就是这个 @Data 了，只需放一个注解在类前面，一切烦恼就都消失了。这么一个注解会帮你生成所有属性的 GET 和 SET 方法（final 属性只生成 GET 方法），帮你实现 hashCode、equals 和 toString 方法以及由必选参数（就是所有 final 属性以及 标识 @NonNull 注解的属性）构成的构造函数。

最棒的是，这些“魔法”是在编译时完成的，和 IDE 自动生成的代码不同，源文件中永远都不会出现这些冗长的代码，一直就是这种短小精悍的形式，这对于开发者实在是太友好了。

```
package dev.local.gtm.api.domain;

import dev.local.gtm.api.domain.enums.Gender;
import lombok.Data;

@Data // <----------这里
public class User {
    private String username;
    private String password;
    private String mobile;
    private String name;
    private String email;
    private Gender gender;
    private String avatar;
}
```

（3）@Builder 让对象创建无烦恼

这个注解也是笔者挚爱之一，在介绍它之前，我们先来看一段代码。

```
User newUser = new User();
newUser.setUsername(userDTO.getUsername());
newUser.setPassword(encryptedPassword);
newUser.setFirstName(userDTO.getFirstName());
newUser.setLastName(userDTO.getLastName());
newUser.setEmail(userDTO.getEmail());
newUser.setImageUrl(userDTO.getImageUrl());
newUser.setLangKey(userDTO.getLangKey());
newUser.setActivated(false);
```

是不是有种很熟悉的感觉，先创建对象，然后一列齐刷刷的 newUser.set×××，笔者感觉是很干扰阅读的，而且在设置属性时，重新输入 newUser.set××× 总有思路被打断的感觉。现代 API 写法中越来越多地使用了 Builder 模式，那么 Lombok 的注解 @Builder 又帮你省去了手写的工作量，可以优雅地写 Builder 模式的代码，何乐而不为呢？

```
User.builder()
  .username(userDTO.getUsername())
  .password(encryptedPassword)
  .firstName(userDTO.getFirstName())
  .lastName(userDTO.getLastName())
  .email(userDTO.getEmail())
  .imageUrl(userDTO.getImageUrl())
  .langKey(userDTO.getLangKey())
  .activated(false)
  .build();
```

此外，和`@Builder`经常在一起使用的有`@Singular`，这个注解标识类中的集合属性，为集合属性生成添加单个元素的方法。详细情况可以参考官网的解释。

（4）别烦我了：try…finally

Java 的异常处理当然是很强大的，但是当你遇到下面这种情况时，是不是也会觉得“生无可恋”啊？

```
import java.io.*;

public class CleanupExample {
  public static void main(String[] args) throws IOException {
    InputStream in = new FileInputStream(args[0]);
    try {
      OutputStream out = new FileOutputStream(args[1]);
      try {
        byte[] b = new byte[10000];
        while (true) {
          int r = in.read(b);
          if (r == -1) break;
          out.write(b, 0, r);
        }
      } finally {
        if (out != null) {
          out.close();
        }
      }
    } finally {
      if (in != null) {
        in.close();
      }
```

```
    }
  }
}
```

Lombok 可以让你这样写。

```
import lombok.Cleanup;
import java.io.*;

public class CleanupExample {
  public static void main(String[] args) throws IOException {
    @Cleanup InputStream in = new FileInputStream(args[0]);
    @Cleanup OutputStream out = new FileOutputStream(args[1]);
    byte[] b = new byte[10000];
    while (true) {
      int r = in.read(b);
      if (r == -1) break;
      out.write(b, 0, r);
    }
  }
}
```

（5）其他几个常用注解

除上面的几个注解，还有本书中会用到的一些 Lombok 提供的工具，它们没有那么强大的功能，但同样会帮你提升效率。

- `val` 和 `var`——用过 Swift 或 Kotlin 的读者肯定知道 `val` 和 `var`，前者是不可变对象，后者是可变对象。有了 Lombok，在 Java 中也可以这么使用了：`val example = new ArrayList<String>();`，而不用写成：`final ArrayList<String> example = new ArrayList<String>();`。`var` 就不举例了，除对象可变外，其他和 `val` 一样。
- `@Log`、`@Slf4j`、`@Log4j`、`@CommonsLog` 等——写日志是我们在编程中经常遇到的，Lombok 提供的这个注解在类上标注后，直接在程序中输入 `log.error()` 即可。

2. 使用 Lombok 改造领域对象

工具介绍完毕，我们回到领域对象中，使用 Lombok 来改造领域对象 `User`。

```
// 省略导入

@Getter
@Setter
@EqualsAndHashCode(of = "id")
@NoArgsConstructor
```

```
@AllArgsConstructor
@Builder
public class User implements Serializable {
    private static final long serialVersionUID = 1L;

    @Id
    private String id;

    @NotNull
    @Pattern(regexp = Constants.LOGIN_REGEX)
    @Size(min = 1, max = 50)
    @Indexed
    private String login;

    @JsonIgnore
    @NotNull
    @Size(min = 60, max = 60)
    private String password;

    @NotNull
    @Pattern(regexp = Constants.MOBILE_REGEX)
    @Size(min = 10, max = 15)
    private String mobile;

    @Size(max = 50)
    private String name;

    @Email
    @Size(min = 5, max = 254)
    @Indexed
    private String email;

    @Size(max = 256)
    private String avatar;

    @Builder.Default
    private boolean activated = false;
}
```

这样看起来一个领域对象的代码就清晰多了。比起之前的 User 对象，我们添加了一些字段的约束，比如 @NotNull、@Pattern、@Size、@Email 这些都属于 JSR 380 Java Bean Validation 提供的注解，用于确保属性满足指定条件，而且这些注解还可以和 Java 8 的 Optional 连用。

```
private LocalDate dateOfBirth;

public Optional<@Past LocalDate> getDateOfBirth() {
    return Optional.of(dateOfBirth);
}
```

`@Indexed` 属于 Spring Data MongoDB 提供的注解，这个注解就是建立索引，如果查看索引的话，则可以登录到 MongoDB 的容器中，使用 MongoDB 的客户端命令进行操作。当然这个索引目前还不是程序规定的样子，直到有数据写入时索引会自动创建。

```
### 先登录到容器
docker exec -it [容器名称] bash
### 然后进入 MongoDB 的客户端
mongo
### 输入下面的命令查看索引
db.user.getIndexes()
```

对于使用的 Lombok 注解，需要指出，`@Builder.Default` 是如果在有属性的默认值时，同时又使用了`@Builder` 的情况下，用来标记有默认值的那个属性。没有使用 `@Data` 的原因是，对于用户的 `equals` 和 `hashCode` 两个方法，我们不想比较所有字段，其实只需比较 `id` 即可，而 `@Data` 默认是全部比较的。指定哪些属性参与到 `equals` 和 `hashCode` 方法中的比较，可以使用 `@EqualsAndHashCode(of = {"属性1", "属性2", ...})` 来实现。

5.1.2 API 的可见控制

有细心的读者可能发现了之前我们的 API 默认暴露了 `repository` 中支持的全部接口。如果启动服务，使用浏览器访问 `http://localhost:8080` 进入 `HAL Browser`，在 `Explorer` 的下方输入框内输入 `/profile/users`，单击“`Go`”按钮，就可以看到 Response Body 中返回的 `JSON`。

```
{
  "alps": {
    "version": "1.0",
    "descriptors": [
      // 省略其他部分
      {
        "id": "create-users",
        "name": "users",
        "type": "UNSAFE",
        "rt": "#user-representation"
      },
      {
        "id": "get-user",
```

```
          "name": "user",
          "type": "SAFE",
          "rt": "#user-representation"
        },
        {
          "id": "update-user",
          "name": "user",
          "type": "IDEMPOTENT",
          "rt": "#user-representation"
        },
        {
          "id": "patch-user",
          "name": "user",
          "type": "UNSAFE",
          "rt": "#user-representation"
        },
        {
          "id": "delete-user",
          "name": "user",
          "type": "IDEMPOTENT",
          "rt": "#user-representation"
        }
      ]
    }
  }
```

从上面的 JSON 中可以看到我们暴露了增、删、改、查等操作，大家可以通过 Postman 进行验证，所有的操作都是允许的。那么问题来了，很多时候我们只想开放某些接口，比如只开放只读接口，不允许写操作。这种情况下怎么办呢？答案非常简单，还是采用注解，但是需要配合适当的策略。

Spring Data Rest 使用 `RepositoryDetectionStrategy` 来决定一个 `repository` 是否作为 Rest 资源暴露出去。需要指出的是，`RepositoryDetectionStrategy` 只决定 API 的外部可见策略，并不涉及权限，也就是说能看到不代表能执行。

- `DEFAULT`——默认情况下会把所有 `public` 的 `respository` 接口暴露成 Rest 资源。但是仍然会尊重 `@RepositoryRestResource` 或 `@RestResource` 的 `exported` 标志位。也就是说，比如某个 `TaskRepo`，加上注解 `@RepositoryRestResource (collectionResourceRel = "tasks", path = "tasks", exported = false)`，那么我们在访问 http://localhost:8080/ api 时将不会把 api/tasks 列出。

- ALL——暴露所有 repository 的接口方法。
- ANNOTATION——只有添加了注解@RepositoryRestResource 和@RestResource 的 repository 方法才会暴露，但仍会尊重 exported 标志位，也就是说如果 exported = false，那么就不暴露该资源。
- VISIBILITY——只暴露 public 添加了注解 @RepositoryRestResource 的 repository 接口。

这个策略的配置有两种方法可以选择，第一种方法是建立一个配置类。

```
package dev.local.gtm.api.config;

import org.springframework.context.annotation.Configuration;
import
org.springframework.data.rest.core.config.RepositoryRestConfiguration;
import
org.springframework.data.rest.core.mapping.RepositoryDetectionStrategy;
import
org.springframework.data.rest.webmvc.config.RepositoryRestConfigurerAdapter;

@Configuration
public class SpringRestConfiguration extends
RepositoryRestConfigurerAdapter {
    @Override
    public void
configureRepositoryRestConfiguration(RepositoryRestConfiguration config) {
        config.setRepositoryDetectionStrategy(RepositoryDetectionStrateg
y.RepositoryDetectionStrategies.DEFAULT);
    }
}
```

第二种方法是通过配置文件，比如 application.yml 或 application. properties 来实现。如果没有复杂的自定义配置，推荐这种方法，因为更简单。

```
## application.yml
spring:
  # 省略其他部分
  data:
    rest:
      detection-strategy: default
```

1. default 策略

我们先来看一下 default（默认）策略，首先把 UserRepo 改造成下面的样子，我们显式地给增、删、改等操作加上了注解 @RestResource(exported = false)，就是说不想暴露成 Rest 资源。

```
@Repository
public interface UserRepo extends MongoRepository<User, String> {


    @RestResource(exported = false)
    @Override
    <S extends User> S insert(S entity);

    @Override
    Page<User> findAll(Pageable pageable);

    @RestResource(exported = false)
    @Override
    <S extends User> S save(S entity);

    @Override
    Optional<User> findById(String s);

    @Override
    boolean existsById(String s);

    @Override
    long count();

    @RestResource(exported = false)
    @Override
    void deleteById(String s);
}
```

现在我们再去 `HAL Browser` 中，输入 `/profile/users`，看看返回的 Response Body，会发现被允许的操作仅剩下了读操作。

```
{
  "alps": {
    "version": "1.0",
    "descriptors": [
      // 省略其他部分
      {
        "id": "get-user",
        "name": "user",
        "type": "SAFE",
        "rt": "#user-representation"
      }
```

```
        ]
    }
}
```

其实，在这种策略下，如果我们还是继承了 MongoRepository，那么只标记不想暴露的操作即可，也就是说根本不用去覆写读操作的方法。

```
@Repository
public interface UserRepo extends MongoRepository<User, String> {

    @RestResource(exported = false)
    @Override
    <S extends User> S insert(S entity);

    @RestResource(exported = false)
    @Override
    <S extends User> S save(S entity);

    @RestResource(exported = false)
    @Override
    void deleteById(String s);
}
```

上面的代码中只保留了对于写操作的 3 个方法，其效果和上面的是等价的，这是因为 default 策略默认开放所有 public 接口，我们的接口继承了 MongoRepository，而 MongoRepository 又继承了 CrudRepository、PagingAndSortingRepository 和 Repository，这些接口中的 public 方法都会默认作为 Rest 资源，当然很多方法其实对应的是一个资源，比如很多的 findAll×××对应的都是 users 这个资源，只不过传递的参数不一样。在这种情况下，我们显式标注哪些不想暴露出去就可以了。

上述方法适合我们需要 MongoRepository 的很多方法，但对外不想暴露成 API，内部的 Service 或 Controller 中还是需要这些未暴露的方法的。如果我们不需要哪些未暴露的方法，做法其实是从继承 MongoRepository 变为继承一个更上层的 Repository 类型，比如 CrudRepository、PagingAndSortingRepository 甚至 Repository。在下面的例子中我们直接使用了 Repository 这个接口，注意这个接口和我们的注解重名，所以注解需要使用全名。这样我们就只暴露这 3 个接口，其他方法无论是内部还是外部都没有提供。

```
// 省略导入

@org.springframework.stereotype.Repository
public interface UserRepo extends Repository<User, String> {
    Page<User> findAll(Pageable pageable);
```

```
    Optional<User> findById(String s);

    long count();
}
```

2. all 策略

首先调整策略为 all，all 和 default 的区别可能很多读者搞不清楚，因为几乎是一样的，只不过对于 public 的界定不同。

```
## application.yml
spring:
  # 省略其他部分
  data:
    rest:
      detection-strategy: all
```

如下代码中，UserRepo 和 default 策略的代码只相差一个 public 的范围限定，但下面的这个代码在 all 策略中就可以把接口暴露出来，但如果改成 default 策略，那么这个 /users 的资源是不存在的。

```
@Repository
interface UserRepo extends MongoRepository<User, String> {

    @RestResource(exported = false)
    @Override
    <S extends User> S insert(S entity);

    @RestResource(exported = false)
    @Override
    <S extends User> S save(S entity);

    @RestResource(exported = false)
    @Override
    void deleteById(String s);
}
```

3. annotation 策略

```
## application.yml
spring:
  # 省略其他部分
  data:
    rest:
      detection-strategy: annotated
```

如果采用 annotation（注释）策略，所有要暴露的 repository 和 repository 方法都需要显式添加注解，不想暴露的方法也需要显式添加注解并设置 exported = false。

```
@RepositoryRestResource
public interface UserRepo extends MongoRepository<User, String> {

    @RestResource(exported = false)
    @Override
    <S extends User> S insert(S entity);

    @RestResource
    @Override
    Page<User> findAll(Pageable pageable);

    @RestResource(exported = false)
    @Override
    <S extends User> S save(S entity);

    @RestResource
    @Override
    Optional<User> findById(String s);

    @RestResource(exported = false)
    @Override
    void deleteById(String s);
}
```

4. visibility 策略

```
## application.yml
spring:
  # 省略其他部分
  data:
    rest:
      detection-strategy: visibility
```

visibility（可见性）策略只暴露 public 添加了注解 @RepositoryRestResource 的 repository 接口。也就是只有类似下面的接口才会暴露为 Rest 资源，缺少 @RepositoryRestResource 或者不是 public 的接口都不能暴露为 Rest 资源。

```
@RepositoryRestResource
public interface UserRepo extends MongoRepository<User, String> {
  // 省略
}
```

Spring Data Rest 在构造 API 方面非常容易，但在目前的使用上还没有普及，大部分项目使用的还是 Level 2 的 API。而且一些周边的开源类库和它的配合还不是太好，比如本书定稿时 Spring 的 Swagger 集成类库 SpringFox 对于 Spring Data Rest 3.× 还是没有支持，笔者感觉在目前阶段还不适合在生产项目中使用，但可以持续关注。所以接下来的项目实践中，我们还是改成经典的 Rest 实现模式。

5.1.3 传统的 API 实现模式

我们先去掉 api/build.gradle 中的 Data Rest 依赖。

```
## 删除下面这个依赖
implementation("org.springframework.boot:spring-boot-starter-data-rest")
```

改造 application.yml 为：

```
spring:
  application:
    name: api-service
  devtools:
    remote:
      secret: thisismysecret
  data:
    mongodb:
      database: gtm-api
```

1. 找回熟悉的 Controller

对于很多习惯了 Spring 开发的读者来讲，Controller、Service、DAO 这些方法突然间都没了会有不适感。其实，这些东西还在，只不过在较简单的情景下，这些都变成了系统背后帮你做的事情。本节我们就先来看看如何将 Controller 再召唤回来。

由于 Controller 其实就是定义 API 资源，所以我们在 rest 包下建立 /api/web/rest/AuthResource.java。这样的语义比较清晰，既然是rest的资源，就叫×××Resource。

如果要让 AuthResource 可以和 UserRepo 配合工作，当然需要在 AuthResource 中引用 TaskRepo。

```
@Log4j2
@RestController
@RequestMapping("/api")
@RequiredArgsConstructor
public class AuthResource {
    private final UserRepo userRepo;
```

```
    //省略其他部分
}
```

Spring 现在鼓励用构造函数来做注入，所以，我们使用 Lombok 提供的注解 @RequiredArgsConstructor 自动生成一个构造函数。为什么叫作 RequiredArgs？因为这个构造是使用必需的参数构成的，它相当于下面的代码，一般来说，如果将成员变量声明成 private final ××××，那么这个××××就是必需的参数，它必须通过构造函数赋值：

```
@Log4j2
@RestController
@RequestMapping("/api")
@RequiredArgsConstructor
public class AuthResource {

    private final UserRepo userRepo;

    @AutoWired
    public AuthResource(UserRepo userRepo){
        this.userRepo = userRepo;
    }
    //省略其他部分
}
```

我们为了可以让 Spring 知道这是一个支持 REST API 的 Controller，还是需要标记其为 @RestController。对 Spring 熟悉的读者可能知道还有另一个注解是 @Controller，那么这个 @RestController 和 @Controller 有什么区别呢？ @RestController 是 Spring 4.× 引入的一个注解，它相当于 @Controller + @RepsonseBody，也就是说使用 @RestController 之后不用给这个 @Controller 的方法加上 @RepsonseBody 了。

拿第 3 章的例子来看，我们在 getAllTasks() 方法的声明里直接返回了 List<Task>，如果使用 Postman 请求，则会发现 Resposne Body 中有了这个数组列表。这个处理其实是通过 @ResponseBody 完成的。如果使用 @Controller，就得声明 public @RepsonseBody List<Task> getAllTasks()，而 @RestController 就是进一步简化了我们的工作。只需在类上使用这个注解，然后在类的方法上就不用使用 @ResponseBody 了。

```
@RestController
public class TaskController {
    // 使用@RequstMapping 指定可以访问的 URL 路径
    @RequestMapping("/tasks")
    public List<Task> getAllTasks() {
        // 省略
    }
}
```

接下来实现登录和注册API，在 `AuthResource` 中添加两个方法 `login` 和 `register`:

```
@Log4j2
@RestController
@RequestMapping("/api")
@RequiredArgsConstructor
public class AuthResource {

    private final UserRepo userRepo;

    @PostMapping(value = "/auth/login")
    public User login(@RequestBody final Auth auth) {
        log.debug("REST 请求 -- 将对用户: {} 执行登录鉴权", auth);
        val user = userRepo.findOneByLogin(auth.getLogin());
        if (!user.isPresent()) {
            throw new LoginNotFoundException();
        }
        if (!user.get().getPassword().equals(auth.getPassword())) {
            throw new InvalidPasswordException();
        }
        return user.get();
    }

    @PostMapping("/auth/register")
    public ResponseEntity<User> register(@RequestBody final User user) {
        log.debug("REST 请求 -- 注册用户: {} ", user);
        if (userRepo.findOneByLogin(user.getLogin()).isPresent()) {
            throw new LoginExistedException();
        }
        if (userRepo.findOneByMobile(user.getMobile()).isPresent()) {
            throw new MobileExistedException();
        }
        if (userRepo.findOneByEmail(user.getEmail()).isPresent()) {
            throw new MobileExistedException();
        }
        return ResponseEntity.ok(userRepo.save(user));
    }
}
```

上面的代码中需要再说明几个要点。

（1）Spring 4.3 之后引入了 `@PostMapping`、`@PutMapping`、`@DeleteMapping`、`@PatchMapping` 和 `@GetMapping` 这几个注解。这些注解和 `@RequestMapping` 的区别在于，它们进一步简化了注解的使用，因为各个注解已经隐含了 HTTP 的方法，比如

`@PostMapping` 就相当于 `@RequestMapping(method = RequestMethod.POST)`。

（2）这些方法接受的参数也使用了各种修饰符，`@PathVariable` 表示参数是从路径中得来的，而 `@RequestBody` 表示参数应该从 HTTP Request 的 `body` 中解析，类似的 `@RequestHeader` 表示参数是 HTTP Request 的 Header 中定义的。

（3）`login` 方法返回的直接是对象，而 `register` 方法返回的是 `ResponseEntity`。那这两种写法有哪些区别？首先 `login` 直接返回对象，其实也是应用了 `@ResponseBody`，Spring 会自动帮我们把返回的对象构建成 json 放到 Response 中。而 `ResponseEntity` 就可以更全面地定制自己想要返回的内容和状态，比如我们在某种情况下希望自己控制返回的状态码和返回的内容。这种时候使用 `ResponseEntity` 就非常方便。也就是说如果你感觉 Spring 默认的处理就很好，那么使用直接的对象返回没有任何问题。但如果你想自己有对状态和内容的控制，那么就可以使用 `ResponseEntity`。

上面的代码中，我们看到有几个方法（`findOneByLogin`、`findOneByMobile` 以及 `findOneByEmail`）在 `UserRepo` 中是没有的。下面我们会来添加这几个方法，也顺便介绍 Spring Data MongoDB 中的查询方式。

5.2 Spring Data 中的查询

Spring Data 是 Spring 的一个子项目（https://projects.spring.io/spring-data），这个项目的目标是要简化对于不同数据库（或者更广泛地叫作不同的数据持久化存储方式）的操作及形成一致的数据访问层的抽象。

5.2.1 基础概念——Repository

Spring Data 的核心是 `Repository`，这个单词译成中文是存储库，感觉还是不太达意，所以后面提到时还是使用英文原文。`Repository` 作为 Spring Data 中最基础的抽象。

```
@Indexed
public interface Repository<T, ID> {
}
```

这个抽象很简单，其中 `T` 是领域对象的类型，`ID` 是领域对象存储在数据库中的 ID。其他的 Repository 都是继承了这个基础接口。当然从这个接口的定义我们也知道，不管使用的是哪种数据库，如果要定义一个 Repository，就要提供领域对象和 ID 的类型。

1．几个内建的 Repository

在 Spring Data 中，我们经常会遇到以下几个 Repository：

- CrudRepository——Crud 就是增、删、改、查的缩写，即 Create、Read、Update、Delete 的首字母连在一起。就是定义了基础的增、删、改、查操作。
- PagingAndSortingRepository——也是命名上就可以看出是定义了分页和排序的操作。它继承了 CrudRepository 之后又提供了分页和排序的操作。
- MongoRepository——和上面两个不同，MongoRepository 是一个和数据库相关的接口，它继承了 PagingAndSortingRepository，同时封装了一些 MongoDB 特有的方法。

需要指出的是，CrudRepository 和 PagingAndSortingRepository 是通用的，也就是说，在 Spring Data Jpa 和 Spring Data MongoDB 中都是一致的，不受限于某一个数据库。

从 CrudRepository 的源码中可以看到基础的增、删、改、查方法都已经在这里定义好了。

```
public interface CrudRepository<T, ID extends Serializable> extends
Repository<T, ID> {

    // 保存
    <S extends T> S save(S entity);
    // 返回指定 ID 对应的对象
    Optional<T> findById(ID primaryKey);
    // 查找所有对象
    Iterable<T> findAll();
    // 返回数量
    long count();
    // 删除实体
    void delete(T entity);
    // 判断指定的 ID 是否存在
    boolean existsById(ID primaryKey);

    // 省略其他部分
}
```

PagingAndSortingRepository 在此之上提供了排序和分页的方法。

```
    public interface PagingAndSortingRepository<T, ID> extends
CrudRepository<T, ID> {

      Iterable<T> findAll(Sort sort);

      Page<T> findAll(Pageable pageable);
    }
```

我们可以从 `MongoRepository` 的源码中看到它增加了 `insert` 方法，这个方法就是不在通用的接口中，而是在 MongoDB 的专属接口中定义的。

```
    public interface MongoRepository<T, ID> extends
PagingAndSortingRepository<T, ID>, QueryByExampleExecutor<T> {

      @Override
      <S extends T> List<S> saveAll(Iterable<S> entites);

      @Override
      List<T> findAll();

      @Override
      List<T> findAll(Sort sort);

      <S extends T> S insert(S entity); // <--- 数据库相关的方法

      <S extends T> List<S> insert(Iterable<S> entities); // <--- 数据库相关的
方法

      @Override
      <S extends T> List<S> findAll(Example<S> example);

      @Override
      <S extends T> List<S> findAll(Example<S> example, Sort sort);
    }
```

5.2.2 查询方式

Spring Data 内建提供了动态依赖方法命名构建查询的方法。这个方法粗听上去感觉可能是个“玩具”，复杂查询可能应付不了，但恰恰相反，这个方法的灵活性非常好，基本可以应对较复杂的查询。

那么我们就来增加一个需求，可以通过查询 Task（任务）描述中的关键字来搜索符合的项目。

显然这个查询不是默认的操作，那么这个需求在 Spring Boot 中怎么实现呢？非常简单，只需在 TaskRepo 中添加一个方法：

```
    ...
    @Repository
    public interface TaskRepo extends MongoRepository<Task, String> {
        Page<Task> findByDescLike(Pageable pageable, @Param("desc") String
desc);
    }
```

新建/api/web/rest/TaskResource.java，我们创建一个 GET 方法 getAllTasks，它对应的 URL 是 /api/tasks，现在让它可以携带参数，也就是 /api/tasks?desc=×××这种形式。在方法中声明@RequestParam 的参数 desc。如果这个参数不存在，那么就返回全部列表，否则就执行 findByDescLike 的查询。

```
    @Log4j2
    @RestController
    @RequestMapping("/api")
    @RequiredArgsConstructor
    public class TaskResource {
        private final TaskRepo taskRepo;

        @GetMapping("/tasks")
        public List<Task> getAllTasks(Pageable pageable,
@RequestParam(required = false) String desc) {
            log.debug("REST 请求 -- 查询所有 Task");
            return desc == null ?
                    taskRepo.findAll(pageable).getContent() :
                    taskRepo.findByDescLike(pageable, desc).getContent();
        }
    }
```

让我们到 Postman 中试一下，看看结果，先试试不带参数，返回了全部列表（见图 5-2）。再试试带上参数 ?desc=you，这个关键字是存在的，返回的是符合条件的列表（见图 5-3）。如果参数改为 ?desc=me，这个关键字就不存在了，所以返回空列表（见图 5-4）。

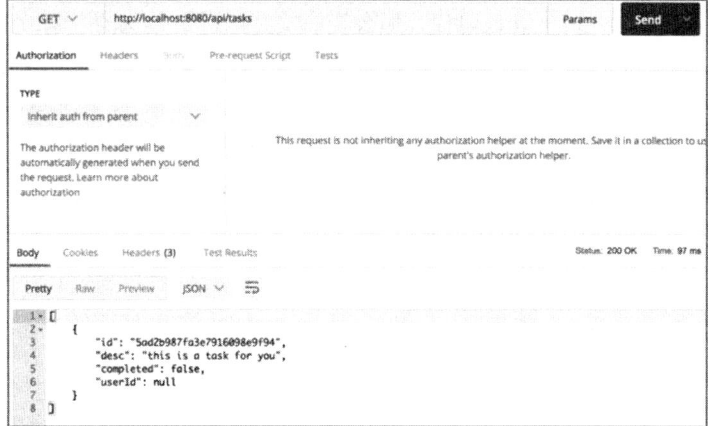

图 5-2

GET　http://localhost:8080/api/tasks?desc=you　Params　Send

Authorization　Headers　Pre-request Script　Tests

TYPE

Inherit auth from parent

The authorization header will be automatically generated when you send the request. Learn more about authorization

This request is not inheriting any authorization helper at the moment. Save it in a collection to u parent's authorization helper.

Body　Cookies　Headers (3)　Test Results　Status: 200 OK　Time: 128 ms

Pretty　Raw　Preview　JSON

```
[
    {
        "id": "5ad2b987fa3e7916098e9f94",
        "desc": "this is a task for you",
        "completed": false,
        "userId": null
    }
]
```

图 5-3

GET　http://localhost:8080/api/tasks?desc=me　Params　Send

Authorization　Headers　Pre-request Script　Tests

TYPE

Inherit auth from parent

The authorization header will be automatically generated when you send the request. Learn more about authorization

This request is not inheriting any authorization helper at the moment. Save it in a collection to u parent's authorization helper.

Body　Cookies　Headers (3)　Test Results　Status: 200 OK　Time: 86 ms

Pretty　Raw　Preview　JSON

```
[]
```

图 5-4

你说这里肯定有“鬼”，笔者同意。那么我们试着把这个方法的名字改为 findDescLike，果然不好用了。为什么呢？这套神奇“疗法”的背后还是那个 Convention over configuration，要神奇的疗效就得遵循 Spring 的“配方”。这个“配方”就是方法的命名是有讲究的：Spring 提供了一套可以通过命名规则进行查询构建的机制。这套机制会首先将方法名过滤一些关键字，比如 find…By、read…By、query…By、count…By 和 get…By 等。系统会根据关键字将命名解析成两个子语句，第一个 By 是区分这两个子语句的关键词。这个 By 之前的子语句是查询子语句（指明返回要查询的对象），后面的部分是条件子语句。如果直接就是 find…By，返回的就是定义 Respository 时指定的领域对象集合（本例中的 Task 组成的集合）。

一般到这里，有的读者可能会问 find…By、read…By、query…By、get…By 到底有什么区别啊？答案是没有区别，就是别名，这种让你不用查文档都可以写对的方式也比较贴近目前流行的自然语言描述风格（类似各种 DSL）。

```
private static final String QUERY_PATTERN = "find|read|get|query|stream";
```

刚刚我们实验了模糊查询，如果是精确查找怎么做呢？比如我们要筛选出已完成或未完成的 Task，也很简单：

```
List<Task> findByCompleted(@Param("completed") boolean completed);
```

5.2.3 复杂类型查询

看到这里你会问，这都是简单类型，如果是复杂类型怎么办？我们还是增加一个需求看一下：现在需求是要这个 API 是多用户的，每个用户看到的 Task 都是他们自己创建的。所以我们先改造一下 Task，让它有一个 User 类型的 owner 属性。

```
@Data @Builder
public class Task {
    @Id
    private String id;
    private String desc;
    private boolean completed;
    private User owner;
}
```

然后给 TaskRepo 添加一个方法：

```
@Repository
public interface TaskRepo extends MongoRepository<Task, String> {
    Page<Task> findByDescLike(Pageable pageable, @Param("desc") String
desc);
    Page<Task> findByOwnerMobile(Pageable pageable, @Param("mobile")
```

```
String mobile);
    }
```

再通过 Postman 使用 /api/tasks 的 POST 方法添加几个数据，比如类似下面的：

```
    {
      "desc": "11 点 30 的午餐",
      "completed": false,
      "owner": {
        "id": "2",
        "login": "lisi",
        "mobile": "13012351235",
        "email": "lisi@local.dev",
        "name": "Li Si"
      }
    }
```

现在，如果执行一个 GET 请求到 http://localhost:8080/api/tasks/search/findBy UserMobile?mobile=13012351235，就会得如图 5-5 所示的结果。

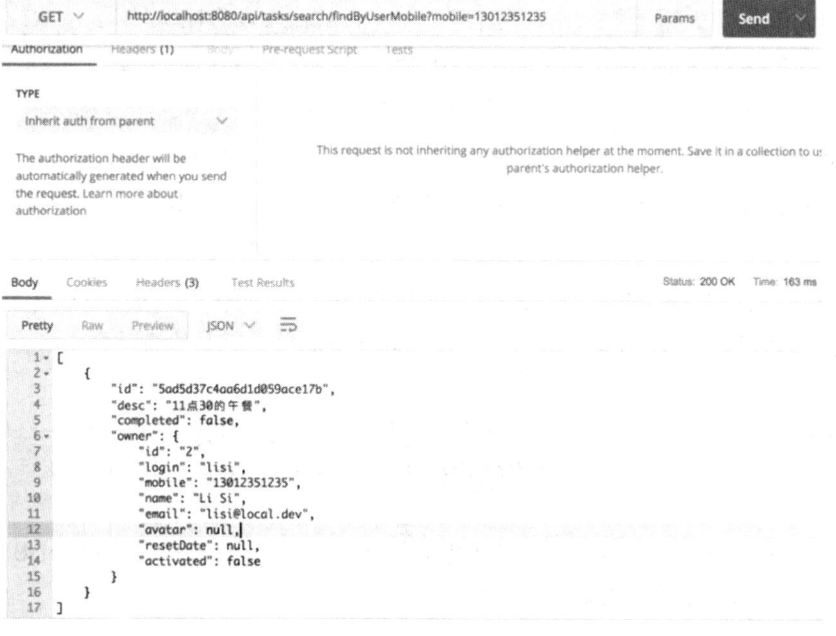

图 5-5

我们来分析这个 findByOwnerMobile 是如何解析的：首先在 By 之后，解析器会按照 camel（每个单词首字母大写）的规则来分词。那么第一个词是 Owner，这个属性在 Task

中有没有呢？有的，但是这个属性是另一个对象类型 User，所以紧跟着这个词的 Mobile 就要在 User 类中去查找是否有 Mobile 这个属性。聪明如你，肯定会想到，那如果在 Task 类中还有一个属性叫 ownerMobile 怎么办？是的，这种情况下 ownerMobile 会被优先匹配，此时请使用 _ 来显性分词处理这种混淆。也就是说，如果我们的 Task 类中同时有 owner 和 ownerMobile 两个属性，如果想要指定的是 owner 的 mobile，那么需要写成 findByOwner_Mobile。

5.2.4 自定义查询

实际开发中需要的查询比上面的要复杂得多，再复杂一些怎么办？还是用例子来说话吧，那么现在我们想要模糊搜索指定用户的 Task 中描述的关键字，返回匹配的集合。这个需求只需改动一行，以命名规则为基础的查询条件是可以加 And、Or 这种关联多个条件的关键字。

```
List<Todo> findByOwnerMobileAndDescLike(@Param("mobile") String mobile,
@Param("desc") String desc);
```

当然，还有其他操作符：Between（值在两者之间）、LessThan（小于）、GreaterThan（大于）、Like（包含）、IgnoreCase（忽略大小写）、AllIgnoreCase（对于多个参数全部忽略大小写）、OrderBy（引导排序子语句）、Asc（升序，仅在 OrderBy 后有效）和 Desc（降序，仅在 OrderBy 后有效）。

刚刚我们谈到的都是对于查询条件子语句的构建，其实在 By 之前，对于要查询的对象也可以有限定的修饰词 Distinct（去重，如有重复取一个值）。比如有可能返回的结果有重复的记录，可以使用 findDistinctTaskByOwnerMobileAndDescLike。

简单是简单了，但若遇到复杂查询，那这个方法名也太长了，可以直接写查询语句吗？几乎所有“码农”都会问的问题。当然可以，同样很简单，就是给方法加上一个元数据修饰符 @Query。

```
@Query("{ 'owner.mobile': ?0, 'desc': { '$regex': ?1} }")
    List<Task> searchTasks(@Param("mobile") String mobile, @Param("desc")
String desc);
```

采用这种方法，就不用按照命名规则起方法名了，可以直接使用 MongoDB 的查询进行。上面的例子中有几个地方需要说明一下。

（1）?0 和 ?1 是参数的占位符，?0 表示第一个参数，也就是 mobile；而 ?1 表示第二个参数，也就是 desc。

（2）MongoDB 中没有关系型数据库的 Like 关键字，需要以正则表达式的方式达成类似的功能。

（3）如果要指定 Id 时，则需要写成 ×××._id，前面的下画线是要有的。

这样的查询已经相当于在 MongoDB 中查询了，有兴趣的读者可以把下面的代码在 MongoDB 的控制台输入，看看是否好用，所以这种程度的支持基本可以让我们写出相对较复杂的查询了。

```
db.task.find(
    {
        "owner.mobile": "13012351235",
        "desc": { "$regex": "11" }
    })
```

但这肯定还是不够的，对于开发人员来讲，如果不给可以自定义的方式基本没人会用的，因为总有这样那样的原因会导致我们希望能完全掌控查询或存储过程。

5.2.5 自定义 Repository

如果想要完全自定义，那就不能只使用这些内建的接口了，需要定义新的接口，并提供具体的实现类，在实现类中就可以采用 MongoTemplate 等进行复杂的逻辑和数据处理。

```
public interface CustomizedMyRepository {
  void someCustomMethod(Param param);
}
public class CustomizedMyRepositoryImpl implements
CustomizedMyRepository {
    public void someCustomMethod(Param param) {
      // 自定义方法的实现
      // 使用 MongoTemplate 进行自定义的复杂操作
      MongoOperations mongoOps = new MongoTemplate(new MongoClient(),
"database");
      mongoOps.insert(new MyDomain("Blablabla", 123));
      mongoOps.findOne(new Query(where("Blablabla").is(123)),
MyDomain.class);
      // 省略其他部分
    }
}
```

然后定义应用中的 repository，除了继承内建的接口（下面的 CrudRepository），也继承自定义接口 CustomizedMyRepository

```
interface MyDomainRepository extends CrudRepository<MyDomain, Long>,
CustomizedMyRepository {
}
```

5.3 Controller 的构建

5.3.1 改造 TaskRepo 和 UserRepo

TaskRepo 目前先保持前面改造后的样子。

```
@Repository
public interface TaskRepo extends MongoRepository<Task, String> {
    Page<Task> findByDescLike(Pageable pageable, @Param("desc") String
desc);
    Page<Task> findByOwnerMobile(Pageable pageable, @Param("mobile")
String mobile);
}
```

UserRepo 需要在用户注册时确保用户名、电子邮件和手机号唯一，因此需要检查这几项是否在数据库中已经存在。对于这几项检查是每个用户注册时都需要做的，而且由于可能会出现同一用户多次提交的情况，比如用户名提示已存在，更改后还是已存在，所以我们给这几个查询分别做缓存 @Cacheable(cacheNames = USERS_BY_×××_CACHE)。

```
package dev.local.gtm.api.repository;

import dev.local.gtm.api.domain.User;
import org.springframework.cache.annotation.Cacheable;
import org.springframework.data.domain.Page;
import org.springframework.data.domain.Pageable;
import org.springframework.data.mongodb.repository.MongoRepository;
import org.springframework.data.repository.query.Param;
import org.springframework.stereotype.Repository;

import java.util.Optional;

@Repository
public interface UserRepo extends MongoRepository<User, String> {
    String USERS_BY_LOGIN_CACHE = "usersByLogin";

    String USERS_BY_MOBILE_CACHE = "usersByMobile";

    String USERS_BY_EMAIL_CACHE = "usersByEmail";

    @Cacheable(cacheNames = USERS_BY_MOBILE_CACHE)
    Optional<User> findOneByMobile(@Param("mobile") String mobile);

    @Cacheable(cacheNames = USERS_BY_EMAIL_CACHE)
    Optional<User> findOneByEmail(@Param("email") String email);
```

```
    @Cacheable(cacheNames = USERS_BY_LOGIN_CACHE)
    Optional<User> findOneByLogin(@Param("login") String login);

    Page<User> findAllByLoginNot(Pageable pageable, @Param("login")
String login);
}
```

这里面还需要指出的是，对于 Java 8 的 `Optional` 不熟悉的读者，可以去看 Oracle 的一个官方教程。

5.3.2 实现 Controller

实现用户的登录、注册和忘记密码，接下来逐一分析它们的业务逻辑。

- 登录：需要先传入用户的用户名和密码，然后根据用户名查找用户。如果系统中不存在该用户，则需要以异常形式告诉客户端。如果存在，则继续比较密码是否相同。如果相同则以 JSON 形式返回该用户信息，否则需要以异常形式提示用户名密码不匹配。
- 注册：由于用户的用户名、电子邮件和手机号都应该是唯一的，所以在注册时，要先去检查是否这些在系统中已经存在，如果存在则需要以异常形式告知用户。如果没有异常，就可以新建用户并返回用户信息了。
- 忘记密码：忘记密码其实是两个步骤的操作——验证手机和重置密码，所以其实我们这个功能会有两个 API。

  验证手机：这里会使用一个第三方的 API——LeanCloud 来做，也顺便学习一下如何在后端访问第三方的 Rest API。验证过程是先看手机号是否存在，如果不存在，则需要以异常形式告知用户。如果存在，则将手机号和短信验证码提交给 LeanCloud 的 Rest API 检查，返回的结果如果是成功的，那么就随机生成一个重置密钥保存到 `User` 的 `resetKey` 属性中。如果不成功，以异常形式告知用户。

  重置密码：客户端发送手机号、密码和重置密钥到此 API。首先检查是否存在手机号对应的用户，如果不存在，则以异常形式告知用户；如果存在，则检查该用户的 `resetKey` 是否和传入的密钥匹配。如果不匹配，以异常形式告知用户；如果匹配，按照传入的密码重置该用户密码，且将 `resetKey` 设置为 null。

```
package dev.local.gtm.api.web.rest;

// 省略导入
/**
 * 用户鉴权资源接口
 *
 * @author Peng Wang (wpcfan@gmail.com)
```

```
 */
@Log4j2
@RestController
@RequestMapping("/api")
@RequiredArgsConstructor
public class AuthResource {

    private final UserRepo userRepo;
    private final RestTemplate restTemplate;
    private final AppProperties appProperties;

    @ApiOperation(value = "用户登录鉴权接口",
            notes = "客户端在 RequestBody 中以 JSON 形式传入用户名、密码，如果成
功，则以 JSON 形式返回该用户信息")
    @PostMapping(value = "/auth/login")
    public User login(@RequestBody final Auth auth) {
        log.debug("REST 请求 -- 将对用户：{} 执行登录鉴权", auth);
        return userRepo.findOneByLogin(auth.getLogin())
                .map(user -> {
                    if (user.getPassword().equals(auth.getPassword())) {
                        throw new InvalidPasswordException();
                    }
                    return user;
                })
                .orElseThrow(LoginNotFoundException::new);
    }

    @PostMapping("/auth/register")
    public ResponseEntity<User> register(@RequestBody User user) {
        log.debug("REST 请求 -- 注册用户：{} ", user);
        if (userRepo.findOneByLogin(user.getLogin()).isPresent()) {
            throw new LoginExistedException();
        }
        if (userRepo.findOneByMobile(user.getMobile()).isPresent()) {
            throw new MobileExistedException();
        }
        if (userRepo.findOneByEmail(user.getEmail()).isPresent()) {
            throw new EmailExistedException();
        }
        return ResponseEntity.ok(userRepo.save(user));
    }

    @PostMapping(value = "/auth/mobile")
    @ResponseStatus(value = HttpStatus.OK)
```

```
    public String verifyMobile(@RequestBody MobileVerification
verification) {
        log.debug("REST 请求 -- 验证手机号 {} 和短信验证码 {}",
verification.getMobile(), verification.getCode());
        return userRepo.findOneByMobile(verification.getMobile())
                .map(user -> {
                    val code = verifySmsCode(verification);
                    if (code.value() != 200) {
                        throw new
MobileVerificationFailedException(code.getReasonPhrase());
                    }
                    user.setResetKey(CredentialUtil.generateResetKey());
                    userRepo.save(user);
                    return "{\"resetKey\": \"" + user.getResetKey() +
"\"}";
                })
                .orElseThrow(MobileNotFoundException::new);
    }

    @PostMapping(value = "/auth/reset")
    public void resetPassword(@RequestBody KeyAndPassword keyAndPassword)
{
        log.debug("REST 请求 -- 重置密码 {}", keyAndPassword);
        userRepo.findOneByMobile(keyAndPassword.getMobile())
                .map(user -> {
                    if
(!user.getResetKey().equals(keyAndPassword.getResetKey())) {
                        throw new ResetKeyNotMatchException();
                    }
                    user.setPassword(keyAndPassword.getPassword());
                    user.setResetKey(null);
                    return userRepo.save(user);
                })
                .orElseThrow(LoginNotFoundException::new);
    }

    @PostMapping("/auth/captcha")
    public Captcha verifyCaptcha(@RequestBody final Captcha captcha) {
        val body = new HashMap<String, String>();
        body.put("captcha_code", captcha.getCode());
        body.put("captcha_token", captcha.getToken());
        val entity = new HttpEntity<>(body);
        try {
            val validateCaptcha =
```

```
restTemplate.postForObject(appProperties.getCaptcha().getVerificationUrl(),
entity, Captcha.class);
                if (validateCaptcha == null) {
                    throw new InternalServerErrorException("返回对象为空，无法进
行验证");
                }
                return Captcha.builder()
                        .code(captcha.getCode())
                        .token(captcha.getToken())
                        .validatedMsg(validateCaptcha.getValidatedMsg())
                        .build();
            } catch (HttpStatusCodeException ex) {
                throw new
CaptchaVerificationFailedException(ex.getStatusCode().getReasonPhrase());
            }
        }

        private HttpStatus verifySmsCode(final MobileVerification
verification) {
            val body = new HashMap<String, String>();
            body.put("mobilePhoneNumber", verification.getMobile());
            val entity = new HttpEntity<>(body);
            try {
                val response = restTemplate.postForEntity(
                        appProperties.getSmsCode().getVerificationUrl()+"/"+
verification.getCode(),
                        entity, Void.class);
                return response.getStatusCode();
            } catch (HttpStatusCodeException ex) {
                return ex.getStatusCode();
            } catch (RestClientException ex) {
                return HttpStatus.INTERNAL_SERVER_ERROR;
            }
        }
    }
```

上面的代码中需要讲解的地方比较多，我们分成几部分来看。

5.3.3 登录

登录对应的 API 路径是 /api/auth/login，我们采用 POST 方法进行。userRepo.findOneByLogin 返回的是一个 Optional<User>，也就是可能为空，Optional 这个 Java 8 带来的新特性有一些函数式编程的感觉（包括 Java 8 提供的 Stream 也是这样）。和传统

的编程方式不太一样，需要一些思维上的转换，一个函数要做的事情很简单，但需要在转换思路时把函数串起来，也就是前一个函数的输出作为后一个的输入。下面代码中…map…orElseThrow 就是这种感觉。map 中的 user -> {...} 是一个 lamda 表达式，也就是一个函数。orElseThrow 中的 LoginNotFoundException::new 是一个语法糖，相当于 _ -> { return new MobileNotFoundException();}，但在只有一个语句时可以简写成 LoginNotFoundException::new 。这里面还有一个自定义的异常 InvalidPasswordException，我们放到后面统一讲异常处理，这里先跳过。

```
@PostMapping(value = "/auth/login")
public User login(@RequestBody final Auth auth) {
    log.debug("REST 请求 -- 将对用户: {} 执行登录鉴权", auth);
    return userRepo.findOneByLogin(auth.getLogin())
            .map(user -> {
                if (user.getPassword().equals(auth.getPassword())) {
                    throw new InvalidPasswordException();
                }
                return user;
            })
            .orElseThrow(LoginNotFoundException::new);
}
```

5.3.4 注册

注册对应的 API 路径是/api/auth/register，同样采用 POST 方法进行。login 方法返回的是一个领域对象类型 User，但 register 方法返回的却是 ResponseEntity<User>，那么这个 ResponseEntity 和前面的 User 有什么区别呢？ResponseEntity 是完整的一个 Response，它可以让你控制返回的状态码、返回对象的形式等，比如 ResponseEntity.ok(userRepo.save(user))相当于 ResponseEntity.ok().body(userRepo.save(user))。所以，如果确定有要对返回的结果做不同于 Spring 默认的操作，则可以使用 ResponseEntity。

```
@PostMapping("/auth/register")
public ResponseEntity<User> register(@RequestBody User user) {
    log.debug("REST 请求 -- 注册用户: {} ", user);
    if (userRepo.findOneByLogin(user.getLogin()).isPresent()) {
        throw new LoginExistedException();
    }
    if (userRepo.findOneByMobile(user.getMobile()).isPresent()) {
        throw new MobileExistedException();
    }
    if (userRepo.findOneByEmail(user.getEmail()).isPresent()) {
```

```
            throw new EmailExistedException();
        }
        return ResponseEntity.ok(userRepo.save(user));
    }
```

5.3.5 忘记密码第一步：验证手机

这个方法前多了一个注解 `@ResponseStatus(value = HttpStatus.OK)` 用来以注解形式提供返回的状态码，但如果是 `HttpStatus.OK` 这种，那么不加也一样，因为默认就是 `OK`，所以在你不想返回 200.OK 时，又不想使用 `ResponseEntity` 的情况下，则可以这样注解达成目的。

另外一个有意思的点是，由于 Java 不是动态语言，所以不可能构建一个临时对象，但是很多时候我们的 API 返回的就是一个简单的字符串，比如：

```
    {
        "message": "hello"
    }
```

在这种情况下，为一个简单的字符串构建一个对象实在有点过于烦琐了，所以我们在下面的代码中 `return "{\"resetKey\": \"" + user.getResetKey() + "\"}";` 返回了一个 `JSON` 字符串。对的，JSON 其实本身就是字符串，这种方式对待简单的 JSON 返回是比较方便的，不用重新封装对象。但是也要注意，不能过度使用这种方式。

```
    @PostMapping(value = "/auth/mobile")
    @ResponseStatus(value = HttpStatus.OK)
    public String verifyMobile(@RequestBody MobileVerification verification)
{
        log.debug("REST 请求 -- 验证手机号 {} 和短信验证码 {}",
verification.getMobile(), verification.getCode());
        return userRepo.findOneByMobile(verification.getMobile())
                .map(user -> {
                    val code = verifySmsCode(verification);
                    if (code.value() != 200) {
                        throw new
MobileVerificationFailedException(code.getReasonPhrase());
                    }
                    user.setResetKey(CredentialUtil.generateResetKey());
                    userRepo.save(user);
                    return "{\"resetKey\": \"" + user.getResetKey() + "\"}";
                })
                .orElseThrow(MobileNotFoundException::new);
    }
```

注意，在判断验证码时，我们使用了一个函数 verifySmsCode(verification)，这不是内建函数，是我们自己封装的。这个函数虽然没几行，但是为它准备的周边内容可是不少呢。

```
    // 还记得这个成员变量吗？如果忘了就翻到前面看看代码
    private final RestTemplate restTemplate;
    // ... 省略其他
    private HttpStatus verifySmsCode(final MobileVerification verification)
{
        val body = new HashMap<String, String>();
        body.put("mobilePhoneNumber", verification.getMobile());
        val entity = new HttpEntity<>(body);
        try {
            val response = restTemplate.postForEntity(
                    appProperties.getSmsCode().getVerificationUrl()+"/"+veri
fication.getCode(),
                    entity, Void.class);
            return response.getStatusCode();
        } catch (HttpStatusCodeException ex) {
            return ex.getStatusCode();
        } catch (RestClientException ex) {
            return HttpStatus.INTERNAL_SERVER_ERROR;
        }
    }
```

下面来认识一位新朋友 RestTemplate。我们之前都是提供 API 给客户端，但很多时候需要自己充当第三方 API 的客户端，这时就需要使用 RestTemplate 。那么有的读者会想为什么不把这些工作交给前端，它们本来就是纯客户端。交给前端会有几个问题。

（1）安全性问题：一般 API 都是有密钥的，而且访问时应该传递这个密钥，如果都放在前端，一旦泄露就是全部的密钥都泄露了，但在后端就安全很多。因为前端相对脆弱，后端可以通过操作系统和网络的各种安全手段保证被攻破的可能性降低。

（2）API 地址过多：一般来说，客户端只需要面对自己的后端，当然这也不绝对，有时客户端也会使用第三方，这个主要看内部的协调。但 API 接口的地址不宜过多是一个基本原则。

（3）影响范围：如果交给前端管理，那么我们日后要替换某个服务时，就需要从前端到后端全面更改。但如果在后端处理，就可以很大程度上保证接口的兼容性。

1. 使用 RestTemplate 发送外部请求

这个 RestTemplate 是怎么得到的呢？我们并没有新建出一个对象来啊。这其实是利用了 Spring 的依赖注入特性将这个 RestTemplate 以单件构造出来提供给应用使用的。所以我们将这个 restTemplate 声明成 private final 也就是必需的参数，只能通过构造提供初始化。然后使用

@RequiredArgsConstructor 提供这样一个构造，这样在 AuthResource 中就得到了实例。

接下来，我们看看是如何提供这个 RestTemplate 实例的，我们在 api 包下新建一个包叫作 config，这个包以后作为我们所有的 Java 配置文件的位置。在下面新建一个 OutboundRestTemplateConfig.java 文件。

```
package dev.local.gtm.api.config;

import dev.local.gtm.api.interceptor.LeanCloudRequestInterceptor;
import lombok.RequiredArgsConstructor;
import org.springframework.boot.web.client.RestTemplateBuilder;
import org.springframework.context.annotation.Bean;
import org.springframework.context.annotation.Configuration;
import org.springframework.web.client.RestTemplate;

/**
 * 为应用访问外部 Http Rest API 提供配置
 *
 * @author Peng Wang (wpcfan@gmail.com)
 */
@RequiredArgsConstructor
@Configuration
public class OutgoingRestTemplateConfig {

    private static final int TIMEOUT = 5000;
    private final AppProperties appProperties;

    @Bean
    public RestTemplate getRestTemplate() {
        return new RestTemplateBuilder()
                .setConnectTimeout(TIMEOUT)
                .interceptors(new
LeanCloudRequestInterceptor(appProperties))
                .build();
    }
}
```

这个文件非常简单，我们利用 RestTemplateBuilder 构造一个 RestTemplate，并加上了 @Bean 注解，这样只要你在应用内声明 RestTemplate 类型的成员变量，并提供了构造函数以供注入，系统就会找到这个实例注入进去。这样提供的好处在于不需要每次都新建一个 RestTemplate 了，避免了一些消耗，同时为 RestTemplate 设置了一些通用的属性，避免团队合作时，每个人使用的参数不统一。

2. 实现 ClientHttpRequestInterceptor 进行外部请求拦截

注意，在构造 RestTemplate 时，我们设置了超时的时间和一个请求拦截器。和 Spring MVC 的拦截器概念类似，只不过 Spring MVC 中的拦截器是拦截进来的请求（incoming），而这里我们要拦截的是出去的 outgoing。那么为什么要提供拦截器呢？因为第三方 API 服务往往要求 HTTP Request 的鉴权，一般是在 HTTP Request 的头设置一些鉴权需要的字段。而这个如果交给使用者自己添加，一是重复代码太多，二是一旦第三方更改了接口鉴权方式，那么代码需要改动时，会影响多个位置。所以使用一个拦截器帮我们统一管理。

我们看看如何实现，在 api 包下面新建一个 interceptor 的包，然后在此包下面新建一个 LeanCloudRequestInterceptor.java 文件。

```
package dev.local.gtm.api.interceptor;

import dev.local.gtm.api.config.AppProperties;
import lombok.RequiredArgsConstructor;
import lombok.val;
import org.springframework.http.HttpRequest;
import org.springframework.http.MediaType;
import org.springframework.http.client.ClientHttpRequestExecution;
import org.springframework.http.client.ClientHttpRequestInterceptor;
import org.springframework.http.client.ClientHttpResponse;

import java.io.IOException;

/**
 * 为 LeanCloud 云服务配置在 Request Header 中写入鉴权信息
 * 关于 LeanCloud 短信服务的鉴权信息可以参考
<a>https://leancloud.cn/docs/rest_sms_api.html</a>
 *
 * @author Peng Wang (wpcfan@gmail.com)
 */
@RequiredArgsConstructor
public class LeanCloudRequestInterceptor implements
ClientHttpRequestInterceptor {

    private final AppProperties appProperties;

    @Override
    public ClientHttpResponse intercept(HttpRequest request, byte[] body,
ClientHttpRequestExecution execution) throws IOException {
        if (!request.getURI().getHost().contains("api.lncld.net")){
            return execution.execute(request, body);
```

```
            }
            val headers = request.getHeaders();
            headers.add("X-LC-Id", appProperties.getLeanCloud().getAppId());
            headers.add("X-LC-Key",
appProperties.getLeanCloud().getAppKey());
            headers.setContentType(MediaType.APPLICATION_JSON);
            return execution.execute(request, body);
        }
    }
```

在这个拦截器中，我们拦截了向外发送的请求，并检查是否是发往 LeanCloud 的请求（`request.getURI().getHost().contains("api.lncld.net")`）。如果是，就写入 LeanCloud 要求的鉴权头 `X-LC-Id` 和 `X-LC-Key`，这部分详细信息可以参考 LeanCloud 的官方文档。这样处理后，以后再访问 LeanCloud 时就无须写鉴权信息了，可以让开发者更聚焦在业务层面。

你可能关注到还有一个地方没讲，这个 `LeanCloudProperties` 是什么？

3. 为应用提供外部可配置属性的能力

前面我们提到过在 `application.yml` 或者 `application.properties` 中可以配置很多属性，但你有没有想过是否可以添加自己的属性呢？这个 Spring 是支持的，而且非常简单，只需要一个类注解 `@ConfigurationProperties("一级属性.二级属性...")` 就行了。

```
    package dev.local.gtm.api.config.propsupport;

    import lombok.Data;
    import
org.springframework.boot.context.properties.ConfigurationProperties;
    import org.springframework.stereotype.Component;

    /**
     * 为应用配置 LeanCloud 外部属性
     *
     * @author Peng Wang (wpcfan@gmail.com)
     */
    @Component
    @ConfigurationProperties("app.leancloud")
    @Data
    public class LeanCloudProperties {
        // 不要复制此处的 appId 和 appKey，这里的是不正确的，请自行申请账号获得自己的
appId 和 appKey
        private String appId = "pqmXXXXXXXXXXXXXXX-yyyyyy";
```

```
    private String appKey = "EUXXXXXXXXXXXXXX";
}
```

那么，我们为 LeanCloud 配置了两个鉴权属性 appId 和 appKey，这里要记得赋初始值，因为如果在 .properties 或 .yml 没有设置时，程序也要能运行。

这样写完之后，需要在根项目的 build.gradle 中添加一个依赖 spring-boot-configuration-processor，它可以对属性注解 @ConfigurationProperties 进行处理。

```
// 省略
buildscript {
    // 省略
}

allprojects {
    // 省略
}

subprojects {
    // 省略
    dependencies {
        optional("org.springframework.boot:spring-boot-devtools")
        optional("org.projectlombok:lombok")
        optional("org.springframework.boot:spring-boot-configuration-
processor") // <--这里
        testImplementation("org.springframework.boot:spring-boot-
starter-test")
    }
    compileJava.dependsOn(processResources)
}
```

在 Intellij IDEA 中试试看，在 application.yml 中输入 app. 的时候还会有智能提示，如图 5-6 所示。

```
app.|
app.leancloud.app-id=pqmaqNUAzliIxr…   String
app.leancloud.app-key=EU4D2VJRUwOx4…   String
app.sms.captcha.request-url=https:/…   String
app.sms.captcha.verification-url=ht…   String
app.sms.code.request-url=https://pq…   String
app.sms.code.verification-url=https…   String
spring.application.admin.enabled=f…   Boolean
spring.application.admin.jmx-name=o…   String
management.endpoint.mappings.cache.time-to…
management.endpoint.mappings.enabled   Boole…
spring.application.index (Applicat…   Integer
Press ^Space to see all configuration keys with different prefix >>
```

图 5-6

这么一个个地添加感觉有点乱，我们干脆为整个应用建立一个自己完整的外部属性配置，在 config 中新建 AppProperties.java，这样就把所有需要配置的属性统一到这个 AppProperties 中管理了。

```
package dev.local.gtm.api.config;

import lombok.Data;
import lombok.Value;
import org.springframework.boot.context.properties.ConfigurationProperties;

/**
 * 为本应用服务提供外部可配置的属性支持
 *
 * @author Peng Wang (wpcfan@gmail.com)
 */
@Value
@Component
@ConfigurationProperties(prefix = "app")
public class AppProperties {

    private final LeanCloud leanCloud = new LeanCloud();
    private final SmsCaptcha captcha = new SmsCaptcha();
    private final SmsCode smsCode = new SmsCode();
    private final Security security = new Security();

    @Data
    public static class LeanCloud {
        // 请注册 LeanCloud 获得自己的应用访问的 appID 和 appKey
        private String appId = "blablabla";
        private String appKey = "nahnahnah";
```

```
        }

        @Data
        public static class SmsCaptcha {
            // 请注册 LeanCloud 获得自己的应用访问的 URL
            private String requestUrl =
"https://xxx.api.lncld.net/1.1/requestCaptcha";
            private String verificationUrl =
"https://xxx.api.lncld.net/1.1/verifyCaptcha";
        }

        @Data
        public static class SmsCode {
            // 请注册 LeanCloud 获得自己的应用访问的 URL
            private String requestUrl =
"https://xxx.api.lncld.net/1.1/requestSmsCode";
            private String verificationUrl =
"https://xxx.api.lncld.net/1.1/verifySmsCode";
        }

        @Data
        public static class Security {
            private Jwt jwt;
            @Data
            public static class Jwt {
                private String secret = "myDefaultSecret";
                private long tokenValidityInSeconds = 7200;
            }
        }
    }
```

对 Spring 熟悉的读者可能还知道另一种属性配置的方式，就是 `@Value` 注解，这两种方式的区别如表 5-1 所示。

表 5-1

特　性	@ConfigurationProperties	@Value
宽松绑定	是	否
元数据支持	是	否
SpEL 表达式支持	否	是

其中，宽松绑定是笔者翻译的名字，英文是 Relaxed Binding，其实就是不严格要求属性的名称和定义的严格一致，比如 `context-path` 可以绑定到 `contextPath`、`port` 可以绑定

到 PORT 等，而 @Value 是不支持这个宽松绑定的，因此在环境设置时，推荐使用 @ConfigurationProperties。此外，@Value 也不支持 IDE 的智能提示，因为没有元数据支持，但在有表达式需求时使用 @Value。

围绕这个 verifyMobile 讲了这么多，这些都是为了更灵活、更方便地开发以后的模块而不得不做的先期工作。

5.3.6 忘记密码第二步：重置密码

讲完了第一步，第二步就简单一些。这里需要指出的是 ResetKey，因为忘记密码是两步操作，现在我们提供两个接口，但存在一个问题，如果用户直接访问第二个接口怎么办？所以需要可以关联第一步和第二步。怎么做呢？具体来说就是第一步验证成功后生成一个随机数，存储到 User 对象中，第二步客户端要将这个随机数和要修改的密码作为参数传递给第二个接口 /auth/reset。

```
    @PostMapping(value = "/auth/mobile")
    @ResponseStatus(value = HttpStatus.OK)
    public String verifyMobile(@RequestBody MobileVerification verification)
{
        // 第一步
        return userRepo.findOneByMobile(verification.getMobile())
                .map(user -> {
                    // 省略部分代码
                    user.setResetKey(CredentialUtil.generateResetKey()); //
<-- 这里
                    // 省略部分代码
                })
                .orElseThrow(MobileNotFoundException::new);
    }

    @PostMapping("/auth/reset")
    public void resetPassword(@RequestBody KeyAndPassword keyAndPassword) {
        // 第二步
        log.debug("REST 请求 -- 重置密码 {}", keyAndPassword);
        userRepo.findOneByMobile(keyAndPassword.getMobile())
                .map(user -> {
                    if
(!user.getResetKey().equals(keyAndPassword.getResetKey())) { // <-- 这里
                        throw new ResetKeyNotMatchException();
                    }
                    user.setPassword(keyAndPassword.getPassword());
                    user.setResetKey(null);
```

```
                return userRepo.save(user);
            })
            .orElseThrow(LoginNotFoundException::new);
}
```

这个随机数的生成我们做成了一个工具类，会放到一个工具类的包中，在 api 包下面新建一个 util 包，然后在此包下新建一个 CredentialUtil.java。这个工具主要生成一个固定长度的随机数，默认是 10 位。

```
package dev.local.gtm.api.util;

import lombok.extern.log4j.Log4j2;

import java.util.Random;

@Log4j2
public class CredentialUtil {
    private static final int COUNT = 10;

    /**
     * 生成一个重置密码的随机数，作为激活密钥
     *
     * @return 生成的密钥
     */
    public static String generateActivationKey() {
        return randomNumeric();
    }

    /**
     * 生成一个重置密码的随机数，作为验证密钥
     *
     * @return 生成的密钥
     */
    public static String generateResetKey() {
        return randomNumeric();
    }

    private static String randomNumeric() {
        return String.valueOf(new Random()
                .nextInt((9 * (int) Math.pow(10, CredentialUtil.COUNT -
1)) - 1)
                + (int) Math.pow(10, CredentialUtil.COUNT - 1));
    }
}
```

5.3.7 API 的异常处理

异常怎么办，上面抛出那么多，也不解释一下吗？嗯，这个异常属于非常重要的一块，所以我们集中在这里讲。

对于一个 API 来说，正常的返回是比较容易定义的，该是个对象就是个对象，该是个数组就是个数组。但是对于异常情况怎么处理是一个比较头疼的问题，如果我们不做任何处理，那么默认出现异常都是返回 500 服务器内部错误。但这个错误的表现形式，一是非常难以理解到底发生了什么，二是对用户非常不友好，如图 5-7 所示。

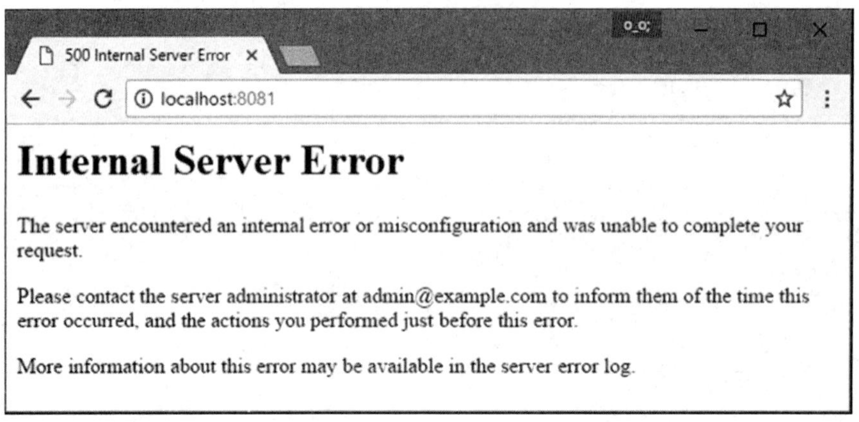

图 5-7

Spring Boot 提供了一个默认的映射：`/error`，当处理中抛出异常之后，会转到该请求中处理，并且该请求有一个全局的错误页面用来展示异常内容（见图 5-8）。当然可以定制化这个界面让它更好看一些，包括信息更明确一些。但是我们要构建的是一个前后端分离的应用，所以更希望让后端可以告知到底发生了什么，而由前端来处理异常。

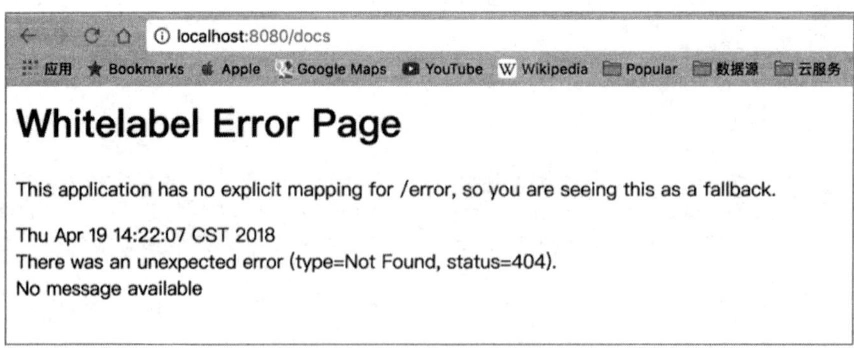

图 5-8

Spring Boot 提供了@ControllerAdvice 以 AOP 的方式进行全局的异常管理。但是如何定义一个错误的对象，都需要什么信息？与其自己定义，不如利用已有的、成熟的问题描述方式。这里我们使用一个开源的项目 Problem[1]来完成问题定义和异常处理。

1. 问题定义

Problem 项目把错误和异常统一名称为“问题”（Problem），一个 Problem 会有几个属性。

- 类型 type：这个 type 是一个 URI 形式，要求指向一个用户可以看懂的问题类型页面。
- 概要 title：使用自然语言简单描述问题类型，一般是给工程师看的，偏技术性语言。
- 状态 status：HTTP 状态码，最小 100，最大 600。具体的 Status Code 列表可以查阅相关资料。
- 具体描述 detail：关于本次发生问题的具体的、特定的描述。要求尽可能使用非技术语言。
- 实例 instance：同样是一个 URI 形式的字符串，用来指向一个可以解释本次发生错误的页面。

有兴趣的读者可以参考 Problem 的 Schema 定义[2]。

```
  Problem:
    type: object
    properties:
      type:
        type: string
        format: uri
        description: |
          An absolute URI that identifies the problem type.  When
dereferenced,
          it SHOULD provide human-readable documentation for the problem
type
          (e.g., using HTML).
        default: 'about:blank'
        example: 'https://zalando.github.io/problem/constraint-violation'
      title:
        type: string
        description: |
          A short, summary of the problem type. Written in english and
```

1 https://zalando.github.io/problem

2 https://zalando.github.io/problem/schema.yaml

```
readable
            for engineers (usually not suited for non technical stakeholders
and
            not localized); example: Service Unavailable
        status:
          type: integer
          format: int32
          description: |
            The HTTP status code generated by the origin server for this
occurrence
            of the problem.
          minimum: 100
          maximum: 600
          exclusiveMaximum: true
          example: 503
        detail:
          type: string
          description: |
            A human readable explanation specific to this occurrence of the
            problem.
          example: Connection to database timed out
        instance:
          type: string
          format: uri
          description: |
            An absolute URI that identifies the specific occurrence of the
problem.
            It may or may not yield further information if dereferenced.
```

2．在工程中集成 Problem 类库

我们使用 zalando 团队提供的在 Spring 中使用 Problem 类库。[1]如果要在我们的工程中使用，则需要在 `api/build.gradle` 中增加如下依赖，由于要使用 Jackson Afterburner 模块作为 `JSON` 的序列化和反序列化的类库配合 Problem，所以也添加了 `jackson-module-afterburner`。在统一异常处理时，我们使用 `@ControllerAdvice`，需要我们添加 `spring-boot-starter-aop`。

```
implementation("org.zalando:problem-spring-web:0.20.1")
implementation("com.fasterxml.jackson.module:jackson-module-afterburner")
implementation("org.springframework.boot:spring-boot-starter-aop")
```

为 Problem 配置 Jackson 模块需要在 `config` 下新建 `JacksonConfig.java` 文件。

1 https://github.com/ zalando/problem-spring-web

```
package dev.local.gtm.api.config;

import com.fasterxml.jackson.module.afterburner.AfterburnerModule;
import org.springframework.context.annotation.Bean;
import org.springframework.context.annotation.Configuration;
import org.zalando.problem.ProblemModule;
import org.zalando.problem.validation.ConstraintViolationProblemModule;

/**
 * 为 zalando problem 配置 Jackson
 *
 * @author Peng Wang (wpcfan@gmail.com)
 */
@Configuration
public class JacksonConfig {
    /*
     * 使用 Jackson Afterburner 模块加速序列化和反序列化过程
     */
    @Bean
    public AfterburnerModule afterburnerModule() {
        return new AfterburnerModule();
    }

    /*
     * 用于序列化和反序列化 RFC7807 Problem 对象的模块。
     */
    @Bean
    ProblemModule problemModule() {
        return new ProblemModule();
    }

    /*
     * 用于序列化和反序列化 ConstraintViolationProblem 的模块
     */
    @Bean
    ConstraintViolationProblemModule constraintViolationProblemModule()
{
        return new ConstraintViolationProblemModule();
    }
}
```

3. 构建统一异常处理

我们需要在 dev/local/gtm/api/web 下新建一个包 exception，在此包下新建一个 ExceptionTranslator.java 文件。

```
package dev.local.gtm.api.web.exception;

// 省略导入

/**
 * 对错误异常进行统一处理，以统一格式输出
 *
 * @author Peng Wang (wpcfan@gmail.com)
 */
@ControllerAdvice
@RequiredArgsConstructor
public class ExceptionTranslator implements ProblemHandling {

    private final HttpServletRequest request;

    @Override
    public ResponseEntity<Problem> process(ResponseEntity<Problem>
entity) {
        if (entity.getBody() == null) {
            return entity;
        }
        Problem problem = entity.getBody();
        if (!(problem instanceof ConstraintViolationProblem || problem
instanceof DefaultProblem)) {
            return entity;
        }
        ProblemBuilder builder = Problem.builder()
                .withType(Problem.DEFAULT_TYPE.equals(problem.getType())
 ? ErrorConstants.DEFAULT_TYPE : problem.getType())
                .withStatus(problem.getStatus())
                .withTitle(problem.getTitle())
                .with("path", request.getRequestURI());

        if (problem instanceof ConstraintViolationProblem) {
            builder
                    .with("violations", ((ConstraintViolationProblem)
problem).getViolations())
                    .with("message", ErrorConstants.ERR_VALIDATION);
            return new ResponseEntity<>(builder.build(),
entity.getHeaders(), entity.getStatusCode());
        } else {
            builder
                    .withCause(((DefaultProblem) problem).getCause())
                    .withDetail(problem.getDetail())
```

```
                    .withInstance(problem.getInstance());
            problem.getParameters().forEach(builder::with);
            if (!problem.getParameters().containsKey("message") &&
problem.getStatus() != null) {
                builder.with("message", "error.http." +
problem.getStatus().getStatusCode());
            }
            return new ResponseEntity<>(builder.build(),
entity.getHeaders(), entity.getStatusCode());
        }
    }

    @Override
    public ResponseEntity<Problem>
handleMethodArgumentNotValid(MethodArgumentNotValidException ex,
NativeWebRequest req) {
        BindingResult result = ex.getBindingResult();
        List<FieldErrorVM> fieldErrors = result.getFieldErrors().stream()
                .map(f -> new FieldErrorVM(f.getObjectName(),
f.getField(), f.getCode()))
                .collect(Collectors.toList());

        Problem problem = Problem.builder()
                .withType(ErrorConstants.CONSTRAINT_VIOLATION_TYPE)
                .withTitle("方法参数不正确")
                .withStatus(defaultConstraintViolationStatus())
                .with("message", ErrorConstants.ERR_VALIDATION)
                .with("fieldErrors", fieldErrors)
                .build();
        return create(ex, problem, req);
    }
}
```

我们实现了 ProblemHandling 接口并覆写了 process 方法和 handleMethodArgumentNotValid 方法。其中 process 方法是通用的问题处理方法，我们在其中根据是验证器错误还是 HTTP 错误而构造不同的 Problem。

在 handleMethodArgumentNotValid 方法中，我们构造了一个更具体针对方法参数错误的 Problem 对象。方法的参数错误属于 Spring 绑定产生的错误——FieldError，这个 FieldError 中的很多属性我们并不关心，所以需要定义一个 FieldErrorVM 用来取出特定的属性，并构建到 Problem 中去——FieldError 中的 objectName 作为 FieldErrorVM 中的 field；FieldError 中的 field 作为 FieldErrorVM 中的 code 作为 FieldErrorVM 中的成员变量 message。

```
@Value
public class FieldErrorVM implements Serializable {

    private static final long serialVersionUID = 1L;

    private final String objectName;

    private final String field;

    private final String message;
}
```

而 ExceptionTranslator 类上的 @ControllerAdvice 注解让这个类成为所有 Controller 错误的全局处理类。

4．构建自定义异常

我们构建了很多自定义异常，但这些异常在实现了统一异常管理之后都变得非常简单，所以下面只分析其中一个。

```
package dev.local.gtm.api.web.exception;

import org.zalando.problem.AbstractThrowableProblem;
import org.zalando.problem.Status;

/**
 * 密码错误异常
 *
 * @author Peng Wang (wpcfan@gmail.com)
 */
public class InvalidPasswordException extends AbstractThrowableProblem {

    public InvalidPasswordException() {
        super(ErrorConstants.INVALID_PASSWORD_TYPE, "密码不正确",
Status.BAD_REQUEST);
    }
}
```

所有的自定义异常均继承 AbstractThrowableProblem 这个抽象类。然后基本需要做的就是实现一个构造函数，在其中调用父类的构造。AbstractThrowableProblem 有很多构造函数，总体目标是要构造出一个 Problem，所以我们就给出问题的类型、概要描述和状态等（具体参考前面提到的问题定义）。

```
public abstract class AbstractThrowableProblem extends ThrowableProblem
{
```

```
    private final URI type;
    private final String title;
    private final StatusType status;
    private final String detail;
    private final URI instance;
    private final Map<String, Object> parameters;

    protected AbstractThrowableProblem() {
        this(null);
    }

    protected AbstractThrowableProblem(@Nullable final URI type) {
        this(type, null);
    }

    protected AbstractThrowableProblem(@Nullable final URI type,
            @Nullable final String title) {
        this(type, title, null);
    }

    protected AbstractThrowableProblem(@Nullable final URI type,
            @Nullable final String title,
            @Nullable final StatusType status) {
        this(type, title, status, null);
    }

    protected AbstractThrowableProblem(@Nullable final URI type,
            @Nullable final String title,
            @Nullable final StatusType status,
            @Nullable final String detail) {
        this(type, title, status, detail, null);
    }

    protected AbstractThrowableProblem(@Nullable final URI type,
            @Nullable final String title,
            @Nullable final StatusType status,
            @Nullable final String detail,
            @Nullable final URI instance) {
        this(type, title, status, detail, instance, null);
    }

    protected AbstractThrowableProblem(@Nullable final URI type,
            @Nullable final String title,
```

```
            @Nullable final StatusType status,
            @Nullable final String detail,
            @Nullable final URI instance,
            @Nullable final ThrowableProblem cause) {
        this(type, title, status, detail, instance, cause, null);
    }

    protected AbstractThrowableProblem(@Nullable final URI type,
            @Nullable final String title,
            @Nullable final StatusType status,
            @Nullable final String detail,
            @Nullable final URI instance,
            @Nullable final ThrowableProblem cause,
            @Nullable final Map<String, Object> parameters) {
        super(cause);
        this.type = Optional.ofNullable(type).orElse(DEFAULT_TYPE);
        this.title = title;
        this.status = status;
        this.detail = detail;
        this.instance = instance;
        this.parameters =
Optional.ofNullable(parameters).orElseGet(LinkedHashMap::new);
    }
    // 省略
}
```

ErrorConstants.java 文件中定义了问题类型的 URI，但这里并不让 URI 真正可访问，在真正的工作中，会有专门的文档团队来配合我们，这里就不花时间去写这些文档了。

```
package dev.local.gtm.api.web.exception;

import java.net.URI;

/**
 * 错误常量定义，使用 URI 形式唯一标识错误类型
 *
 * @author Peng Wang (wpcfan@gmail.com)
 */
public final class ErrorConstants {

    public static final String ERR_CONCURRENCY_FAILURE =
"error.concurrencyFailure";
    static final String ERR_VALIDATION = "error.validation";
    private static final String PROBLEM_BASE_URL =
"http://www.twigcodes.com/problem";
```

```
        static final URI DEFAULT_TYPE = URI.create(PROBLEM_BASE_URL +
"/problem-with-message");
        static final URI CONSTRAINT_VIOLATION_TYPE =
URI.create(PROBLEM_BASE_URL + "/constraint-violation");
        public static final URI PARAMETERIZED_TYPE =
URI.create(PROBLEM_BASE_URL + "/parameterized");
        static final URI INVALID_PASSWORD_TYPE = URI.create(PROBLEM_BASE_URL
+ "/invalid-password");
        static final URI EMAIL_EXISTED_TYPE = URI.create(PROBLEM_BASE_URL +
"/email-already-used");
        static final URI LOGIN_EXISTED_TYPE = URI.create(PROBLEM_BASE_URL +
"/login-existed");
        static final URI LOGIN_NOT_FOUND_TYPE = URI.create(PROBLEM_BASE_URL
+ "/login-not-found");
        static final URI MOBILE_EXISTED_TYPE = URI.create(PROBLEM_BASE_URL +
"/mobile-already-used");
        static final URI MOBILE_NOT_FOUND_TYPE = URI.create(PROBLEM_BASE_URL
+ "/mobile-not-found");
        static final URI MOBILE_VERIFICATION_FAILED_TYPE =
URI.create(PROBLEM_BASE_URL + "/mobile-verification-failed");
        static final URI RESET_KEY_NOT_MATCH_TYPE =
URI.create(PROBLEM_BASE_URL + "/reset-key-not-match");
        static final URI EMAIL_NOT_FOUND_TYPE = URI.create(PROBLEM_BASE_URL
+ "/email-not-found");

        private ErrorConstants() {
        }
    }
```

5.4 构建安全的 API 接口

估计很多读者看了之前的登录和注册之后会吐槽，这个实现方式实在是不怎么样。是的，前面实现的鉴权方式的缺点主要如下。

（1）密码明文存储：这个导致的安全性问题就不用多说了。

（2）并未实现对 API 的保护：换句话说，登录与否和能否访问 API 没有关系，所以这是一个假登录。

（3）没有角色的划分：所有的接口都是一视同仁的，但实际项目中肯定要有各种角色允许访问的接口是不一样的限制。

那么，本节我们就一起学习如何使用 JWT 实现一个基于 token 的 API 鉴权方式以及使用 Spring Security 实现给予角色的权限控制。

5.4.1 为什么要保护 API

通常情况下，把 API 直接暴露出去风险是很大的，那么，对 API 要划分出一定的权限级别，然后做一个用户的鉴权，依据鉴权结果给予用户开放对应的 API。目前，比较主流的方案有如下几种。

（1）用户名和密码鉴权，使用 Session 保存用户鉴权结果。

（2）自行采用 token 进行鉴权，自己设计的 token 往往由于设计时没有考虑周全，后期存在各种兼容性问题。

（3）使用 OAuth/OAuth2 进行鉴权（其实 OAuth 也是一种基于 token 的鉴权，只是没有规定 token 的生成方式）。

（4）使用 JWT 作为 token。

第一种方案就不介绍了，由于依赖 Session 来维护状态，也不太适合移动时代，新的项目就不要采用了。第二种方案的兼容性和可维护性较差，而 OAuth 其实对于不做开放平台的公司有些过于复杂。我们主要介绍第四种方案：JWT。

5.4.2 什么是 JWT

图 5-9 是一个 JWT 的工作流程图。模拟一下实际的流程是这样的（假设受保护的 API 在 /protected 中）。

（1）用户导航到登录页，输入用户名、密码，进行登录。

（2）服务器验证登录鉴权，如果该用户合法，则根据用户的信息和服务器的规则生成 JWT token。

（3）服务器将该 token 以 JSON 形式返回（其实不一定要 JSON 形式，这里说的是一种常见的做法）

（4）用户得到 token，存在 localStorage、Cookie、IndexDB 或其他数据存储形式中。

（5）以后用户请求/protected 中的 API 时，在请求的 header 中加入 Authorization: Bearer ××××(token)。此处注意 token 之前有一个 7 字符长度的 Bearer，注意 Bearer 后应该有一个空格。

（6）服务器端对此 token 进行检验，如果合法就解析其中内容，根据其拥有的权限和自己的业务逻辑给出对应的响应结果。

（7）用户取得结果。

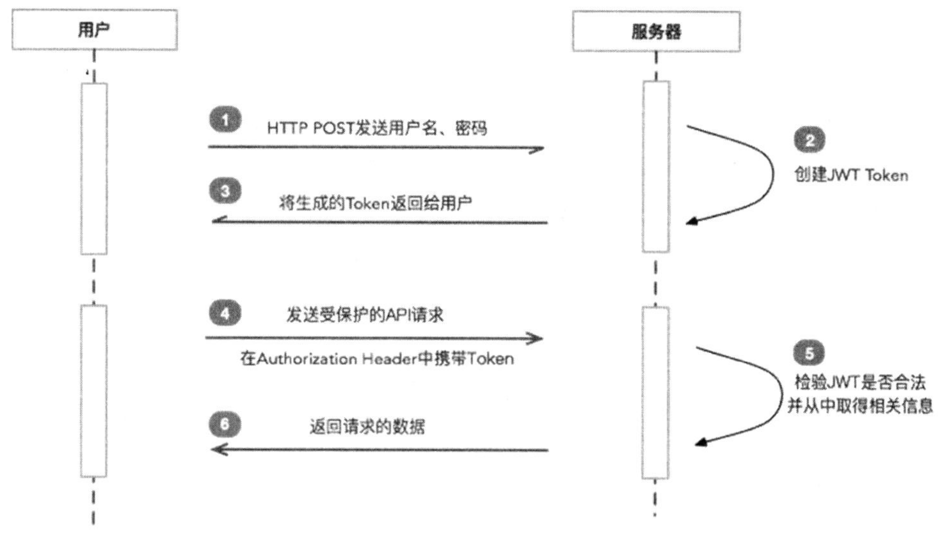

图 5-9

为了更好地理解这个 token 是什么，我们先来看一个 token 生成后的样子：

```
eyJhbGciOiJIUzUxMiJ9.eyJzdWIiOiJ3YW5nIiwiY3JlYXRlZCI6MTQ4OTA3OTk4MTM5Myw
iZXhwIjoxNDg5Njg0NzgxfQ.RC-BYCe_UZ2URtWddUpWXIp4NMsoeq2O6UF-8tVplqXY1-CI9u1-
a-9DAAJGfNWkHE81mpnR3gXzfrBAB3WUAg
```

仔细看还是可以看到这个 `token` 分成了三部分，每部分用 `.` 分隔，每段都是用 `Base64` 编码的。如果我们用一个 Base64 的解码器，则可以看到第一部分 `eyJhbGciOiJIUzUxMiJ9` 被解析成了：

```
{
    "alg":"HS512"
}
```

这是告诉我们 HMAC 采用 HS512 算法对 JWT 进行的签名。

第二部分 `eyJzdWIiOiJ3YW5nIiwiY3JlYXRlZCI6MTQ4OTA3OTk4MTM5MywiZXhwIjoxNDg5Njg0NzgxfQ` 被解码之后是：

```
{
    "sub":"wang",
    "created":1489079981393,
```

```
    "exp":1489684781
}
```

上述代码告诉我们这个 token 中含有的数据声明（Claim），这个例子里面有 3 个声明：sub、created 和 exp。在这个例子中，分别代表着用户名、创建时间和过期时间，当然你可以把任意数据声明在这里。

看到这里，你可能会想这是个什么 token，所有信息都透明，安全怎么保障？别急，我们看看 token 的第三段 RC-BYCe_UZ2URtWddUpWXIp4NMsoeq2O6UF-8tVplqXY1-CI9u1-a-9DAAJGfNWkHE81mpnR3gXzfrBAB3WUAg。同样使用 Base64 解码之后：

```
D X •DmYTeàL•UZcPZ0$gZAY•_7•wY@
```

最后一段代码其实是签名，这个签名必须知道密钥才能计算。这个也是 JWT 的安全保障。这里提一点注意事项，由于数据声明（Claim）是公开的，千万不要把密码等敏感字段放进去，否则就等于是公开给别人了。

也就是说，JWT 是由三段组成的，按官方的叫法分别是 header（头）、payload（负载）和 signature（签名）：

```
header.payload.signature
```

header 中的数据通常包含两部分：一部分是我们刚刚看到的 alg，这个词是 algorithm 的缩写，就是指明算法。另一部分可以添加的字段是 token 的类型（按 RFC 7519 实现的 token 机制可不只 JWT 一种），但如果我们采用的是 JWT，指定这个就多余了。

```
{
  "alg": "HS512",
  "typ": "JWT"
}
```

payload 中可以放置三类数据：系统保留的、公共的和私有的。

- 系统保留的声明（Reserved claim）：这类声明不是必需的，但是是建议使用的，包括 iss（签发者）、exp（过期时间）、sub（主题）、aud（目标受众）等。这里我们发现都用的是缩写的三个字符，这是由于 JWT 的目标就是尽可能小巧。
- 公共声明：这类声明需要在 IANA JSON Web Token Registry 中定义或者提供一个 URI，因为要避免重名等冲突。
- 私有声明：这个就是你根据业务需要自己定义的数据了。

签名的过程是这样的：采用 header 中声明的算法，接受 3 个参数：Base64 编码的 header、base64 编码的 payload 和密钥（secret）进行运算。签名这一部分如果安全性要求高，则可以采用 RSASHA256 的方式设置公钥、私钥对。

```
HMACSHA256(
  base64UrlEncode(header) + "." +
  base64UrlEncode(payload),
  secret)
```

5.4.3 JWT 的生成和解析

为了简化工作，这里引入一个比较成熟的 JWT 类库，叫 jjwt[1]。这个类库可以用于 Java 和 Android 的 JWT token 的生成和验证。

需要在 api/build.gradle 中添加依赖：

```
implementation("io.jsonwebtoken:jjwt:0.9.0")
```

JWT 的生成可以使用下面这样的代码完成：

```
String generateToken(Map<String, Object> claims) {
    return Jwts.builder()
            .setClaims(claims)
            .setExpiration(generateExpirationDate())
            .signWith(SignatureAlgorithm.HS512, secret) //采用什么算法是可
以自己选择的，不一定非要采用 HS512
            .compact();
}
```

数据声明（claim）其实就是一个 Map，比如我们想放入用户名，简单地创建一个 Map 然后放进去就可以了。

```
Map<String, Object> claims = new HashMap<>();
claims.put(CLAIM_KEY_USERNAME, username());
```

解析也很简单，利用 jjwt 提供的解析器（parser）传入密钥，然后就可以解析 token 了。

```
Claims getClaimsFromToken(String token) {
    Claims claims;
    try {
        claims = Jwts.parser()
                .setSigningKey(secret)
                .parseClaimsJws(token)
                .getBody();
    } catch (Exception e) {
        claims = null;
    }
    return claims;
```

1 https://github. com/jwtk/ jjwt

```
}
```

JWT 本身没什么难度，但安全是一个比较复杂的事情，JWT 只不过提供了一种基于 token 的请求验证机制。但我们的用户权限、对于 API 的权限划分、资源的权限划分、用户的验证等都不是 JWT 负责的。也就是说，请求验证后，你是否有权限查看对应的内容是由你的用户角色决定的。所以这里要利用 Spring 的一个子项目 Spring Security 来简化工作。在开始 Spring Security 的工作前，让我们对基于角色的权限做一个简单的背景知识介绍。

5.4.4 权限的设计

1. ACL 权限模型

ACL 是英文 Access Control Lists 的缩写，指的是对于某个数据对象的权限列表，一个 ACL 会指出哪些用户或系统进程被授予了对数据对象的访问权限，以及允许什么样的操作。比如文件的 ACL 通常类似 (ZhangSan: read, write; LiSi: read)。

2. 基于 RBAC 的权限模型

RBAC 是 Role-Based Access Control 的英文缩写，翻译过来就是基于角色的访问控制。RBAC 认为权限授权实际上是由 Who，What，How 决定的。在 RBAC 模型中，Who，What，How 构成了访问权限三元组，即 Who 对 What 进行 How 的操作。其中 Who 是权限的拥有者或主体（如 User，Role），What 是资源或对象（如 Resource，Class），如图 5-10 所示。

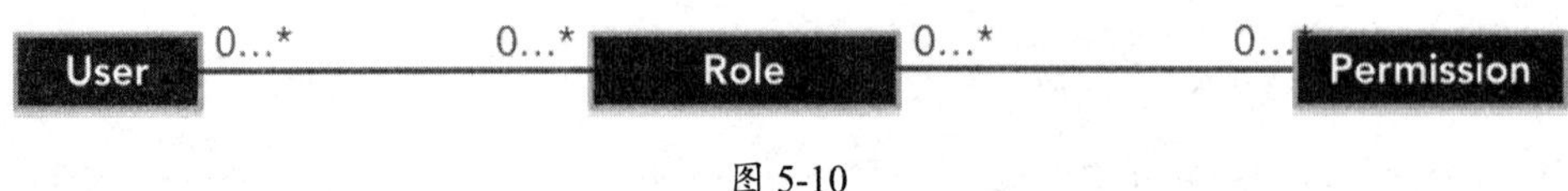

图 5-10

RBAC 主要分为 4 种变化形式：

- 核心模型 RBAC-0（Core RBAC）
- 角色分层模型 RBAC-1（Hierarchal RBAC）
- 角色限制模型 RBAC-2（Constraint RBAC）
- 统一模型 RBAC-3（Combines RBAC）

最重要也是最基本的是核心模型 RBAC-0，因为这个模型是最小化实现 RBAC 权限思想的方式，其他模型都是在此基础上的补充和变化。

RBAC 和 ACL 的区别在于 RBAC 将权限分配到对组织有意义的特定操作上而不是分配到底层的数据对象上。举个小例子，ACL 可以用来授予或拒绝某个系统文件的写访问请求，但它不能判断这个文件是怎样被更改的。

在 RBAC 系统中，一个操作可以是在一个财务系统中“创建一个账簿”或者在一个医疗系统中“执行一个血糖测试”。这些操作的权限分配对组织来讲是有意义的，因为组织内的这些操作是一个基本单位的流程。

RBAC 非常适合职责分离的需求，这种需求下经常会要求确保至少两个或两个以上的人员参与到授权的操作中去。其实一个最小化的 RBAC 模型和 ACLg（带分组的 ACL 模型）是等效的。

3. 前后端的权限划分

一个系统的前后端都会涉及权限的处理，但后端的权限更为重要一些，因为这涉及系统的数据，也就是说后端如果不设防，那么前端的安全做得再好也没有用。相反如果后端的安全性设计得较好，那么即使前端有些漏洞，也不会影响后端的数据。

通常一个前后端分离的系统，我们需要对后端的 API 进行保护，这种保护有多个层次。

- token：这是最基础的，就是验证你的请求是一个授权请求。
- 角色：即使有了合法的 token，还要看你的角色是否可以访问某些 API。打个比方，token 类似于小区的大门钥匙，但进入你的家，还得有家门钥匙。这个角色相当于是你是否有权进入这个房间。
- 访问频次（rate limit）：API 既然开放出来就是让用户使用的，但是我们不欢迎恶意的使用，比如有人写程序对你的鉴权接口进行字典攻击去得到用户的密码，如果没有限制，他们这么试下去，既增加了系统的安全风险，又消耗了系统的大量资源。但是机器和人是有区别的，比如人是不可能在非常短的时间内（比如 100ms 内）发送多个请求的，或者正常用户不会在一个短时间内，比如 1h 尝试 3000 次同一个接口的请求。
- 加固请求来源的信任：对于安全性高的系统会要求客户端（iOS、Android 等）上传自身应用的签名，避免有人反编译程序或者通过抓包等进行分析，重新伪造请求。对于 Web 应用可以通过 CORS 指定某些地址的请求才可以访问。
- 黑名单：这个就是根据各种不正常的用户行为或者已经查明的攻击来源进行拉黑处理。

还有很多其他的手段，安全是一个大问题，但不是本书的主要目的，所以就不展开讨论了。

5.4.5 使用 Spring Security 规划角色安全

Spring Security 是一个基于 Spring 的通用安全框架，里面内容太多了，我们的主要目的也不是展开讲这个框架，而是如何利用 Spring Security 和 JWT 一起来完成 API 保护。所以关于 Spring Secruity 的基础内容或展开内容，请自行去官网学习。

如果你的系统有用户的概念，一般来说，应该有一个用户表，最简单的用户表应该有三列：ID，Username 和 Password，类似表 5-2 所示。

表 5-2

ID	Username	Password
10	wang	abcdefg

不是所有用户都是一种角色，比如网站管理员、供应商、财务等，这些角色和网站的直接用户需要的权限可能是不一样的。那么我们就需要一个角色表，Spring Security 中的角色叫 Authority，如表 5-3 所示。

表 5-3

ID	Authority
10	User
20	Admin

我们还需要一个可以将用户和角色关联起来建立映射关系的表，如表 5-4 所示。

表 5-4

User_ID	Authority_Id
10	10
20	20

这是典型的一个关系型数据库的用户角色的设计，由于我们要使用的 MongoDB 是一个文档型数据库，所以重新审视一下这个结构。

这个数据结构的优点在于它避免了数据冗余，每个表负责自己的数据，通过关联表进行关系的描述，同时也保证数据的完整性：比如当你修改角色名称后，没有脏数据的产生。

1. 在 MongoDB 中如何实现

在 MongoDB 中，我们可以将其简化为有两个文档集合 User 和 Authority，这两个集合各自是独立的，MongoDB 在数据库层面是没有提供用来定义两个集合关系的操作，那么我们怎么体现这种关系呢？

一般来说，可以在 User 对象中设置一个属性，这个属性是一个集合，由 Authority 的 ID 组成。当然在 Authority 中我们也可以有一个集合，由 User 的 ID 构成。这样手动维护了这种关系，但是这个集合最好不要太大，否则会导致文档的体积过大，不光是查询性能，增、删、改的性能都会下降。那么我们再来想一下，是否有必要在两个对象中都维护这种关系呢？角色相对的数量是比较少的，而且应该也不会增长到很大数字，估计到 100 系统维护人员就得发疯了吧。反过来，有同样角色的用户，这个数量可就不好估计了，如果是面向消费者的系统，那么这个数字就得十万、百万、千万……，所以我们应该只在 User 中维护一个 Authority 的 ID 组成的集合。

看到这里，估计很多读者会疑惑，这是什么数据库啊，还要手动维护关系。但是大家要注意的是，没有十全十美的方案，所有的方案都是要根据具体场景来做取舍和妥协的。大家一定还记得大学里学习数据结构的时候，数组适合查找，却不适合插入和删除。而链表更适合插入和删除，却不适合查找就很麻烦。如果存在一个万能的数据结构，我们就不用学习多种数据结构了。

如果业务系统中有大量的多对多关系，而且对数据完整性有要求，那最好还是使用关系型数据库。另外，一个业务中其实可以有多个数据库，互为补充，在适合的场景使用擅长这个方向的数据结构。

再回到我们的数据模型上，现在我们先建立两个 collection，一个是 User，一个是 Authority。

```
## User
{
  _id: <id_generated>
  username: 'zhangsan',
  password: 'pass',
  authorityIds: ['10', '20']
}
{
  _id: <id_generated>
  username: 'lisi',
  password: 'pass',
  authorityIds: ['10']
}
## Authority
{
  _id: '10',
  name: 'USER'
}
{
  _id: '20',
  name: 'ADMIN'
}
```

如果我们希望列出角色为 User 的用户，可以通过以下查询得到：

```
db.user.find({
    authorityIds: "10"
})
```

在 MongoDB 3.2 以上版本中，增加了类似 join 的功能，比如可以在 aggregate 的 pipeline 中使用 $lookup 关联两个 collection，为什么说是类似呢，因为实际上 MongoDB 做的动作是关联了两个大文件，做了两层循环，外层是 db.××× 的这个 ××× collection，内层是要关联的那个 collection，一旦发现符合关联条件的数据就丢到外层的数组中。所以如果是两个非常大数据量的 collection 做关联，请首先一定要做索引，其次尽可能利用 $match 和 $project 减数据量。而且不要因为有了这个类似 join 的功能就按照关系型数据库的经验来做，关于 $lookup 推荐大家看 MongoDB 官方团队的解释。

我们在 MongoDB 中可以根据下面的 JSON 建立 User 和 Authority。

```
// Authority
{
    "_id" : ObjectId("5ad88524b78ca780560ac3aa"),
    "name" : "USER"
}
{
    "_id" : ObjectId("5ad8852fb78ca780560ac3ab"),
    "name" : "ADMIN"
}
// User
{
    "_id" : ObjectId("5ad8853db78ca780560ac3ac"),
    "username" : "zhangsan",
    "password" : "pass",
    "authorityIds" : [
        ObjectId("5ad88524b78ca780560ac3aa")
    ]
}
{
    "_id" : ObjectId("5ad885c2b78ca780560ac3ad"),
    "username" : "lisi",
    "password" : "pass",
    "authorityIds" : [
        ObjectId("5ad8852fb78ca780560ac3ab"),
        ObjectId("5ad88524b78ca780560ac3aa")
    ]
}
```

然后，使用 MongoDB 的 $lookup 将所有角色名为 ADMIN 的用户列出来。

```
db.user.aggregate(
    [
        {
            "$unwind" : {
                "path" : "$authorityIds"
            }
        },
        {
            "$lookup" : {
                "from" : "authority",
                "localField" : "authorityIds",
                "foreignField" : "_id",
                "as" : "authorities"
            }
        },
        {
            "$match" : {
                "authorities.name" : "ADMIN"
            }
        }
    ],
    {
        "allowDiskUse" : false
    }
);
```

5.4.6 在 Spring Boot 中启用 Spring Security

要在 Spring Boot 中引入 Spring Security 非常简单，修改 `build.gradle`，增加一个引用 `org.springframework.boot:spring-boot-starter-security`:

```
dependencies {
    implementation("org.springframework.boot:spring-boot-starter-web")
    implementation("io.jsonwebtoken:jjwt:0.9.0")
    implementation("org.springframework.boot:spring-boot-starter-
security")
    implementation("org.springframework.boot:spring-boot-starter-aop")
    implementation("org.zalando:problem-spring-web:0.20.1")
    implementation("com.fasterxml.jackson.module:jackson-module-
afterburner")
    implementation("org.springframework.boot:spring-boot-starter-data-
mongodb")
}
```

5.4.7 改造用户对象

为什么需要对用户对象进行改造？其一是我们需要把用户拥有的角色放入 JWT token 中，传递给客户端，这样客户端可以根据需要决定那些界面功能需要开放给用户还是隐藏起来。其二是 Spring Security 需要实现 `UserDetails` 类，这个接口中规定了用户的几个必须要有的方法，所以我们要让 `User` 类添加几个属性以满足这个接口。为什么不直接实现 `UserDetails` 类？因为 `UserDetails` 类完全是为了安全服务的，它和我们的领域类可能有部分属性重叠，但很多接口其实是为了 Spring Security 的方式定制的，如果以后采用别的安全框架，那这样的紧耦合是不太好的。因此需要做一些隔离，但不管采用哪个框架，User 类确实还需要添加几个属性：

- `Activated`——标识用户是否激活
- `resetKey`——重置密码密钥
- `resetDate`——重置密码时间
- `authorities`——用户的角色集合

```
package dev.local.gtm.api.domain;

// 省略导入
@Data
@Builder
@EqualsAndHashCode(callSuper = false, of = {"id"})
@ToString(exclude = "authorities")
@NoArgsConstructor
@AllArgsConstructor
@Document(collection = "api_users")
public class User implements Serializable {
    private static final long serialVersionUID = 1L;

    @Id
    private String id;

    @NotNull
    @Pattern(regexp = Constants.LOGIN_REGEX)
    @Size(min = 1, max = 50)
    @Indexed
    private String login;

    @JsonIgnore
    @NotNull
    @Size(min = 60, max = 60)
    private String password;
```

```
    @NotNull
    @Pattern(regexp = Constants.MOBILE_REGEX)
    @Size(min = 10, max = 15)
    private String mobile;

    @Size(max = 50)
    private String name;

    @Email
    @Size(min = 5, max = 254)
    @Indexed
    private String email;

    @Size(max = 256)
    private String avatar;

    @Size(max = 20)
    @Field("reset_key")
    @JsonIgnore
    private String resetKey;

    @Field("reset_date") @Builder.Default
    private Instant resetDate = null;

    @Builder.Default
    private boolean activated = false;

    @JsonIgnore
    @Singular("authority")
    @DBRef(lazy = true)
    @Field("authority_ids")
    private Set<Authority> authorities;
}
```

@Singular("authority") 是和 Lombok 提供的 @Builder 配合为 User 类增加一个可以单独添加一个角色到集合的方法。而括号中的 authority 就是在 Builder 中这个方法的名称，一般集合起的属性名都是复数形式（authorities），@Singular ("authority") 就是我们给添加单个元素的方法起个名称，这里就是 authority 了。具体的用法如下面代码所示：

```
User.builder().authority(new Authority("Blablabla")).build();
```

我们在 User 的 authorities 属性上还应用了一个注解 @DBRef(lazy = true)，之

前我们提过在 MongoDB 3.2 以上支持类似关系型数据库的多个 collection 的查询，那么这个 `@DBRef` 就类似于标识“外键”的感觉。

- 手动引用：我们手动保存某个文档的 `_id` 到另一个文档中作为引用。应用可以根据这些引用执行新的查询得到想要的数据，一般情况下，这种方式是完全够用的。
- `DBRef` 是把文档的 `_id` 和 `collection` 名称以及可选的数据库名称一起保存起来。和手动方式类似，这种方式也需要查询两次，很多 MongoDB 的驱动提供了一些工具类，但是这些驱动并不会自动解析 `DBRef`。可以这么理解，`DBRef` 提供了一种通用的格式来表示文档之间的关系。

值得指出的是，MongoDB 官方并不推荐使用 `DBRef`，而是推荐尽可能地手动关联。因为通过手动方式有很多情况可以避免多次查询，比如除了 `_id`，我们还可以冗余某些字段到文档中，而 `DBRef` 只是保存了 `_id`。但在这个例子中，角色的数量不会很多，而且我们后面想要提供角色名称的更改功能，所以这里还是使用了 `DBRef`。在实际工作中，请慎重使用 `DBRef`。

Spring 对于 `DBRef` 也做了一些优化，比如注解中的 `lazy = true` 就是给 DBRef 字段懒加载的特性，只有使用时才会去查询，避免无谓的二次查询。

5.4.8 构建 JWT token 工具类

改造完用户这个领域对象之后，需要根据这个用户的信息生成 JWT token，也需要在请求中得到 token 的时候检验这个 token 是否合法，还需要得到这个合法的 token 之后可以解析它。所以，我们在 `api` 包下面新建一个 `security` 包，然后在 `security` 下面再建立一个 `jwt` 包，在这个包下面新建 `TokenProvider.java` 文件。

```
package dev.local.gtm.api.security.jwt;

// 省略导入
/**
 * Jwt token 的工具类
 *
 * @author Peng Wang (wpcfan@gmail.com)
 */
@Log4j2
@RequiredArgsConstructor
@Component
public class TokenProvider {

    private static final String AUTHORITIES_KEY = "auth";
```

```
        private String secretKey;

        private long tokenValidityInMilliseconds;

        private final AppProperties appProperties;

        @PostConstruct
        public void init() {
            this.secretKey =
appProperties.getSecurity().getJwt().getSecret();
            this.tokenValidityInMilliseconds = 1000 *
appProperties.getSecurity().getJwt().getTokenValidityInSeconds();
        }

        public String createToken(Authentication authentication) {
            val authorities = authentication.getAuthorities().stream()
                    .map(GrantedAuthority::getAuthority)
                    .collect(Collectors.joining(","));

            val now = (new Date()).getTime();
            val validity = new Date(now + this.tokenValidityInMilliseconds);

            return Jwts.builder()
                    .setSubject(authentication.getName())
                    .claim(AUTHORITIES_KEY, authorities)
                    .signWith(SignatureAlgorithm.HS512, secretKey)
                    .setExpiration(validity)
                    .compact();
        }

        public Authentication getAuthentication(String token) {
            val claims = Jwts.parser()
                    .setSigningKey(secretKey)
                    .parseClaimsJws(token)
                    .getBody();

            val authorities =
                    Arrays.stream(claims.get(AUTHORITIES_KEY).toString().spl
it(","))
                            .map(SimpleGrantedAuthority::new)
                            .collect(Collectors.toList());

            val principal = new User(claims.getSubject(), "", authorities);
```

```
            return new UsernamePasswordAuthenticationToken(principal, token,
authorities);
        }

        public boolean validateToken(String authToken) {
            try {
                Jwts.parser().setSigningKey(secretKey).parseClaimsJws(authTo
ken);
                return true;
            } catch (SignatureException e) {
                log.info("非法 JWT 签名");
                log.trace("非法 JWT 签名的 trace: {}", e);
            } catch (MalformedJwtException e) {
                log.info("非法 JWT token.");
                log.trace("非法 JWT token 的 trace: {}", e);
            } catch (ExpiredJwtException e) {
                log.info("过期 JWT token");
                log.trace("过期 JWT token 的 trace: {}", e);
            } catch (UnsupportedJwtException e) {
                log.info("系统不支持的 JWT token");
                log.trace("系统不支持的 JWT token 的 trace: {}", e);
            } catch (IllegalArgumentException e) {
                log.info("JWT token 压缩处理不正确");
                log.trace("JWT token 压缩处理不正确的 trace: {}", e);
            }
            return false;
        }
    }
```

这个文件提供了 3 种方法。

- createToken——设置用户的登录名为 sub，其拥有的角色（数组形式）为 auth，当然我们可以再放一些其他信息进去，比如用户的姓名等。
  - token 中的信息如果是客户端常用的信息就可以减少客户端对 API 的请求频次和数量。但不要在 token 中放入敏感的信息，因为即使没有签名也能解开 token 中的其他信息。
  - 在这个方法中，我们还设置了过期时间，一般的规则是 token 的生命周期不要太长，比如一般不要给一个几个月或一年都有效的 token，测试目的除外。实际工作中都是设一个几个小时有效的 token。
- getAuthentication——使用签名解析 JWT token，取出相关信息形成 Spring Security 中的 UsernamePasswordAuthenticationToken，Spring Security 根据这个去判断用户是否已经鉴权。

  - 值得注意的是，虽然叫作 UsernamePasswordAuthenticationToken，但我们构造时使用了 token 作为 password 参数。换句话说，如果 token 合法，就认为这个用户是已登录的用户。
  - 为什么方法定义中写返回应该是 Authentication，但实际返回的是 UsernamePasswordAuthenticationToken？因为 UsernamePasswordAuthenticationToken 实现了 Authentication 接口。
- validateToken——验证 token 是否合法。

读者大家可能留意到了，我们使用了 AppProperties，而且似乎为它添加了几个属性，是的，我们为应用属性添加了 app.security.authorization 和 app.security.jwt。

- app.security.authorization——用于设置授权相关的属性。
  - app.security.authorization.header——在 Http Request 的头当中写的授权信息的键值，也就是 Authorization: Bearer ××××(token)中 Authorization 位置的这个字符串，默认的标准是 Authorization，但也可以允许设置成别的字符串。
- app.security.jwt——设置 JWT 相关属性。
  - app.security.jwt.secret——设置加密的密钥，这个千万要保存好，否则就是大门洞开了。
  - app.security.jwt.tokenValidityInSeconds——设置有效期，默认给出值是 7200 s，也就是 2h。
  - app.security.jwt.tokenPrefix——就是 Authorization: Bearer ××××(token) 中 Bearer 和空格这个字符串。

```
@Value
@Component
@ConfigurationProperties(prefix = "app")
public class AppProperties {

    // 省略其他部分
    private final Security security = new Security();
    // 省略其他部分
    @Data
    public static class Security {
        private final Jwt jwt = new Jwt();
        private final Authorization authorization = new Authorization();

        @Data
        public static class Authorization {
```

```
            private String header = "Authorization";
        }

        @Data
        public static class Jwt {
            private String secret = "myDefaultSecret";
            private long tokenValidityInSeconds = 7200;
            private String tokenPrefix = "Bearer ";
        }
    }
}
```

5.4.9 如何检查任何请求的授权信息

有了工具类，接下来怎么办？在所有的 API Controller 中应用这个工具进行检查吗？显然不行，重复代码太多了，我们还是利用 Filter，这样所有的请求进来时就都被 Filter 检查和处理了。

```
package dev.local.gtm.api.security.jwt;

// 省略导入

@RequiredArgsConstructor
public class JWTFilter extends GenericFilterBean {

    private final TokenProvider tokenProvider;
    private final AppProperties appProperties;

    @Override
    public void doFilter(ServletRequest servletRequest, ServletResponse
servletResponse, FilterChain filterChain)
            throws IOException, ServletException {
        val httpServletRequest = (HttpServletRequest) servletRequest;
        val jwt = getToken(httpServletRequest);
        if (StringUtils.hasText(jwt) &&
this.tokenProvider.validateToken(jwt)) {
            val authentication =
this.tokenProvider.getAuthentication(jwt);
            SecurityContextHolder.getContext().setAuthentication(authent
ication);
        }
        filterChain.doFilter(servletRequest, servletResponse);
    }
```

```
        private String getToken(HttpServletRequest request){
            val bearerToken =
request.getHeader(appProperties.getSecurity().getAuthorization().getHeader()
);
            val prefix =
appProperties.getSecurity().getJwt().getTokenPrefix();
            if (StringUtils.hasText(bearerToken) &&
bearerToken.startsWith(prefix)) {
                return bearerToken.substring(prefix.length(),
bearerToken.length());
            }
            val jwt =
request.getParameter(appProperties.getSecurity().getAuthorization().getHeade
r());
            if (StringUtils.hasText(jwt)) {
                return jwt;
            }
            return null;
        }
    }
```

Filter 做的事情就是当请求进来后，先从 Request 的头部把授权信息取出来，然后返回解析后的 JWT token。如果这个 token 是合法的，那么我们就使用工具类得到 Spring Security 中的鉴权 `val authentication = this.tokenProvider.getAuthentication(jwt)`。最后在 Spring Security 的安全上下文中设置这个鉴权信息 `SecurityContextHolder.getContext().setAuthentication(authentication)`。

多说两句，`SecurityContextHolder` 是一个关联 `SecurityContext`（安全上下文）和当前线程的类，一般常用就是使用 `SecurityContextHolder` 的 `getContext()` 得到 `SecurityContext`。那么这个 `SecurityContext` 又是个什么呢？看一下 Spring Security 的源码，原来就是存和取 `Authentication` 鉴权对象的。

```
// 摘自 Spring Security 项目，为阅读方便，只保留部分代码
public interface SecurityContext extends Serializable {

    Authentication getAuthentication();

    void setAuthentication(Authentication authentication);
}
```

那么，究竟这个 `Authentication` 里面有什么？又可以做到什么呢？还是看源码更清晰。

```
// 摘自 Spring Security 项目，为阅读方便，只保留部分代码
```

```
public interface Authentication extends Principal, Serializable {
    // 得到角色集合
    Collection<? extends GrantedAuthority> getAuthorities();

    // 得到密码或者起到密码作用的值，比如 token
    Object getCredentials();

    // 简单理解就是得到用户信息
    Object getDetails();

    // 简单理解就是得到用户名，实际 Principal 的概念要更抽象一些
    Object getPrincipal();

    // 返回是否已鉴权
    boolean isAuthenticated();

    // 设置是否已鉴权
    void setAuthenticated(boolean isAuthenticated) throws
IllegalArgumentException;
}
```

这些知识对于更好地理解 Spring Security 是很有帮助的，希望大家可以细致做一些功课，但这里我们就只是略做介绍了。

我们需要回头接着说 Filter，Filter 建好之后，还得配置到 Spring 中去，否则 Spring 也不知道应该怎么应用这个 Filter。在 Spring 中 Filter 是有顺序的，而且这个顺序很重要，因为如果你的 Filter 在不适合的位置，那么有可能你的 Filter 本来预期的效果会被其他 Filter 更改掉了。

所以我们在 `jwt` 包中新建一个 `JWTConfigurer.java`，这个类里面只做了一件事，就是把刚刚的 Filter 加到 Spring Security 内建的 `UsernamePasswordAuthenticationFilter` 之前。也就是说在 Spring Security 内建的鉴权 Filter 开始之前就完成了我们自己的鉴权。

```
package dev.local.gtm.api.security.jwt;

// 省略导入

@RequiredArgsConstructor
@Component
public class JWTConfigurer extends
SecurityConfigurerAdapter<DefaultSecurityFilterChain, HttpSecurity> {

    private final TokenProvider tokenProvider;
    private final AppProperties appProperties;
```

```
    @Override
    public void configure(HttpSecurity http) {
        val customFilter = new JWTFilter(tokenProvider, appProperties);
        http.addFilterBefore(customFilter,
UsernamePasswordAuthenticationFilter.class);
    }
}
```

5.4.10 得到用户信息

刚刚，我们提到过在 Authentication 中可以得到用户信息，这个信息其实是我们自己可以定义的信息，于是，接下来要实现的是 UserDetailsService，这个接口只定义了一个方法 loadUserByUsername，顾名思义，就是提供一种从用户名可以查到用户并返回的方法。注意，不一定是数据库，文本文件、XML 文件等都可能成为数据源，这也是为什么 Spring 提供这样一个接口的原因：保证你可以采用灵活的数据源。接下来我们建立一个 UserDetailsServiceImpl 来实现这个接口。

```
package dev.local.gtm.api.security;

// 省略导入

/**
 * 用户信息服务的具体实现
 *
 * @author Peng Wang (wpcfan@gmail.com)
 */
@Log4j2
@RequiredArgsConstructor
@Component
public class UserDetailsServiceImpl implements UserDetailsService {

    private final UserRepo userRepo;

    /**
     * 通过数据库加载用户信息
     * @param login 用户名
     * @return 返回 Spring Security User
     */
    @Override
    public UserDetails loadUserByUsername(final String login) {
        log.debug("正在对用户名为 {} 的用户进行鉴权", login);

        if (new EmailValidator().isValid(login, null)) {
```

```
                val userByEmailFromDatabase =
userRepo.findOneByEmailIgnoreCase(login);
                return userByEmailFromDatabase.map(user ->
createSpringSecurityUser(login, user))
                        .orElseThrow(() -> new UsernameNotFoundException("系
统中不存在 email 为 " + login + " 的用户"));
            }

            if (Pattern.matches(Constants.MOBILE_REGEX, login)) {
                val userByMobileFromDatabase =
userRepo.findOneByMobile(login);
                return userByMobileFromDatabase.map(user ->
createSpringSecurityUser(login, user))
                        .orElseThrow(() -> new UsernameNotFoundException("系
统中不存在手机号为 " + login + " 的用户"));
            }

            String lowercaseLogin = login.toLowerCase(Locale.ENGLISH);
            val userByLoginFromDatabase =
userRepo.findOneByLogin(lowercaseLogin);
            return userByLoginFromDatabase.map(user ->
createSpringSecurityUser(lowercaseLogin, user))
                    .orElseThrow(() -> new UsernameNotFoundException("User "
+ lowercaseLogin + " was not found in the database"));

        }

        /**
         * 通过应用的用户领域对象创建 Spring Security 的用户
         *
         * 这里有两个 User，为避免混淆，对于 Spring Security 的 User 采用 Full
Qualified Name
         * @param lowercaseLogin 小写的用户登录名
         * @param user 领域对象
         * @return Spring Security User
         * @see org.springframework.security.core.userdetails.User
         */
        private org.springframework.security.core.userdetails.User
createSpringSecurityUser(String lowercaseLogin, User user) {
            if (!user.isActivated()) {
                throw new UserNotActivatedException("用户 " + lowercaseLogin
+ " 没有激活");
            }
            val grantedAuthorities = user.getAuthorities().stream()
```

```
                    .map(authority -> new
SimpleGrantedAuthority(authority.getName()))
                    .collect(Collectors.toList());
            return new
org.springframework.security.core.userdetails.User(user.getLogin(),
                    user.getPassword(),
                    grantedAuthorities);
        }
    }
```

5.4.11 配置 Spring Security

做了这么多前期工作，最后还是得让 Spring Security 配置安全，我们还需要建立一个安全配置类 SecurityConfig，这里把 PasswordEncoder 和 AuthenticationManager 以 Bean 的形式提供。在 void configure(WebSecurity web) 方法中指定哪些资源是 Spring Security 忽略的，也就是不做鉴权，一般是静态资源。在 void configure (HttpSecurity http) 中设置具体的安全策略，包括是否需要鉴权，需要的角色等。

```
    package dev.local.gtm.api.config;

    // 省略导入

    @RequiredArgsConstructor
    @Configuration
    @EnableWebSecurity
    @EnableGlobalMethodSecurity(prePostEnabled = true, securedEnabled = true)
    @Import(SecurityProblemSupport.class)
    public class SecurityConfig extends WebSecurityConfigurerAdapter {

        private final AuthenticationManagerBuilder
authenticationManagerBuilder;

        private final UserDetailsService userDetailsService;

        private final SecurityProblemSupport problemSupport;

        private final JWTConfigurer jwtConfigurer;

        @PostConstruct
        public void init() {
            try {
                authenticationManagerBuilder
                        .userDetailsService(userDetailsService)
```

```
                    .passwordEncoder(passwordEncoder());
            } catch (Exception e) {
                throw new BeanInitializationException("安全配置失败", e);
            }
        }

        @Override
        @Bean
        public AuthenticationManager authenticationManagerBean() throws
Exception {
            return super.authenticationManagerBean();
        }

        @Bean
        public PasswordEncoder passwordEncoder() {
            return new BCryptPasswordEncoder();
        }

        @Override
        public void configure(WebSecurity web) {
            // 以下资源 Spring Security 会忽略，一般用于忽略静态资源
            web.ignoring()
                    .antMatchers(HttpMethod.OPTIONS, "/**")
                    .antMatchers("/app/**/*.{js,html}")
                    .antMatchers("/i18n/**")
                    .antMatchers("/content/**")
                    .antMatchers("/test/**");
        }

        @Override
        protected void configure(HttpSecurity http) throws Exception {
            http
                    .exceptionHandling()
                    .authenticationEntryPoint(problemSupport)
                    .accessDeniedHandler(problemSupport)
                    .and()
                        .csrf()
                        .disable()
                        .headers()
                        .frameOptions()
                        .disable()
                    .and()
                        .sessionManagement()
                        .sessionCreationPolicy(SessionCreationPolicy.STATELE
```

```
SS)
                    .and()
                        .authorizeRequests()
// 限制 Request 需要认证，以下是不同路径需要的认证方式
                            .antMatchers("/api/auth/**").permitAll()
// 无需鉴权
                            .antMatchers("/api/**").authenticated()
// 其他路径均需认证
                            .antMatchers("/websocket/tracker").hasAuthority(
AuthoritiesConstants.ADMIN) // 不仅需要鉴权而且需要角色为 ADMIN
                            .antMatchers("/management/health").permitAll()
// 服务状态接口
                            .antMatchers("/management/**").hasAuthority(Auth
oritiesConstants.ADMIN) //服务管理接口
                    .and()
                        .apply(jwtConfigurer);
        }
    }
```

此时，如果启动应用，访问 `http://localhost:8080/api`，可以看到如下输出，可见我们的安全配置生效了。

```
{
    "type":"http://www.twigcodes.com/problem/problem-with-message",
    "title":"Unauthorized",
    "status":401,
    "detail":"Full authentication is required to access this resource",
    "path":"/api/",
    "message":"error.http.401"
}
```

1．重构 AuthResource

之前的 AuthResource 是直接使用 UserRepo 的，但是现在我们还需要使用 AuthorityRepo 和访问外部的 API（LeanCloud 短信验证和 Captcha 验证），这就增加了业务的复杂度，我们应该把业务抽离出来，放在 Service 中。所以现在 Controller 就变成了下面的样子：

```
package dev.local.gtm.api.web.rest;

// 省略导入

/**
 * 用户鉴权资源接口
 *
 * @author Peng Wang (wpcfan@gmail.com)
```

```
    */
   @Log4j2
   @RestController
   @RequestMapping("/api")
   @RequiredArgsConstructor
   public class AuthResource {

       private final AuthService authService;
       private final AppProperties appProperties;

       @PostMapping(value = "/auth/login")
       public ResponseEntity<JWTToken> login(@RequestBody final Auth auth)
{
           log.debug("REST 请求 -- 将对用户: {} 执行登录鉴权", auth);
           return generateJWTHeader(auth.getLogin(), auth.getPassword());
       }

       @PostMapping("/auth/register")
       public ResponseEntity<JWTToken> register(@Valid @RequestBody UserVM
userVM) {
           log.debug("REST 请求 -- 注册用户: {} ", userVM);
           if (!checkPasswordLength(userVM.getPassword())) {
               throw new InvalidPasswordException();
           }
           authService.registerUser(userVM, userVM.getPassword());
           return generateJWTHeader(userVM.getLogin(),
userVM.getPassword());
       }

       @GetMapping(value = "/auth/mobile")
       public void requestSmsCode(@RequestParam String mobile,
@RequestParam String token) {
           log.debug("REST 请求 -- 请求为手机号 {} 发送验证码，Captcha 验证 token
为 {} ", mobile, token);
           authService.requestSmsCode(mobile, token);
       }

       @PostMapping(value = "/auth/mobile")
       public ResetKey verifyMobile(@RequestBody MobileVerification
verification) {
           log.debug("REST 请求 -- 验证手机号 {} 和短信验证码 {}",
verification.getMobile(), verification.getCode());
           val key = authService.verifyMobile(verification.getMobile(),
verification.getCode());
```

```
            return new ResetKey(key);
        }

        @PostMapping(value = "/auth/reset")
        public void resetPassword(@RequestBody KeyAndPasswordVM
keyAndPasswordVM) {
            log.debug("REST 请求 -- 重置密码 {}", keyAndPasswordVM);
            authService.resetPassword(keyAndPasswordVM.getResetKey(),
keyAndPasswordVM.getMobile(), keyAndPasswordVM.getPassword());
        }

        @GetMapping(value = "/auth/captcha")
        public Captcha requestCaptcha() {
            log.debug("REST 请求 -- 请求发送图形验证码 Captcha");
            return authService.requestCaptcha();
        }

        @PostMapping("/auth/captcha")
        public CaptchaResult verifyCaptcha(@RequestBody final
CaptchaVerification verification) {
            log.debug("REST 请求 -- 验证 Captcha {}", verification);
            val result = authService.verifyCaptcha(verification.getCode(),
verification.getToken());
            log.debug("Captcha 验证返回结果 {}", verification);
            return new CaptchaResult(result);
        }

        @GetMapping("/auth/search/username")
        public ExistCheck usernameExisted(@RequestParam("username") String
username) {
            log.debug("REST 请求 -- 用户名是否存在 {}", username);
            return new ExistCheck(authService.usernameExisted(username));
        }

        @GetMapping("/auth/search/email")
        public ExistCheck emailExisted(@RequestParam("email") String email)
{
            log.debug("REST 请求 -- email 是否存在 {}", email);
            return new ExistCheck(authService.emailExisted(email));
        }

        @GetMapping("/auth/search/mobile")
        public ExistCheck mobileExisted(@RequestParam("mobile") String
mobile) {
```

```
            log.debug("REST 请求 -- email 是否存在 {}", mobile);
            return new ExistCheck(authService.mobileExisted(mobile));
        }

        private static boolean checkPasswordLength(String password) {
            return !StringUtils.isEmpty(password) &&
                    password.length() >= UserVM.PASSWORD_MIN_LENGTH &&
                    password.length() <= UserVM.PASSWORD_MAX_LENGTH;
        }

        private ResponseEntity<JWTToken> generateJWTHeader(String login,
String password) {
            val jwt = authService.login(login, password);
            val headers = new HttpHeaders();
            headers.add(
                    appProperties.getSecurity().getAuthorization().getHeader
(),
                    appProperties.getSecurity().getJwt().getTokenPrefix() +
jwt);
            log.debug("JWT token {} 加入 HTTP 头", jwt);
            return new ResponseEntity<>(new JWTToken(jwt), headers,
HttpStatus.OK);
        }

        /**
         * 简单返回 JWT token
         * 对于非常简单的需要封装成 JSON 的类，可以直接定义在 Controller 中
         */
        @Getter
        @Setter
        @AllArgsConstructor
        private static class JWTToken {
            @JsonProperty("id_token")
            private String idToken;
        }

        @Getter
        @Setter
        @Builder
        @NoArgsConstructor
        @AllArgsConstructor
        private static class CaptchaVerification {
            @JsonProperty("captcha_token")
            private String token;
```

```
        @JsonProperty("captcha_code")
        private String code;
    }

    @Getter
    @Setter
    @AllArgsConstructor
    @NoArgsConstructor
    public static class MobileVerification {
        private String mobile;
        private String code;
    }

    @Getter
    @Setter
    @AllArgsConstructor
    private static class ResetKey {
        @JsonProperty("reset_key")
        private String resetKey;
    }

    @Getter
    @Setter
    @AllArgsConstructor
    private static class CaptchaResult {
        @JsonProperty("validate_token")
        private String validatedToken;
    }

    @Getter
    @Setter
    @AllArgsConstructor
    private static class ExistCheck {
        private boolean existed;
    }
}
```

我们增加了几个接口，包括“请求图形验证码”“验证图形验证码”“验证用户名是否存在”、“验证电子邮件是否存在”和“验证手机号是否存在”等。

由于使用的临时对象比较多，包括接受的参数及返回的对象，都需要封装成对象以便做 JSON 的序列化和反序列化，所以我们建立了很多私有的类。在 Lombok 的帮助下这些类也不是很复杂。

2. 鉴权的服务层

在我们将业务逻辑剥离到 Service 中时，需要注意对于第三方 API 的调用，虽然我们还是使用了前面提到的 RestTemplate，但在这里做了一些改动。

```
package dev.local.gtm.api.service.impl;

// 省略包的导入信息

@Log4j2
@RequiredArgsConstructor
@Service
public class AuthServiceImpl implements AuthService {

    private final UserRepo userRepo;
    private final AuthorityRepo authorityRepo;
    private final PasswordEncoder passwordEncoder;
    private final TokenProvider tokenProvider;
    private final AuthenticationManager authenticationManager;
    @Qualifier("leanCloudTemplate")
    private final RestTemplate leanCloudTemplate;
    private final AppProperties appProperties;

    @Override
    public void registerUser(UserDTO userDTO, String password) {
        if (userRepo.findOneByLogin(userDTO.getLogin()).isPresent()) {
            throw new LoginExistedException();
        }
        if (userRepo.findOneByMobile(userDTO.getMobile()).isPresent()) {
            throw new MobileExistedException();
        }
        if
(userRepo.findOneByEmailIgnoreCase(userDTO.getEmail()).isPresent()) {
            throw new EmailExistedException();
        }
        val newUser = User.builder()
                .password(passwordEncoder.encode(password))
                .login(userDTO.getLogin())
                .mobile(userDTO.getMobile())
                .email(userDTO.getEmail())
                .name(userDTO.getName())
                .avatar(userDTO.getAvatar())
                .activated(true)
                .authority(authorityRepo.findOneByName(AuthoritiesConsta
nts.USER).orElseThrow(AuthorityNotFoundException::new))
```

```
                    .build();
            log.debug("user to be saved {} ", newUser);
            userRepo.save(newUser);
            log.debug("用户 {} 创建成功", newUser);
        }

        @Override
        public String login(String login, String password) {
            val authenticationToken = new
UsernamePasswordAuthenticationToken(login, password);
            val authentication =
this.authenticationManager.authenticate(authenticationToken);
            SecurityContextHolder.getContext().setAuthentication(authenticat
ion);
            return tokenProvider.createToken(authentication);
        }

        @Override
        public void requestSmsCode(String mobile, String validateToken) {
            sendSmsCode(mobile, validateToken);
        }

        @Override
        public String verifyMobile(String mobile, String code) {
            return userRepo.findOneByMobile(mobile)
                    .map(user -> {
                        verifySmsCode(mobile, code);
                        user.setResetKey(CredentialUtil.generateResetKey());
                        userRepo.save(user);
                        return user.getResetKey();
                    })
                    .orElseThrow(MobileNotFoundException::new);
        }

        @Override
        public Captcha requestCaptcha() {
            val captcha =
leanCloudTemplate.getForObject(appProperties.getCaptcha().getRequestUrl(),
Captcha.class);
            if (captcha == null) {
                log.debug("由于某种原因，远程返回的是 200，但得到的对象为空");
                throw new InternalServerErrorException("请求 Captcha 返回对象为
空");
            }
```

```
        return captcha;
    }

    @Override
    public String verifyCaptcha(String code, String token) {
        val body = new HashMap<String, String>();
        body.put("captcha_code", code);
        body.put("captcha_token", token);
        val entity = new HttpEntity<>(body);
        val validateToken =
leanCloudTemplate.postForObject(appProperties.getCaptcha().getVerificationUr
l(), entity, ValidateToken.class);
        if (validateToken == null) {
            log.debug("由于某种原因，远程返回的是 200，但得到的对象为空");
            throw new InternalServerErrorException("验证 Captcha 返回对象为
空，无法进行验证");
        }
        return validateToken.getValidatedToken();
    }

    @Override
    public void resetPassword(String key, String mobile, String password)
{
        userRepo.findOneByMobile(mobile)
                .map(user -> {
                    if (!user.getResetKey().equals(key)) {
                        log.debug("ResetKey 不匹配，客户端传递的 key 为：{}，
期待值为 {} ", key, user.getResetKey());
                        throw new ResetKeyNotMatchException();
                    }
                    user.setPassword(passwordEncoder.encode(password));
                    user.setResetKey(null);
                    user.setResetDate(Instant.now());
                    return userRepo.save(user);
                })
                .orElseThrow(LoginNotFoundException::new);
    }

    @Override
    public boolean usernameExisted(String username) {
        return userRepo.findOneByLogin(username).isPresent();
    }

    @Override
```

```
    public boolean emailExisted(String email) {
        return userRepo.findOneByEmailIgnoreCase(email).isPresent();
    }

    @Override
    public boolean mobileExisted(String mobile) {
        return userRepo.findOneByMobile(mobile).isPresent();
    }

    private void verifySmsCode(final String mobile, final  String code)
{
        val body = new HashMap<String, String>();
        body.put("mobilePhoneNumber", mobile);
        val entity = new HttpEntity<>(body);
        leanCloudTemplate.postForEntity(
                appProperties.getSmsCode().getVerificationUrl() + "/" +
code,
                entity, Void.class);
    }

    private void sendSmsCode(String mobile, String validateToken) {
        val body = new HashMap<String, String>();
        body.put("mobilePhoneNumber", mobile);
        body.put("validate_token", validateToken);
        val entity = new HttpEntity<>(body);
        leanCloudTemplate.postForEntity(appProperties.getSmsCode().getRe
questUrl(), entity, Void.class);
    }

    @Getter
    @Setter
    private static class ValidateToken {
        @JsonProperty("validate_token")
        private String validatedToken;
    }
  }
```

你可能注意到了，`@Qualifier("leanCloudTemplate")`，为什么要加这个 `@Qualifier` 呢？因为随着业务的拓展，我们感觉到似乎不只会单单使用 LeanCloud 一个第三方服务，但目前的 RestTemplate 似乎是为了 LeanCloud 打造的（比如，我们的认证头截断器就是专门处理 LeanCloud 认证的），这不太好，所以需要有所区分。

因此，在 `OutgoingRestTemplateConfig` 中，我们也标识了 `@Bean ("leanCloudTemplate")`，这样以后可以为每个第三方服务提供一个专门的 Bean，用于注入适合该服务的 RestTemplate。

比如在 getLeanCloudRestTemplate 中，除原来的认证头截断器 LeanCloudAuthHeaderInterceptor 外，还为它定制了专门的错误处理器 LeanCloudRequestErrorHandler。

之所以要有这个错误处理器，是由于在访问第三方服务时，如果出现错误，那么对方服务器会返回错误信息。而我们如果不处理，则在自己的客户端调用时统一都是 500 内部服务器错误，这显然是不对的。我们需要将对方的出错信息根据不同情况转换成自己的错误信息，而这个就是 LeanCloudRequestErrorHandler 的作用。

除此之外，我们还可以指定不同的 HTTP 类库，现在我们使用的是 JDK 内建的 HTTP 类库，但如果你想使用 Apache 的 HttpComponent，就可以使用 new RestTemplateBuilder().requestFactory(new HttpComponentsAsyncClientHttpRequestFactory()) 去设置，当然，别忘了在 build.gradle 中添加依赖。

```
package dev.local.gtm.api.config;

// 省略包导入部分
@Log4j2
@RequiredArgsConstructor
@Configuration
public class OutgoingRestTemplateConfig {

    private static final int TIMEOUT = 10000;
    private final AppProperties appProperties;

    @Bean("leanCloudTemplate")
    public RestTemplate getLeanCloudRestTemplate() {
        return new RestTemplateBuilder()
                .setConnectTimeout(TIMEOUT)
                .setReadTimeout(TIMEOUT)
                .errorHandler(new LeanCloudRequestErrorHandler())
//                 .requestFactory(new
HttpComponentsAsyncClientHttpRequestFactory()) // Use Apache HttpComponent
                .interceptors(new
LeanCloudAuthHeaderInterceptor(appProperties))
                .build();
    }

    public class LeanCloudRequestErrorHandler implements
ResponseErrorHandler {

        @Override
        public boolean hasError(ClientHttpResponse response) throws
IOException {
```

```
            return response.getStatusCode() != HttpStatus.OK;
        }

        @Override
        public void handleError(ClientHttpResponse response) throws
IOException {
            if (response.getStatusCode() == HttpStatus.FORBIDDEN) {
                throw new OutgoingBadRequestException("403 Forbidden");
            }
            val err = extractErrorFromResponse(response);
            if (err == null) {
                throw new InternalServerErrorException("从 Response 中提取
json 返回 null");
            }
            throw new OutgoingBadRequestException(err.getError());
        }

        private LeanCloudError
extractErrorFromResponse(ClientHttpResponse response) throws IOException {
            String json = readResponseJson(response);
            try {
                ObjectMapper mapper = new ObjectMapper(new
JsonFactory());
                JsonNode jsonNode = mapper.readValue(json,
JsonNode.class);
                Integer code = jsonNode.has("code") ?
jsonNode.get("code").intValue() : null;
                String err = jsonNode.has("error") ?
jsonNode.get("error").asText() : null;

                val error = new LeanCloudError(code, err);
                log.debug("LeanCloud error: ");
                log.debug(" CODE: " + error.getCode());
                log.debug(" ERROR: " + error.getError());
                return error;
            } catch (JsonParseException e) {
                return null;
            }
        }

        private String readResponseJson(ClientHttpResponse response)
throws IOException {
            val json = readFully(response.getBody());
            log.debug("LeanCloud 返回的错误: " + json);
```

```
            return json;
        }

        private String readFully(InputStream in) throws IOException {
            val reader = new BufferedReader(new InputStreamReader(in));
            val sb = new StringBuilder();
            while (reader.ready()) {
                sb.append(reader.readLine());
            }
            return sb.toString();
        }
    }

    @Data
    @AllArgsConstructor
    private class LeanCloudError {
        private Integer code;
        private String error;
    }
}
```

3．验证 JWT Token

在开始我们的实验之前，需要先插入两个角色记录，这是因为验证鉴权接口时需要几个现成的角色。可以使用 MongoDB 的 shell 或者任一 MongoDB 的图形化管理工具，比如 Studo 3T。

```
db.api_authority.insert({
    "name": "ROLE_USER"
});
db.api_authority.insert({
    "name": "ROLE_ADMIN"
});
```

如果使用 Postman 做个实验，那以 POST 方法访问 http://localhost:8080/api/auth/ register 并在 body 中给出 json 的用户注册信息，如图 5-11 所示。

我们成功地得到了 JWT token。如果得到的 token 在 JWT 的网站上进行验证，就可以看到 token 的信息中包含了在 TokenProvider 中放进去的元素。但是请留意下方红色按钮，上面写着 Invalid Signature（非法签名），如图 5-12 所示。

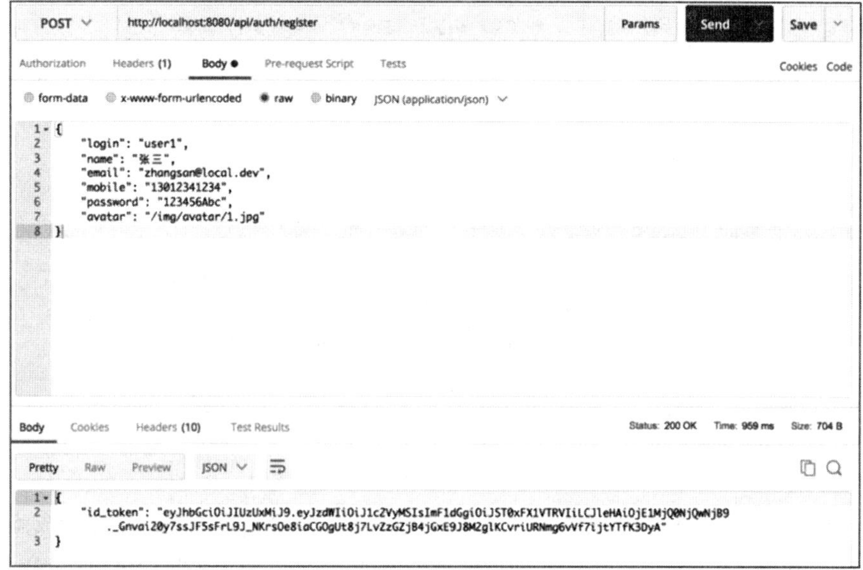

图 5-11

图 5-12

如果你还记得前面在 AppProperties 中定义的默认值，那这个密钥的默认值是 myDefaultSecret。

```
// 省略其他代码
public class AppProperties {
    // 省略其他代码
```

```
    private final Security security = new Security();
    // 省略其他代码
    @Data
    public static class Security {
        private final Jwt jwt = new Jwt();
        // 省略其他代码
        @Data
        public static class Jwt {
            private String secret = "myDefaultSecret";
            private long tokenValidityInSeconds = 7200;
            private String tokenPrefix = "Bearer ";
        }
    }
    // 省略其他代码
}
```

所以，我们把这个值粘贴到写着 `secret` 的输入框，就会发现红色的按钮变成蓝色，文字也变成了 `Signature Verified`（签名已验证）。好的，这也验证了我们生成的 JWT token 是正确的。

5.4.12 使用 JWT 进行 API 访问

由于我们在 `SecurityConfig` 中的设置，`/api/auth` 路径下是不需要鉴权的，如果想验证鉴权访问的结果，则需要新建一个 `TaskResource`，其基础路径为 `/api/tasks`。

```
package dev.local.gtm.api.web.rest;

// 省略导入

@Log4j2
@RestController
@RequestMapping("/api")
@RequiredArgsConstructor
public class TaskResource {
    private final TaskRepo taskRepo;
    private final UserRepo userRepo;

    @GetMapping("/tasks")
    public List<Task> getAllTasks(Pageable pageable,
@RequestParam(required = false) String desc) {
        log.debug("REST 请求 -- 查询所有 Task");
        return desc == null ?
                taskRepo.findAll(pageable).getContent() :
                taskRepo.findByDescLike(pageable, desc).getContent();
```

```
    }

    @GetMapping("/tasks/search/findByUserMobile")
    public List<Task> findByUserMobile(Pageable pageable, @RequestParam
String mobile) {
        log.debug("REST 请求 -- 查询所有手机号为 {} Task", mobile);
        return taskRepo.findByOwnerMobile(pageable, mobile).getContent();
    }

    @PostMapping("/tasks")
    Task addTask(@RequestBody Task task) {
        log.debug("REST 请求 -- 新增 Task {}", task);
        return userRepo.findById(task.getOwner().getId())
                .map(user -> {
                    task.setOwner(user);
                    return taskRepo.save(task);
                })
                .orElseThrow(() -> new ResourceNotFoundException("Task 中
id 为 " + task.getOwner().getId() + " 的 owner 不存在"));
    }

    @PutMapping("/tasks/{id}")
    public Task updateTask(@PathVariable String id, @RequestBody Task
toUpdate) {
        log.debug("REST 请求 -- 更新 id: {} 的 Task {}", id, toUpdate);
        val task = taskRepo.findById(id);
        return task.map(res -> {
            res.setDesc(toUpdate.getDesc());
            res.setCompleted(toUpdate.isCompleted());
            res.setOwner(toUpdate.getOwner());
            return taskRepo.save(res);
        }).orElseThrow(() -> new ResourceNotFoundException("id 为 "+ id +
" 的 Task 没有找到"));
    }

    @GetMapping("/tasks/{id}")
    public Task getTask(@PathVariable String id) {
        log.debug("REST 请求 -- 取得 id: {} 的 Task", id);
        val task = taskRepo.findById(id);
        return task.orElseThrow(() -> new ResourceNotFoundException("id
为 "+ id + " 的 Task 没有找到"));
    }

    @DeleteMapping("/tasks/{id}")
```

```
    @ResponseStatus(HttpStatus.OK)
    public void deleteTask(@PathVariable String id) {
        log.debug("REST 请求 -- 删除 id 为 {} 的 Task", id);
        val task = taskRepo.findById(id);
        task.ifPresent(taskRepo::delete);
    }
}
```

这个 Controller 简单地实现了任务的增、删、改、查操作，为什么没有实现 Service 隔离，而是直接使用 Repository ？一般情况下如果只是简单增、删、改、查操作，笔者不建议增加 Service 层，单纯地为这种领域对象增加 Service 接口和实现，除了增加了代码的臃肿，没有其他好处。只有当业务逻辑复杂起来之后，我们才需要 Service 的封装。

然后，我们就可以在 Postman 中发一个 GET 请求访问 http://localhost:8080/api/tasks，如图 5-13 所示。

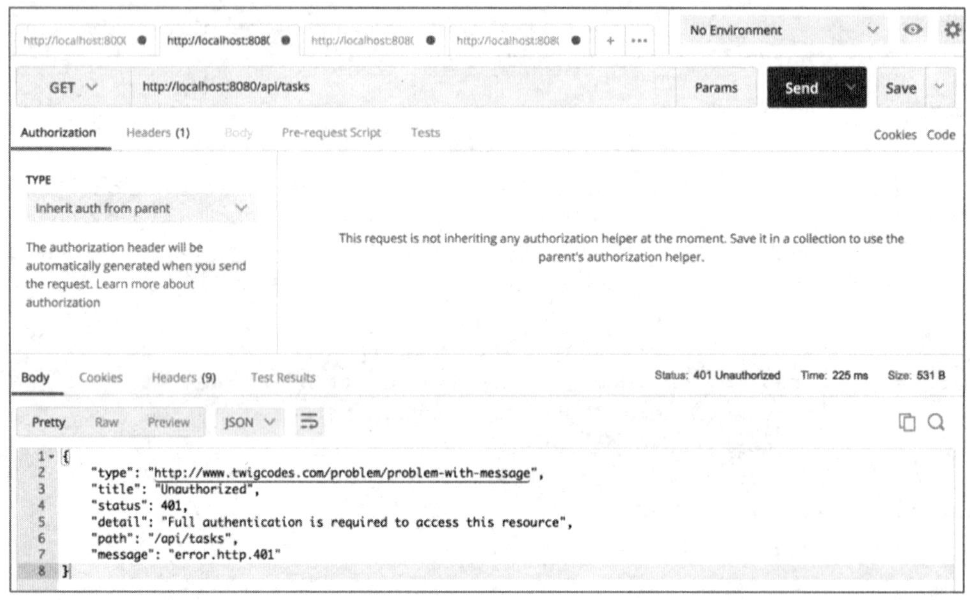

图 5-13

我们会得到一个 401 的未授权的返回结果，这验证了访问 API 需要鉴权：

```
{
    "type": "http://www.twigcodes.com/problem/problem-with-message",
    "title": "Unauthorized",
    "status": 401,
    "detail": "Full authentication is required to access this resource",
```

```
        "path": "/api/tasks",
        "message": "error.http.401"
    }
```

如果把鉴权的 Header 在 Postman 中加进去，就可以正常得到返回结果了，注意，得到的是一个 200 响应码和一个空数组，这很正常，因为我们并未添加新的 Task，但这个 API 经过鉴权是可以访问的了，如图 5-14 所示。

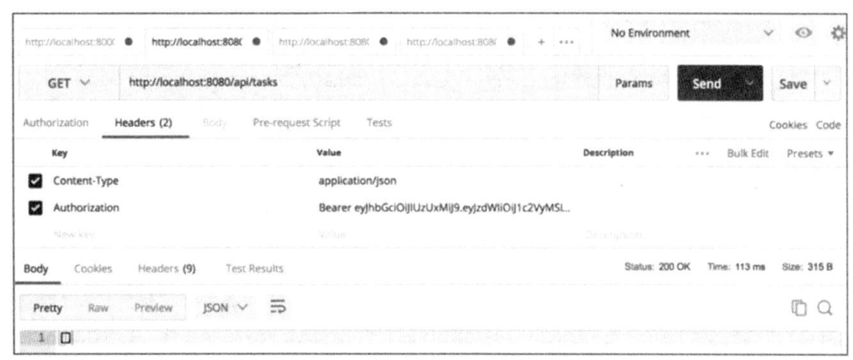

图 5-14

下面的问题就是如果同样是系统用户，那如何限制不同角色的用户可以访问不同的资源？

5.4.13 使用角色

接下来，我们要规定一下哪些资源需要什么样的角色可以访问，首先将 `TaskResource` 改造一下，给方法加上注解，现在只利用两个注解 `@PreAuthorize` 和 `@PostAuthorize`。简单说一下下面代码的逻辑。

- `getAllTasks`——查询任务，默认情况下每个人只能得到自己的任务列表，而管理员可以得到所有的任务列表。这个权限是数据的查询方式不一样，虽然也可以使用注解，比如 `@PostFilter`，但这种过滤是全部查出来之后过滤，浪费资源。我们这里采用的是根据角色给出不同的查询。
- `findByUserMobile`——给管理员一个可以通过手机号过滤任务的方法。
- `addTask`——每个人创建任务的所有者只能是自己（这个需求后面有变化，暂时这里先这么做）。
- `updateTask`——每个人只能更新自己的任务，当然管理员除外，管理员可以更新所有人的任务。
- `getTask`——取得某一个任务，还是每个人只能取自己创建的任务，但管理员除外。
- `deleteTask`——删除任务，这个我们先不加权限，后面再处理。

```
public class TaskResource {
    private final TaskRepo taskRepo;
    private final UserRepo userRepo;

    @GetMapping("/tasks")
    public Page<Task> getAllTasks(Pageable pageable) {
        log.debug("REST 请求 -- 查询所有 Task");
        return
SecurityUtils.isCurrentUserInRole(AuthoritiesConstants.ADMIN) ?
                taskRepo.findAll(pageable) :
                SecurityUtils.getCurrentUserLogin().map(login ->
taskRepo.findByOwnerLogin(pageable, login))
                        .orElseThrow(() -> new
InternalServerErrorException("未找到当前登录用户的登录名"));
    }

    @PreAuthorize("hasRole('ADMIN')")
    @GetMapping("/tasks/search/findByUserMobile")
    public List<Task> findByUserMobile(Pageable pageable, @RequestParam
String mobile) {
        log.debug("REST 请求 -- 查询所有手机号为 {} Task", mobile);
        return taskRepo.findByOwnerMobile(pageable, mobile).getContent();
    }

    @PostMapping("/tasks")
    Task addTask(@RequestBody Task task) {
        log.debug("REST 请求 -- 新增 Task {}", task);
        return SecurityUtils.getCurrentUserLogin()
                .flatMap(userRepo::findOneByLogin)
                .map(user -> {
                    task.setOwner(user);
                    return taskRepo.save(task);
                })
                .orElseThrow(() -> new ResourceNotFoundException("未找到用
户鉴权信息"));
    }

    @PreAuthorize("#toUpdate.owner.login == principal.username or
hasRole('ADMIN')")
    @PutMapping("/tasks/{id}")
    public Task updateTask(@PathVariable String id, @RequestBody Task
toUpdate) {
        log.debug("REST 请求 -- 更新 id: {} 的 Task {}", id, toUpdate);
        val task = taskRepo.findById(id);
```

```
        return task.map(res -> {
            res.setDesc(toUpdate.getDesc());
            res.setCompleted(toUpdate.isCompleted());
            val user = userRepo.findById(toUpdate.getOwner().getId());
            if (user.isPresent()) {
                res.setOwner(user.get());
            } else {
                throw new ResourceNotFoundException("id 为 "+ id + " 的
User 没有找到");
            }
            return taskRepo.save(res);
        }).orElseThrow(() -> new ResourceNotFoundException("id 为 "+ id +
" 的 Task 没有找到"));
    }

    @PostAuthorize("returnObject.owner.login == principal.username or
hasRole('ADMIN')")
    @GetMapping("/tasks/{id}")
    public Task getTask(@PathVariable String id) {
        log.debug("REST 请求 -- 取得 id: {} 的 Task", id);
        val task = taskRepo.findById(id);
        return task.orElseThrow(() -> new ResourceNotFoundException("id
为 "+ id + " 的 Task 没有找到"));
    }

    @DeleteMapping("/tasks/{id}")
    @ResponseStatus(HttpStatus.OK)
    public void deleteTask(@PathVariable String id) {
        log.debug("REST 请求 -- 删除 id 为 {} 的Task", id);
        val task = taskRepo.findById(id);
        task.ifPresent(taskRepo::delete);
    }
}
```

Spring Security 提供 4 个表达“之前”或“之后”意义的注解，这 4 个注解在 `SecurityConfig` 中以`@EnableGlobalMethodSecurity(prePostEnabled = true)`形式启用，用途分别如下。

- `@PreAuthorize`——在方法调用之前检验其表达式，以确定是否有权调用方法，这个注解也是 4 个中最常用的一个。
- `@PostAuthorize`——在方法执行之后检验其表达式，以决定是否有权得到返回值。一般的使用场景是，在方法使用期无法得到某些数据，执行之后才可以，而权限表达式又需要这个值时，就可以使用这个注解。但要注意，一般如果方法是要改变某些数据时，则要慎重使用这个注解，因为方法实际上已经执行，无论有没有权限，限制的只是返回结果而已。通常会使用到内建的变量 returnObject（方法的返回值）。

- @PreFilter——在方法执行前对数据进行筛选，但使用的场景较少，一般是在得到数据之后进行筛选，而不是之前。
- @PostFilter——在方法执行后对得到的数据进行筛选。需要注意这个筛选对于大数量级的数据来说是不适合的。

所以，@PreAuthorize("hasRole('ADMIN')") 的意思就是只能有 ROLE_ADMIN 角色的用户有权调用，需注意的一点是 hasRole 表达式认为每个角色名字前都有一个前缀 ROLE_。所以这里的 ADMIN 其实在数据库中存储的是 ROLE_ADMIN。而 hasRole 就是一个内建表达式，Spring Security 提供的常用的内建表达式如表 5-5 所示。

表 5-5

内建表达式	功能描述
hasRole(role)	如果当前用户有该角色，则返回 true，否则 false。角色名称默认是不加 ROLE_ 前缀的
hasAnyRole([role1,role2])	和上面的区别就是这个是只要有多个角色中的一个就返回 true
principal	当前用户的 Principal 对象，很多时候我们会在表达式中使用用户名信息 principal.username 或激活状态 principal.enabled 等
authentication	Authentication 对象
permitAll	永远返回 true
denyAll	永远返回 false
isAnonymous()	如果是未鉴权用户就返回 true
isAuthenticated()	是否已鉴权
hasPermission(Object target, Object permission)	如果用户对目标对象有访问权限，则返回 true
hasPermission(Object targetId, String targetType, Object permission)	和上面类似，只不过多了一个参数 id

这种表达式中如果想引用方法的参数，则可以使用 #，比如下面表达式中的 #toUpdate 就是方法的参数 @RequestBody Task toUpdate。

```
    @PreAuthorize("#toUpdate.owner.login == principal.username or
hasRole('ADMIN')")
        @PutMapping("/tasks/{id}")
        public Task updateTask(@PathVariable String id, @RequestBody Task
toUpdate) {
            // 省略方法
        }
```

前面的 TaskResource 的代码中，我们还使用了一个工具类 SecurityUtils，这个工具类比较简单，就是进一步封装了 Spring Security 的一些方法，让我们使用起来更方便。

```
    package dev.local.gtm.api.security;

    // 省略导入

    /**
     * Spring Security 的工具类
     */
    public final class SecurityUtils {

        private SecurityUtils() {
        }

        /**
         * 得到当前用户的登录名
         *
         * @return 返回当前用户的登录名
         */
        public static Optional<String> getCurrentUserLogin() {
            SecurityContext securityContext =
SecurityContextHolder.getContext();
            return Optional.ofNullable(securityContext.getAuthentication())
                    .map(authentication -> {
                        if (authentication.getPrincipal() instanceof
UserDetails) {
                            UserDetails springSecurityUser = (UserDetails)
authentication.getPrincipal();
                            return springSecurityUser.getUsername();
                        } else if (authentication.getPrincipal() instanceof
String) {
                            return (String) authentication.getPrincipal();
                        }
                        return null;
                    });
        }
```

```
    /**
     * 得到当前用户的 JWT token
     *
     * @return 返回当前用户的 JWT toekn
     */
    public static Optional<String> getCurrentUserJWT() {
        SecurityContext securityContext =
SecurityContextHolder.getContext();
        return Optional.ofNullable(securityContext.getAuthentication())
                .filter(authentication -> authentication.getCredentials()
instanceof String)
                .map(authentication -> (String)
authentication.getCredentials());
    }

    /**
     * 用户是否已鉴权
     *
     * @return 已鉴权返回 true，否则返回 false
     */
    public static boolean isAuthenticated() {
        SecurityContext securityContext =
SecurityContextHolder.getContext();
        return Optional.ofNullable(securityContext.getAuthentication())
                .map(authentication ->
authentication.getAuthorities().stream()
                        .noneMatch(grantedAuthority ->
grantedAuthority.getAuthority().equals(AuthoritiesConstants.ANONYMOUS)))
                .orElse(false);
    }

    /**
     * 当前用户是否有指定角色
     *
     * @param authority 要检查的角色
     * @return 如果有该角色，则返回 true，否则返回 false
     */
    public static boolean isCurrentUserInRole(String authority) {
        SecurityContext securityContext =
SecurityContextHolder.getContext();
        return Optional.ofNullable(securityContext.getAuthentication())
                .map(authentication ->
authentication.getAuthorities().stream()
```

```
                           .anyMatch(grantedAuthority ->
grantedAuthority.getAuthority().equals(authority)))
                   .orElse(false);
        }
    }
```

现在大家可以去试着通过 /api/auth/register 接口注册几个新用户，其中 ADMIN 角色的用户可以在 MongoDB 中手动修改一下，然后使用它们的 token 去创建各自的任务，体验一下角色对于资源的限制。

下面，我们来做一个验证，首先创建两个用户 admin（ROLE_ADMIN 和 ROLE_USER 两个角色）和 user2（ROLE_USER 角色）。

```
   db.api_users.insert({
       "_id" : ObjectId("5add712330ccb9a9e7f95f7b"),
       "login" : "admin",
       "name" : "admin user",
       "email" : "admin@local.dev",
       "mobile" : "13012340001",
       "activated" : true,
       "authority_ids" : [
           DBRef("api_authority", ObjectId("5add5d2c301b322e28e288f3")),
           DBRef("api_authority", ObjectId("5add5d48301b322e28e288f4"))
       ],
       "password" :
"$2a$10$.DN8bpMCQxkyvLjhfq5ht.6v8U6cICyW3aU57nizkBsuPHcYyF.4y",
       "avatar" : "/img/avatar/2.jpg"
   });
   db.api_users.insert({
       "_id" : ObjectId("5adda1eaf68cb3a9a905dc55"),
       "login" : "user2",
       "password" :
"$2a$10$TiGrPs11iL3rqmVclg2mhOIiviXuB6NbkwFRXiCjc2gxaGJBpjuUe",
       "mobile" : "13012340002",
       "name" : "李四",
       "email" : "lisi@local.dev",
       "avatar" : "/img/avatar/3.jpg",
       "activated" : true,
       "authority_ids" : [
           DBRef("api_authority", ObjectId("5add5d2c301b322e28e288f3"))
       ],
       "created_date" : ISODate("2018-04-23T09:05:46.922+0000"),
       "last_modified_date" : ISODate("2018-04-23T09:05:46.922+0000"),
       "_class" : "dev.local.gtm.api.domain.User"
   });
```

然后，可以利用 /api/login 得到每个用户的 token。分别使用几个用户的 token 创建任务，再试一下各个接口的权限。

1. 安全日志

很多时候，如果 Spring Security 出现问题，那么我们希望可以看到日志来帮助分析问题，日志怎么开启呢？只需要在 application.yml 中配置 org.springframework. security: DEBUG 即可。

```
logging:
  level:
    org.apache.http: DEBUG
    org.springframework:
          web:
            client.RestTemplate: DEBUG
          data: DEBUG
          security: DEBUG
    org.springframework.data.mongodb.core.MongoTemplate: DEBUG
    dev.local.gtm.api: DEBUG
```

其他日志也可以采用这种方式开启，比如上面设置了 org.springframework. web. client.RestTemplate: DEBUG，这个是开启 RestTemplate 的请求日志，可以看到每个对外请求的细节信息。

而 org.springframework.data.mongodb.core.MongoTemplate: DEBUG 可以看到组装的 MongoDB 的查询语句。

2. 多环境的配置

上面在做各种实验时，笔者不知道大家是否留意到了，我们操作的其实是真正的数据库，在平时做一些个人的探索项目时倒也无所谓。但如果在商业项目中，如果直接去访问生产库，则是绝对不允许的。当然，除了数据库，比如有很多第三方 API，或者其他的配置，在生产环境和开发环境中也是很不一样的。这就引出了一个现代开发中的重要概念——多环境配置。

在 Spring Boot 中对于多环境配置有良好的支持，拿.yml 配置文件来说，有两种配置方式。

- 单文件形式：在 application.yml 中以 --- 区隔不同的环境配置。
- 多文件形式：以 applicaiton-环境配置名.yml 形式提供多个文件。

```
## 单文件形式
spring:
  application:
    name: api-service
  profiles:
```

```
    active: dev
  data:
    mongodb:
      database: gtm-api
server:
  port: 8080
logging:
  level:
    org.apache.http: ERROR
    org.springframework:
      web: ERROR
      data: ERROR
      security: ERROR
    org.springframework.data.mongodb.core.MongoTemplate: ERROR
    dev.local.gtm.api: ERROR

---

spring:
  profiles: dev
  devtools:
    remote:
      secret: thisismysecret
  data:
    mongodb:
      database: gtm-api-dev
logging:
  level:
    org.apache.http: DEBUG
    org.springframework:
          web:
            client.RestTemplate: DEBUG
          data: DEBUG
          security: DEBUG
    org.springframework.data.mongodb.core.MongoTemplate: DEBUG
    dev.local.gtm.api: DEBUG

---

spring:
  profiles: test
  data:
    mongodb:
      database: gtm-api-test
```

```
logging:
  level:
    org.apache.http: DEBUG
    org.springframework:
          web:
            client.RestTemplate: DEBUG
          data: DEBUG
          security: DEBUG
    org.springframework.data.mongodb.core.MongoTemplate: DEBUG
    dev.local.gtm.api: DEBUG

---

spring:
  profiles: prod
```

如果配置项不是很多，那么采用单文件，.yml 支持环境配置的“继承”，比如文件开始的那段就是公共配置，每个环境只需要改动在自己这里发生变化的即可，不写的自动使用公共配置。大家可以看到我们对于不同环境使用了不同的数据库和不同的日志。具体的环境配置名称使用类似 `profiles: test` 形式进行命名。

（1）使用不同的环境

在 IDEA 中，我们可以在 `Run/Debug Configurations` 中设置 `Active profiles`，比如填 `dev`、`test` 或 `prod` 等，如图 5-15 所示。

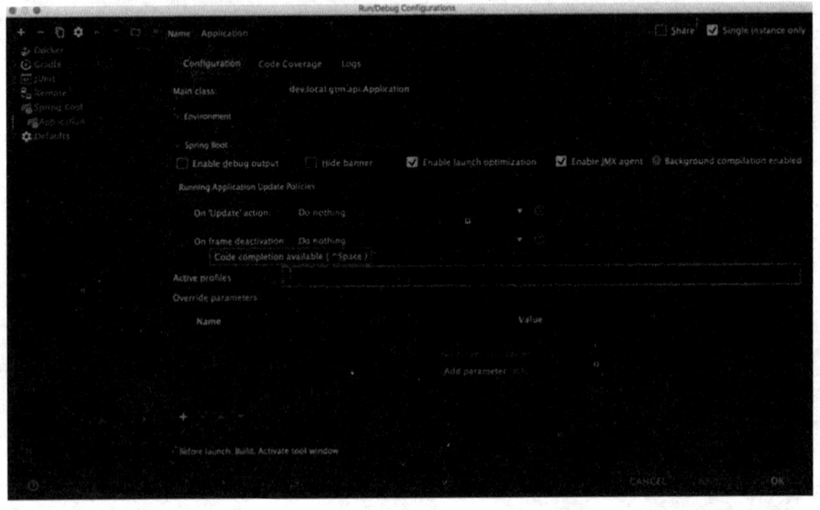

图 5-15

如果使用 Gradle 命令行，则需要在 api/build.gradle 中增加 bootRun 区块。

```
    apply plugin: 'org.springframework.boot'
    repositories {
        maven { setUrl('http://oss.jfrog.org/artifactory/oss-snapshot-
local/') }
    }
    configurations {
        // 省略
    }
    bootRun {
        systemProperties = System.properties as Map<String, ?>
    }
    test {
        systemProperties['spring.profiles.active'] = 'test'
    }
    dependencies {
    // 省略
    }
```

然后就可以使用下面的命令运行 prod 环境了，当然你还可以试试其他的环境配置，并在不同的环境配置下使用 API 创建任务或注册用户，可以看看 MongoDB 中是不是不同的配置下数据库是不一样的。

```
    ./gradlew :api:bootRun -Dspring.profiles.active=prod
```

5.5 跨域和 API 文档

5.5.1 跨域解决方案——CORS

前面我们初步做出了一个可以实现受保护的 REST API，但是没有涉及一个前端领域很重要的问题，那就是跨域请求（cross-origin HTTP request）。先来回顾一些背景知识。

1. 什么是跨域

定义：当我们从本身站点请求不同域名或端口的服务所提供的资源时，就会发起跨域请求。

例如最常见的很多 CSS 样式文件是会链接到某个公共 CDN 服务器上，而不是在本身的服务器上，这其实就是一个典型的跨域请求。但浏览器由于安全原因限制了在脚本（script）中发起的跨域 HTTP 请求。也就是说，XMLHttpRequest 和 Fetch 等是遵循“同源规则”的，即只能访问自己服务器的指定端口的资源（同一服务器不同端口也会视为跨域）。但这种限制在今天，我们的应用需要访问多种外部 API 或资源时就不能满足开发者的需求了，因此就产生了若干对于跨域的解决方案，JSONP 是其中一种，但在今天来看主流的更彻底的解决方案是 CORS（Cross-Origin Resource Sharing）。

2. 跨域资源共享

这种机制将跨域的访问控制权交给服务器，这样可以保证安全的跨域数据传输。现代浏览器一般会将 CORS 的支持封装在 HTTP API 之中（比如 XMLHttpRequest 和 Fetch），这样可以有效控制使用跨域请求的风险。

概括来说，这个机制是增加一系列的 HTTP 头来让服务器可以描述哪些源是允许使用浏览器来访问资源的。而且对于简单的请求和复杂请求，处理机制是不一样的。

简单请求仅允许 3 个 HTTP 方法：GET、POST 及 HEAD，另外只能支持若干 Header 参数：Accept 、 Accept-Language 、 Content-Language 、 Content-Type （ 值只能是 application/x-www-form-urlencoded、multipart/form-data 和 text/plain）、DPR、Downlink、Save-Data、Viewport-Width 和 Width。

对于简单请求，比如下面这样一个简单的 GET 请求：从 http://me.domain 发起到 http://another.domain/data/×××的资源请求。

```
GET /data/×××/ HTTP/1.1
// 请求的域名
Host: another.domain
...//省略其他部分，重点是下面这句，说明了发起请求者的来源
Origin: http://me.domain
```

应用了 CORS 的对方服务器返回的响应应该像下面这个样子，当然这里 Access-Control-Allow-Origin: * 中的 * 表示任何网站都可以访问该资源，如果要限制只能从 me.domain 访问，那么需要改成 Access-Control-Allow-Origin: http://me. Domain。

```
HTTP/1.1 200 OK
...//省略其他部分
Access-Control-Allow-Origin: *
...//省略其他部分
Content-Type: application/json
```

对于复杂请求怎么办呢？这需要一次预检请求和一次实际的请求，也就是说需要两次和对方服务器的请求/响应。预检请求是以 OPTION 方法进行的，因为 OPTION 方法不会改变任何资源，所以这个预检请求是安全的，它的职责在于发送实际请求将会使用的 HTTP 方法以及将要发送的 Header 中将携带哪些内容，这样对方服务器可以根据预检请求的信息决定是否接受。

```
// 预检请求
OPTIONS /resources/post/ HTTP/1.1
```

```
Host: another.domain
...// 省略其他部分
Origin: http://me.domain
Access-Control-Request-Method: POST
Access-Control-Request-Headers: Content-Type
```

服务器对预检请求的响应如下：

```
HTTP/1.1 200 OK
// 省略其他部分
Access-Control-Allow-Origin: http://me.domain
Access-Control-Allow-Methods: POST, GET, OPTIONS
Access-Control-Allow-Headers: Content-Type
Access-Control-Max-Age: 86400
// 省略其他部分
Content-Type: text/plain
```

接下来的正式请求上面的简单请求差不多，就不赘述了。

3. Spring Boot 中如何启用 CORS

讲了这么多，终于进入正题，加入 CORS 的支持在 Spring Boot 中简单到不忍直视，添加一个配置类即可：

```
package dev.local.gtm.api.config;

// 省略导入

@Log4j2
@RequiredArgsConstructor
@Configuration
public class CorsConfig {

    private final AppProperties appProperties;

    @Bean
    public CorsFilter corsFilter() {
        val source = new UrlBasedCorsConfigurationSource();
        val config = appProperties.getCors();
        if (config.getAllowedOrigins() != null
&& !config.getAllowedOrigins().isEmpty()) {
            log.debug("注册 CORS 过滤器");
            config.addAllowedOrigin("*");
            config.addAllowedMethod("*");
            config.addAllowedHeader("*");
            source.registerCorsConfiguration("/api/**", config);
```

```
                source.registerCorsConfiguration("/management/**", config);
                source.registerCorsConfiguration("/v2/api-docs", config);
                source.registerCorsConfiguration("/*/api/**", config);
                source.registerCorsConfiguration("/*/management/**", config);
            }
            return new CorsFilter(source);
        }
    }
```

在 SecurityConfig 中配置我们建立的 CorsFilter 就可以了，这样前端就可以访问 API 了。

```
    @RequiredArgsConstructor
    @Configuration
    @EnableWebSecurity
    @EnableGlobalMethodSecurity(prePostEnabled = true, securedEnabled = true)
    @Import(SecurityProblemSupport.class)
    public class SecurityConfig extends WebSecurityConfigurerAdapter {

        // 省略其他成员变量
        private final CorsFilter corsFilter;
        // 省略其他方法
        @Override
        protected void configure(HttpSecurity http) throws Exception {
            http
                    .addFilterBefore(corsFilter,
UsernamePasswordAuthenticationFilter.class) // <--- 将 corsFilter 配置在
UsernamePasswordAuthenticationFilter 之前
                    .exceptionHandling()
                    .authenticationEntryPoint(problemSupport)
                    .accessDeniedHandler(problemSupport)
                    .and()
                        .csrf()
                        .disable()
                        .headers()
                        .frameOptions()
                        .disable()
                    .and()
                        .sessionManagement()
                        .sessionCreationPolicy(SessionCreationPolicy.STATELE
SS)
                    .and()
                        .authorizeRequests()
                            .antMatchers("/api/auth/**").permitAll()
                            .antMatchers("/api/**").authenticated()
```

```
                        .antMatchers("/websocket/tracker").hasAuthority(
AuthoritiesConstants.ADMIN)
                        .antMatchers("/websocket/**").permitAll()
                        .antMatchers("/management/health").permitAll()
                        .antMatchers("/management/**").hasAuthority(Auth
oritiesConstants.ADMIN)
                 .and()
                     .apply(jwtConfigurer);
        }
    }
```

5.5.2 API 文档

大家前面使用 Postman 时是不是感觉各种 API 的构造有点麻烦？想象一下，如果有 4 个团队：后端、前端、iOS 客户端和 Android 客户端。后端开发完 API 之后，其他几个团队怎么知道 API 应该怎么使用呢？当然可以使用传统的文档，但传统文档有几个明显弱点。

（1）没有交互性：API 文档阅读时很少会引发人提出什么问题，因为 API 是前端或者客户端和后端交互的手段，很多问题也是交互时才发现的。

（2）更新不及时：API 的重构和修正在系统联调时是很频繁的，文档的更新往往不够及时。

（3）错误率较高：写文档时可能会有笔误或有遗漏，这可能导致后端认为很简单的事情，前端调试很久才发现是文档错了。

因此，在这个敏捷开发盛行的时代，我们需要一种敏捷的、可交互的文档来描述 API。Swagger（https://swagger.io）便是这种文档类型中的佼佼者。Swagger 是一种 Rest API 的简单但强大的表现形式，它是标准的和编程语言无关的，不但人可以阅读，而且机器也可读，也就是可以直接进行 API 交互。所以它既可以作为 Rest API 的交互式文档，也可以作为 Rest API 的形式化的接口描述，甚至可以直接生成客户端和服务端的代码。

而能将 Swagger 和 Spring Boot 平滑地对接起来的开源类库就是 SpringFox[1]。

1. 配置 Swagger

首先，需要在 `api/build.gradle` 中配置 SpringFox 的依赖项，共有 3 个依赖 `springfox-swagger2`、`springfox-bean-validators` 和 `springfox-swagger-ui`。

```
// api/build.gradle
// 省略其他部分
```

1 https://github. com/springfox/springfox

```
dependencies {
        implementation("io.springfox:springfox-swagger2:${springFoxVersion}")
       implementation("io.springfox:springfox-bean-
validators:${springFoxVersion}")
       implementation("io.springfox:springfox-swagger-
ui:${springFoxVersion}")
       implementation("org.springframework.boot:spring-boot-starter-
undertow")
       implementation("org.springframework.boot:spring-boot-starter-
actuator")
       implementation("org.springframework.boot:spring-boot-starter-web")
       implementation("io.jsonwebtoken:jjwt:0.9.0")
       implementation("org.springframework.boot:spring-boot-starter-
security")
       implementation("org.springframework.boot:spring-boot-starter-aop")
       implementation("org.zalando:problem-spring-web:0.20.1")
       implementation("com.fasterxml.jackson.module:jackson-module-
afterburner")
       implementation("org.springframework.boot:spring-boot-starter-data-
mongodb")
       testImplementation("org.springframework.security:spring-security-
test")
}
```

当然，也需要在根项目的 build.gradle 中的 ext 中指定版本号。

```
buildscript {
    ext {
        // 省略
        springFoxVersion = '2.8.1-SNAPSHOT'
    }
    // 省略
}
```

配置完类库依赖之后，就可以在 config 包下新建一个 SwaggerConfig.java 文件。

```
package dev.local.gtm.api.config;

// 省略导入

/**
 * 配置 Swagger 以提供 API 文档
 *
 * @author Peng Wang (wpcfan@gmail.com)
 */
@EnableSwagger2
```

```
@ComponentScan(basePackages = "dev.local.gtm.api.web.rest")
@Import({
        springfox.bean.validators.configuration.BeanValidatorPluginsConf
iguration.class
})
@Configuration
public class SwaggerConfig {
    /**
     * 配置 Swagger 扫描哪些 API（不列出那些监控 API）
     *
     * @return Docket
     */
    @Bean
    public Docket apiDoc() {
        return new Docket(DocumentationType.SWAGGER_2)
                .select()
                    .apis(RequestHandlerSelectors.basePackage("dev.local
.gtm.api.web.rest"))
                    .paths(PathSelectors.any())
                    .build()
                .pathMapping("/")
                .directModelSubstitute(LocalDate.class, String.class)
                .genericModelSubstitutes(ResponseEntity.class)
                .apiInfo(apiInfo());
    }

    /**
     * 对 API 的概要信息进行定制
     *
     * @return ApiInfo
     */
    private ApiInfo apiInfo() {
        return new ApiInfo(
                "GTM API 文档",
                "所有 GTM 开放的 API 接口，供 Android，iOS 和 Web 客户端调用",
                "1.0",
                "http://www.twigcodes.com/gtm/tos.html",
                new Contact("Peng Wang", "http://www.twigcodes.com",
"wpcfan@gmail.com"),
                "API 授权协议", "http://www.twigcodes.com/gtm/api-
license.html", Collections.emptyList());
    }
}
```

这段配置代码要解释几个地方。

- `@EnableSwagger2`——这个注解就可以将 Swagger 集成进来。
- `@ComponentScan(basePackages = "...")`——指明我们需要扫描哪些包下的 Controller。
- `@Import`——如果要使用 Bean 验证，需要导入 SpringFox 的验证类。
- `Docket`——这个是 SpringFox 的基础配置类。
- `ApiInfo`——对于 API 文档基础信息，比如版本号，团队信息等进行配置。

现在，为了可以看到文档效果，还需要在 `SecurityConfig` 中加入 Swagger UI 的支持，默认情况下，Swagger 集成了一个自己的 UI，它的入口 `URL` 一般是 `/swagger-ui.html`，我们暂时允许匿名访问该 `URL`。

```
    public class SecurityConfig extends WebSecurityConfigurerAdapter {

        // 省略

        @Override
        protected void configure(HttpSecurity http) throws Exception {
            http
                    .addFilterBefore(corsFilter,
UsernamePasswordAuthenticationFilter.class)
                    .exceptionHandling()
                    .authenticationEntryPoint(problemSupport)
                    .accessDeniedHandler(problemSupport)
                    .and()
                        .csrf()
                        .disable()
                        .headers()
                        .frameOptions()
                        .disable()
                    .and()
                        .sessionManagement()
                        .sessionCreationPolicy(SessionCreationPolicy.STATELE
SS)
                    .and()
                        .authorizeRequests()
                            .antMatchers("/api/auth/**").permitAll()
                            .antMatchers("/api/**").authenticated()
                            .antMatchers("/websocket/tracker").hasAuthority(
AuthoritiesConstants.ADMIN)
                            .antMatchers("/websocket/**").permitAll()
                            .antMatchers("/management/health").permitAll()
```

```
                    .antMatchers("/management/**").hasAuthority(Auth
oritiesConstants.ADMIN)
                    .antMatchers("/v2/api-docs/**").permitAll()
                    .antMatchers("/swagger-
resources/configuration/ui").permitAll()
                     // 这里暂时允许匿名访问 API 文档
                    .antMatchers("/swagger-
ui/index.html").permitAll()
              .and()
                 .apply(jwtConfigurer);
      }
   }
```

现在，我们看看文档的效果，访问 http://localhost:8080/swagger-ui/index.html，可以看到文档首页，里面现在列出了现有的两个资源（Controller），如图 5-16 所示。

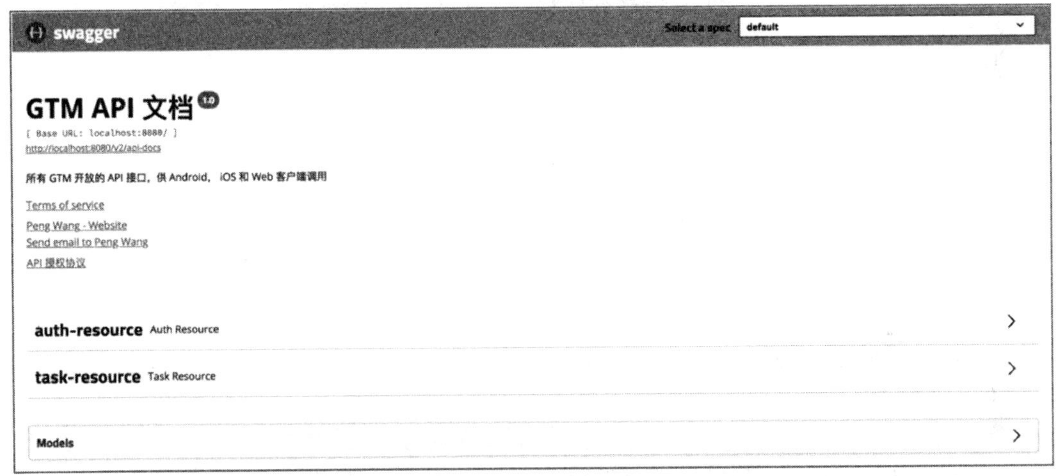

图 5-16

那么，如果我们展开其中的一项，比如 GET /api/auth/captcha，可以看到如图 5-17 所示的界面，包括这个 API 需要的参数和可能的响应码。

如果我们点击 Try it out 按钮，然后点击 Execute，那么就可以发送该 API 请求，并得到返回结果，如图 5-18 所示。

这样我们可以把 Postman 抛弃了，对于前端或客户端开发的同事来说，直接看这个文档的同时就可以进行验证和调试了。

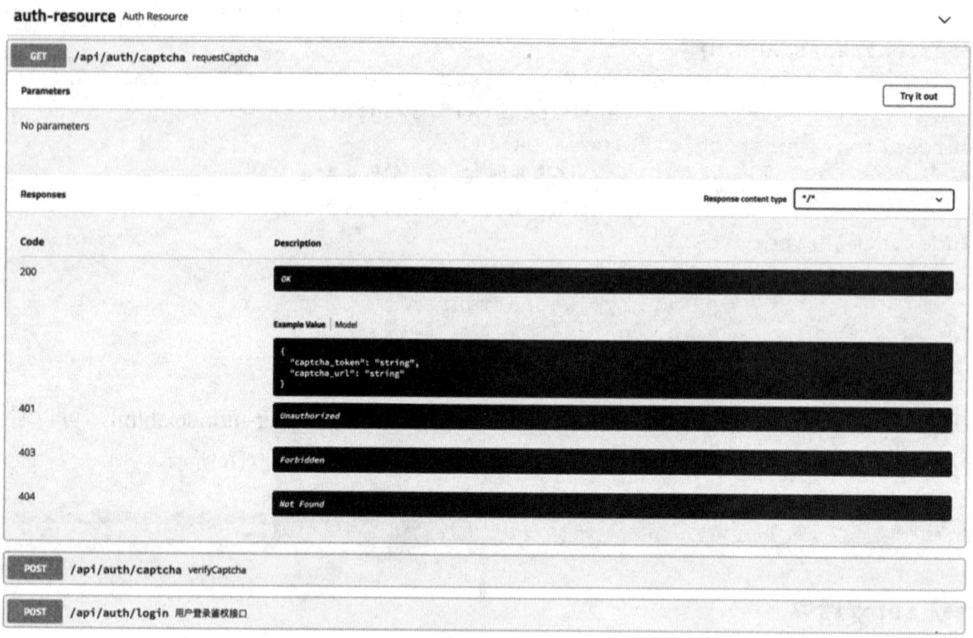
auth-resource Auth Resource
GET
/api/auth/captcha requestCaptcha
Parameters
Try it out
No parameters
Responses
Response content type
/
Code
Description
200
OK
Example Value
Model
{
"captcha_token": "string",
"captcha_url": "string"
}
401
Unauthorized
403
Forbidden
404
Not Found
POST
/api/auth/captcha verifyCaptcha
POST
/api/auth/login 用户登录鉴权接口

图 5-17

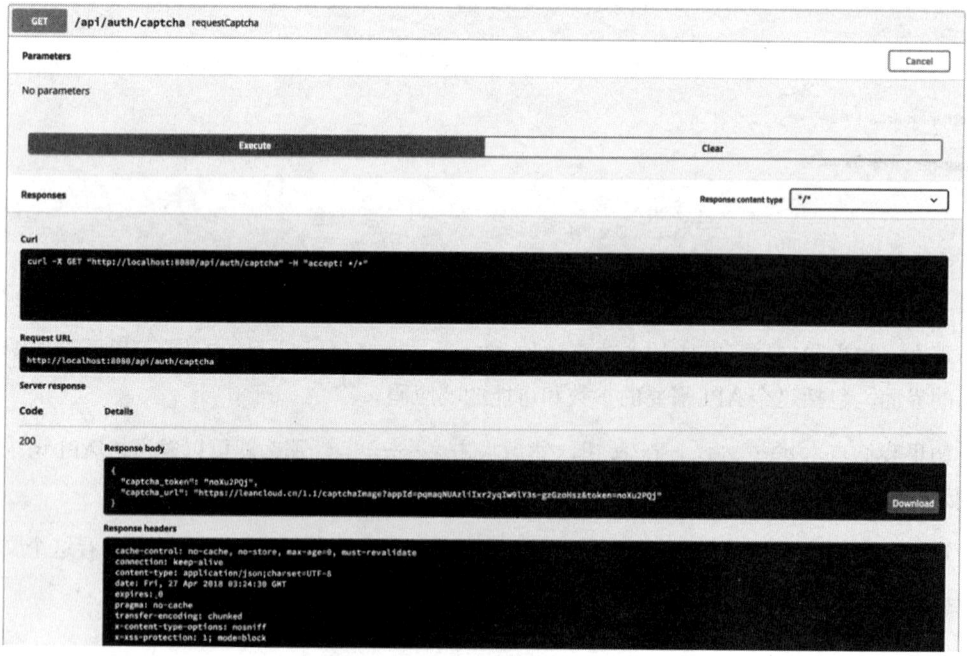
GET
/api/auth/captcha requestCaptcha
Parameters
Cancel
No parameters
Execute
Clear
Responses
Response content type
/
Curl
curl -X GET "http://localhost:8080/api/auth/captcha" -H "accept: */*"
Request URL
http://localhost:8080/api/auth/captcha
Server response
Code
Details
200
Response body
Download
Response headers
cache-control: no-cache, no-store, max-age=0, must-revalidate
connection: keep-alive
content-type: application/json;charset=UTF-8
date: Fri, 27 Apr 2018 03:24:30 GMT
expires: 0
pragma: no-cache
transfer-encoding: chunked
x-content-type-options: nosniff
x-xss-protection: 1; mode=block

图 5-18

2．对于受保护的 API 的访问

刚刚我们都是对匿名可以访问的 API 进行验证，但如果是刚刚建立的受保护的 API，比如需要 HTTP 中的 Authorization 请求消息头的这种怎么办呢？SpringFox 提供了 SecurityContext 和 SecurityScheme 来实现这个访问安全 API 的目的。

```
public class SwaggerConfig {
    /**
     * 配置 Swagger 扫描哪些 API（不列出那些监控 API）
     *
     * @return Docket
     */
    @Bean
    public Docket apiDoc() {
        return new Docket(DocumentationType.SWAGGER_2)
                .select()
                    .apis(RequestHandlerSelectors.basePackage("dev.local
.gtm.api.web.rest"))
                    .paths(PathSelectors.any())
                    .build()
                .pathMapping("/")
                .securitySchemes(newArrayList(apiKey()))
                .securityContexts(newArrayList(securityContext()))
                .directModelSubstitute(LocalDate.class, String.class)
                .genericModelSubstitutes(ResponseEntity.class)
                .apiInfo(apiInfo());

    }

    private ApiKey apiKey() {
        // 用于 Swagger UI 测试时添加 Bearer token
        return new ApiKey("Bearer", HttpHeaders.AUTHORIZATION,
In.HEADER.name());
    }

    private SecurityContext securityContext() {
        return SecurityContext.builder()
                .securityReferences(defaultAuth())
                .forPaths(PathSelectors.regex("/api/((?!auth).)*"))
                .build();
    }

    private List<SecurityReference> defaultAuth() {
        AuthorizationScope authorizationScope = new
```

```
AuthorizationScope("global", "accessEverything");
            AuthorizationScope[] authorizationScopes = new
AuthorizationScope[1];
            authorizationScopes[0] = authorizationScope;
            return newArrayList(new SecurityReference("Bearer",
authorizationScopes));
        }
        // 省略
    }
```

- SecurityScheme——提供一种保护 API 的安全策略，目前内建支持 ApiKey、BasicAuth 和 OAuth。由于之前实现的是 JWT token，属于 ApiKey，所以下面提供的也是 ApiKey。其他实现形式请查阅官网文档。
- ApiKey——在下面的配置中，我们使用了 ApiKey("Bearer", HttpHeaders.AUTHORIZATION, In.HEADER.name()) 构造一个 ApiKey。
  - — ApiKey 构造函数的第 1 个参数是如何命名 ApiKey，后面在 SecurityReference ("Bearer", authorizationScopes) 引用的名字需要和这里定义的一致。
  - — 第 2 个参数指出 ApiKey 的 key 是什么。举例来说，我们的 API 鉴权是 Authorization: Bearer ×××× 的形式，这个 key 就是 Authorization，所以使用 HttpHeaders.AUTHORIZATION 作为第 2 个参数。
  - — 第 3 个参数是这个 key 和我们传入的 value（value 就是 Bearer ×××× 部分）要在请求的什么地方添加，我们需要的是在 Request 头部添加，所以是 In.HEADER.name()。
- SecurityContext——提供一种全局方式用来选择实施哪种安全策略（安全策略就是上面的 SecurityScheme），在下面的配置中以正则表达式的形式 .forPaths(PathSelectors.regex("/api/((?!auth).)*")) 对不是/api/auth**的这种路径实施 ApiKey 的安全策略。也就是说，在 API 文档中进行交互式测试时，对于非 auth 路径下的 API 添加 ApiKey 中定义的 Header。

现在如果重新访问 http://localhost/swagger-ui/index.html，那么我们会看到出现了一个 Authorize 按钮，如图 5-19 所示。

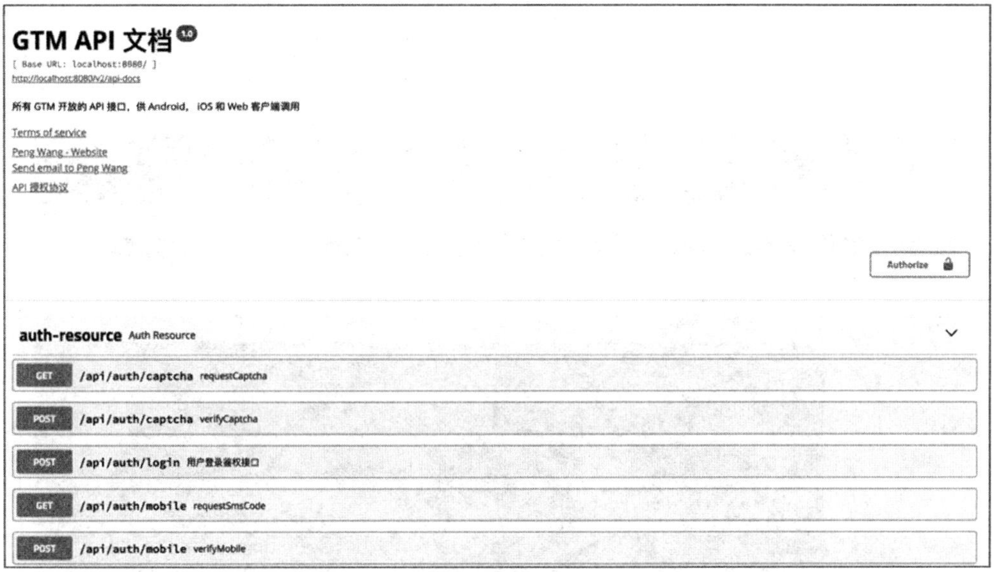

图 5-19

点击这个按钮，就会出现下面的界面，让你输入 Authorization 的值，也就是 Bearer ××××，这个值可以通过调用注册或登录的 API 得到，把得到的值前面加上 Bearer 和一个空格，粘贴到 value 的输入框中，点击 Authorize 按钮，然后点击 Close 按钮关闭对话框，如图 5-20 所示。

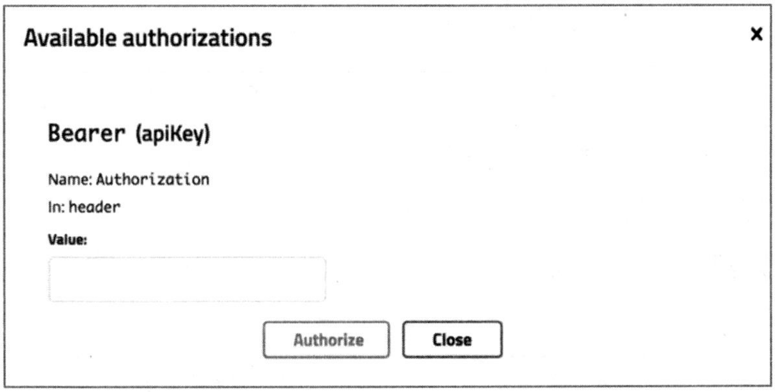

图 5-20

现在，如果访问某个受保护的 API，就会得到类似图 5-21 所示的效果。

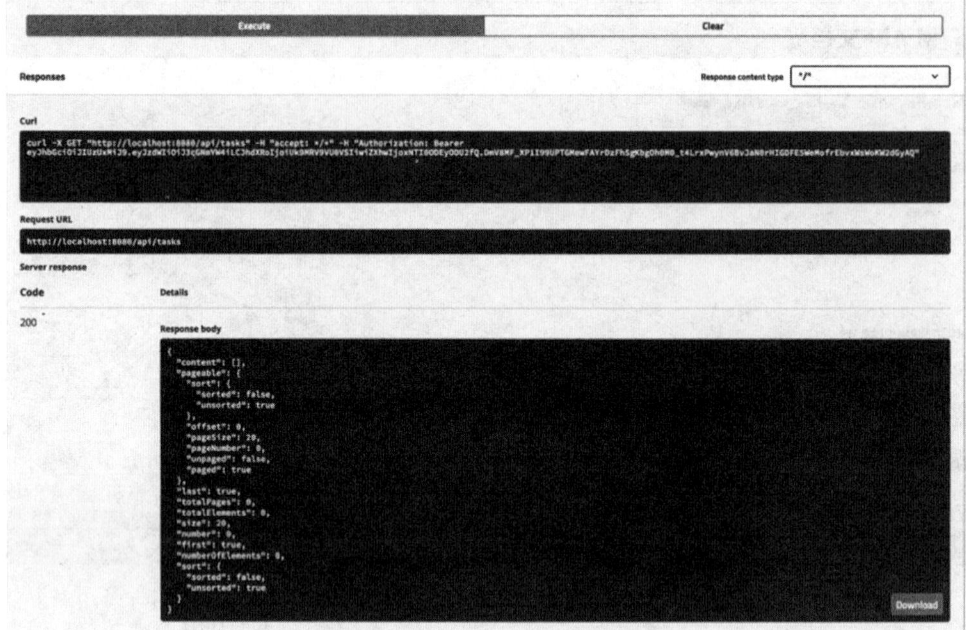

图 5-21

我们可以看到在文档生成的 curl 命令中，使用 -H 添加了一个 Authorization 的 Header 键值对。有兴趣的读者在 *nix 系统上可以开一个 terminal 实验下面的命令，得到的输出和我们界面上看到的应该一样。

```
curl -X GET "http://localhost:8080/api/tasks" -H "accept: */*" -H "Authorization: Bearer eyJhbGciOiJIUzUxMiJ9.eyJzdWIiOiJ3cGNmYW4iLCJhdXRoIjoiUk9MRV9VU0VSIiwiZXhwIjoxNTI0ODEyODU2fQ.DmV8MF_XP1I99UPTGMewFAYrDzFhSgKbgOh0M0_t4LrxPwynV6BvJaN0rHIGDFE5WeMofrEbvxWsWoKW2dGyAQ"
```

3. 使用注解完善文档

SpringFox 提供了一系列注解来帮助我们完善文档，常用的注解如表 5-6 所示。

表 5-6

常用注解	应用对象	举例说明
@ApiModelProperty	领域对象的属性	@ApiModelProperty(value="姓名")
@ApiParam	Controller 的参数	@ApiParam(value = "用户名") @RequestParam ("username") String username

续表

常用注解	应用对象	举例说明
@ApiOperation	Controller 方法	@ApiOperation(value = "用户登录鉴权接口",notes = "客户端在 RequestBody 中以 JSON 形式传入用户名、密码，如果成功以 JSON 形式返回该用户信息")

我们具体看一下在代码中如何使用，下面的代码体现了 ApiOperation 和 @ApiParam。

```
// 省略
public class AuthResource {

    // 省略
    // 对这个 API 的描述
    @ApiOperation(value = "用户登录鉴权接口",
            notes = "客户端在 RequestBody 中以 JSON 形式传入用户名、密码，如果
成功以 JSON 形式返回该用户信息")
    @PostMapping(value = "/auth/login")
    public ResponseEntity<JWTToken> login(@RequestBody final Auth auth)
{
        log.debug("REST 请求 -- 将对用户: {} 执行登录鉴权", auth);
        return generateJWTHeader(auth.getLogin(), auth.getPassword());
    }

    // 省略

    @GetMapping("/auth/search/username")
    public ExistCheck usernameExisted(
            //对参数的描述
            @ApiParam(value = "用户名") @RequestParam("username") String
username) {
        log.debug("REST 请求 -- 用户名是否存在 {}", username);
        return new ExistCheck(authService.usernameExisted(username));
    }
    // 省略
}
```

由图 5-22 可以看出，@ApiOperation 的 value 指定的是一个简短描述，而 notes 可以较详细地说明这个 API 的用途。

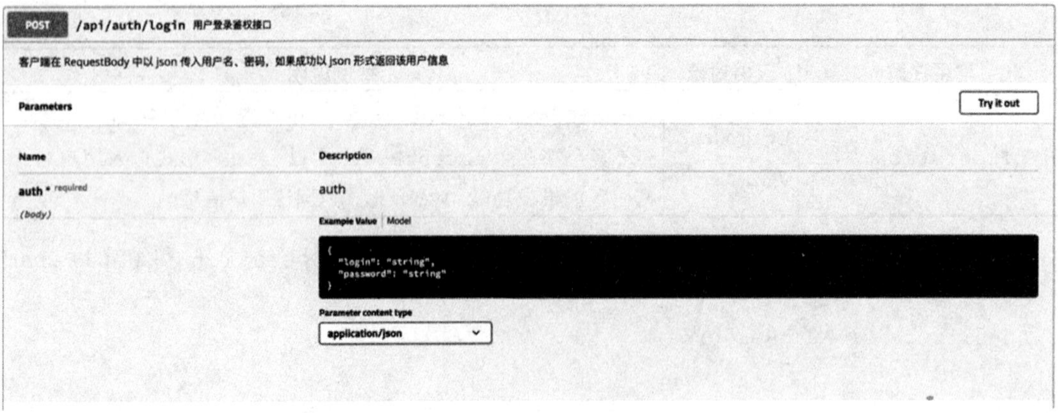

图 5-22

而图 5-23 展现了`@ApiParam`的效果，就是对 Controller 方法参数的描述。

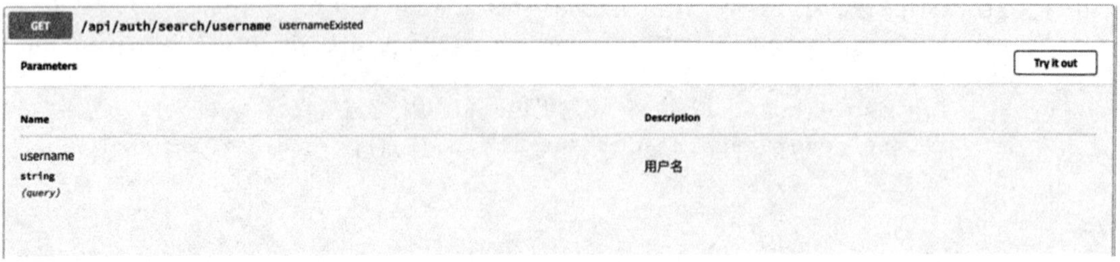

图 5-23

如果使用 `@ApiModelProperty` 对领域对象进行描述，那我们可以试一下把 `Auth` 改造成下面的样子。

```
@Data
public class Auth {
    @ApiModelProperty(value = "登录名")
    private String login;
    @ApiModelProperty(value = "密码")
    private String password;
}
```

我们在文档的 `Parameters` 区块下的 `Description` 中，点击 `Example Value` 右边的 `Model`，就可以看到效果了，是对于领域对象属性的说明，如图 5-24 所示。

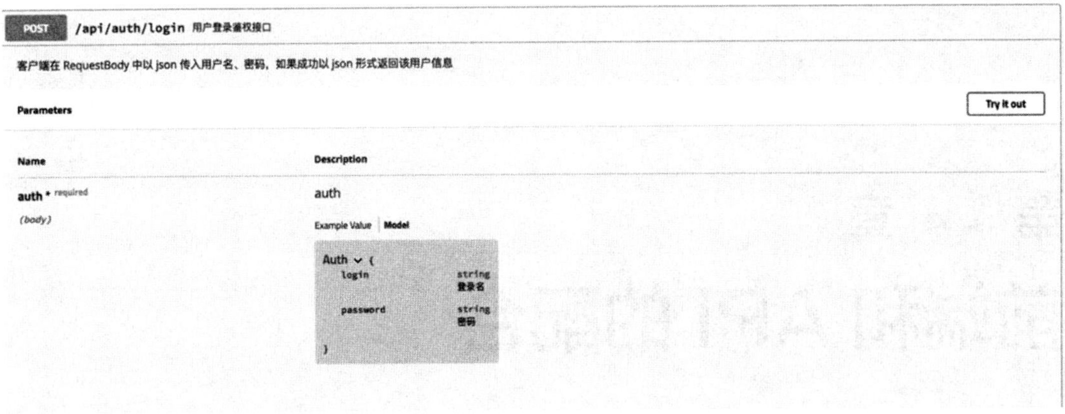

图 5-24

API 的文档注解功能还是很强大的，但在实际的使用中笔者不是很推荐每个 API 都用注解做详细说明，因为这样会导致代码的注解部分太长，可读性下降。一般我们只对关键 API 或容易产生误解的部分进行注解就可以了。

第 6 章 前端和 API 的配合

前面的章节中，我们是前端和后端分开介绍的，但在实际应用中，这两者是要配合起来的。后端作为内容资源的提供者，而前端作为这些资源的消费者，将资源以需要的形式展现给使用者，也就是我们的最终用户。

6.1 响应式的 HTTP API 处理

后端的 API 既然已经搭好了，那么前端就需要调用了。但这时问题往往也伴随而来，由于 HTTP 访问一般要使用异步方式，而采用异步方式之后，很多时候会产生逻辑错误或者回调地狱。这时也是体验 RxJS 优越性的时候了。

6.1.1 Angular 中的 HTTP 服务

要使用 Angular 中的 HTTP 服务，一般有两个选择，如果在 Angular 4.× 时可以使用传统的位于@angular/http 中的 HttpModule。但在 Angular 4.3 之后，官方增加了位于@angular/common/http 中的 HttpClientModule，并且推荐使用 HttpClient。我们这里采用的是 HttpClientModule。注意，之所以把这个模块放在 CoreModule 中导入，是因为 HttpClient 是以依赖形式提供的，在同一个应用没有必要初始化多个 HttpClient。

```
// 省略其他导入
import { HttpClientModule } from '@angular/common/http'; // <-- 这里

@NgModule({
  declarations: [
    // 省略组件声明
```

```
  ],
  imports: [
    SharedModule,
    HttpClientModule, // <--- 这里
    AuthModule,
    AppRoutingModule,
    BrowserAnimationsModule
  ]
})
export class CoreModule {
  // 省略
}
```

还记得我们的后端的 API 地址吗？是 http://localhost:8080/api/×××，如果直接这么使用，没有后端代码怎么办？以后正式部署了怎么办？种种问题都会接踵而来。这个其实和第 5 章在后端要解决不同环境下使用不同数据库的问题类似，都是一个开发环境配置问题。

6.1.2 Angular 的开发环境配置

在使用 Angular CLI 生成工程时，如果你留意就会发现在 `src` 下面有一个 `environments` 子目录，如图 6-1 所示。

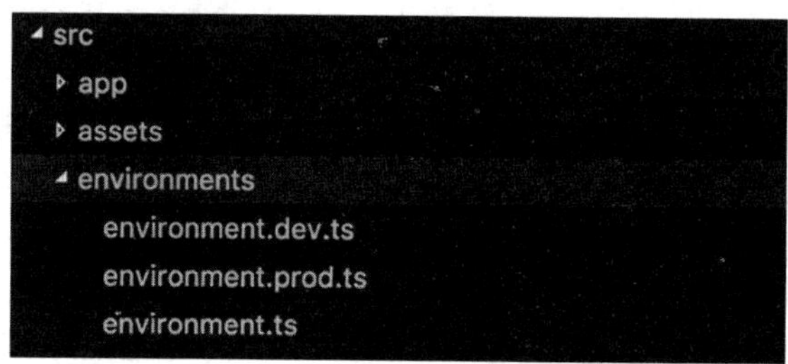

图 6-1

`environment.ts` 是默认的环境配置，如果在没有指定使用哪些配置时，Angular 会默认取这个文件中的值。

```
export const environment = {
  production: false,
  apiBaseUrl: 'http://localhost:8080/api/'
};
```

在上面的环境配置中，我们指定了一个环境变量 `apiBaseUrl`，现在，如果我们在 `environments/environment.prod.ts` 中设置一个不同的 `apiBaseUrl`，那应用中使用这个环境变量时会发生什么：

```
export const environment = {
  production: true,
  apiBaseUrl: 'http://local.dev/api/'
};
```

1. 使用环境变量

为方便起见，我们直接在 CoreModule 的构造函数中打印这个变量。

```
// 省略
import { environment } from '../../environments/environment'; // <--- 这里

@NgModule({
  declarations: [
    // 省略
  ],
  imports: [
    // 省略
  ]
})
export class CoreModule {
  constructor(
    @Optional()
    @SkipSelf()
    parentModule: CoreModule,
    ir: MatIconRegistry,
    ds: DomSanitizer
  ) {
    // 省略其他
    console.log(environment.apiBaseUrl); // <--- 这里
  }
}
```

然后使用 `ng serve` 启动应用，打开 Chrome 的开发者工具，观察控制台，可以看到现在打印出来的是 http://localhost:8080/api/，如图 6-2 所示。

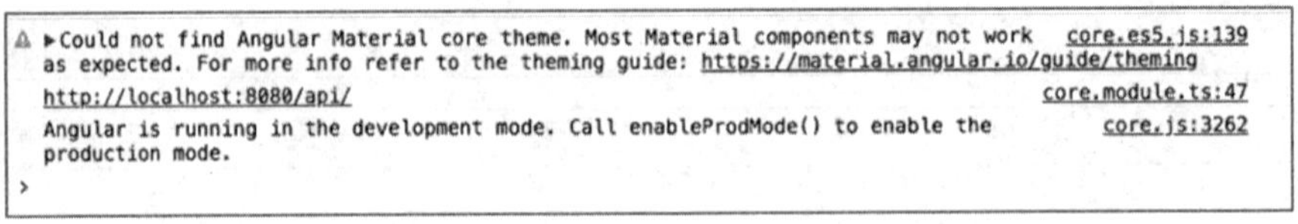

```
⚠ ▶Could not find Angular Material core theme. Most Material components may not work   core.es5.js:139
  as expected. For more info refer to the theming guide: https://material.angular.io/guide/theming
  http://localhost:8080/api/                                                           core.module.ts:47
  Angular is running in the development mode. Call enableProdMode() to enable the       core.js:3262
  production mode.
>
```

图 6-2

接下来使用 `ng serve -c production` 或者 `ng serve --prod`，可以看到生产环境

下的输出是 http://local.dev/api/，如图 6-3 所示。

```
http://local.dev/api/                                          main.26e8666….js:1
>
```

图 6-3

命令中的 production 是怎么回事？看一下项目根目录的 angular.json，就会发现 build 和 serve 中都有 configuration，里面定义了 production 这个配置。所以如果想有更多配置，除要新建对应的 environment.×××.ts 外，还需要在这个 angular.json 中也定义一下配置的名称。而 ng serve --prod 其实是 Angular CLI 在生产模式时自动切换到了 production 这个配置。

```
{
  "$schema": "./node_modules/@angular-devkit/core/src/workspace/workspace-schema.json",
  "version": 1,
  "newProjectRoot": "projects",
  "projects": {
    "frontend": {
      "root": "",
      "projectType": "application",
      "prefix": "app",
      "architect": {
        "build": {
          // 省略其他设置
          "configurations": {
            "production": {
              "fileReplacements": [
                {
                  "replace": "src/environments/environment.ts",
                  "with": "src/environments/environment.prod.ts"
                }
              ],
              "optimization": true,
              "outputHashing": "all",
              "sourceMap": false,
              "extractCss": true,
              "namedChunks": false,
              "aot": true,
              "extractLicenses": true,
              "vendorChunk": false,
              "buildOptimizer": true
            }
```

```
          }
        },
        "serve": {
          "builder": "@angular-devkit/build-angular:dev-server",
          "options": {
            "browserTarget": "frontend:build"
          },
          "configurations": {
            "production": {
              "browserTarget": "frontend:build:production"
            }
          }
        },
        // 省略其他
      }
    },
    // 省略其他部分
  },
  "schematics": {
    "@schematics/angular:component": {
      "styleext": "scss"
    }
  }
}
```

2. 让容器支持环境配置

其实在第 2 章，我们给出前端的容器镜像文件时已经支持了环境配置，在 RUN npm run build --prod --configuration $env 中已经指定了环境配置。

```
#### STAGE 1: Build

## We label our stage as 'builder'
FROM node:8-alpine as builder

### 省略其他部分

### Build the angular app in production mode and store the artifacts in
dist folder
ARG env=production
RUN npm run build --prod --configuration $env

### 省略其他部分
```

我们在 Dockerfile 中定义了一个参数 env，而这个配置名称可以通过参数形式传入

Docker 制作过程，再回顾我们的 docker-compose.yml，我们在构建镜像时使用 args 传入了 env

```
version: '3'
services:
  nginx:
    build:
      context: .
      dockerfile: ./docker/nginx/Dockerfile
      args:
        - env=production # <--- 这里以参数形式传入
    container_name: nginx
    ports:
      - 80:80
```

6.1.3 在前端服务中使用 HttpClient

现在我们可以更改 frontend/src/app/auth/services/auth.service.ts 为下面的代码。代码中可以看到 Angular 消费 Rest API 是非常简单的，因为 Angular 中将 HttpClient 的方法也分成了 GET、PUT、PATCH、POST 和 DELETE，对应不同的 Rest API 操作，我们使用不同的方法。我们可以通过在构造函数中参数注入的方式得到 HttpClient 的实例 constructor(private http: HttpClient)。

```
// 省略导入

@Injectable()
export class AuthService {
  private headers = new HttpHeaders().append(
    'Content-Type',
    'application/json'
  );
  constructor(private http: HttpClient) {}
  /**
   * 用于用户的登录鉴权
   * @param auth 用户的登录信息，一般是登录名（目前是用户名，以后会允许手机号）和密码
   */
  login(auth: Auth): Observable<string> {
    return this.http
      .post<{ id_token: string }>(
        `${environment.apiBaseUrl}auth/login`,
        JSON.stringify(auth),
        { headers: this.headers }
      )
```

```
      .pipe(pluck('id_token'));
  }
  /**
   * 用于用户的注册
   * @param user 用户注册信息
   */
  register(user: User): Observable<string> {
    return this.http
      .post<{ id_token: string }>(
        `${environment.apiBaseUrl}auth/register`,
        JSON.stringify(user),
        { headers: this.headers }
      )
      .pipe(pluck('id_token'));
  }
  /**
   * 请求发送短信验证码到待验证手机，成功返回空对象 {}
   * @param mobile 待验证的手机号
   */
  requestSmsCode(mobile: string, token: string): Observable<void> {
    const params = new HttpParams()
      .append('mobile', mobile)
      .append('token', token);
    return this.http.get<void>(`${environment.apiBaseUrl}auth/mobile`, {
      headers: this.headers,
      params: params
    });
  }
  /**
   * 验证手机号和短信验证码是否匹配
   * @param mobile 待验证手机号
   * @param code 收到的短信验证码
   */
  verifySmsCode(mobile: string, code: string): Observable<void> {
    return this.http.post<void>(
      `${environment.apiBaseUrl}auth/mobile`,
      JSON.stringify({ mobile: mobile, code: code }),
      { headers: this.headers }
    );
  }
  /**
   * 请求发送短信验证码到待验证手机，成功返回空对象 {}
   * @param mobile 待验证的手机号
   */
```

```
    requestCaptcha() {
      return
this.http.get<Captcha>(`${environment.apiBaseUrl}auth/captcha`, {
        headers: this.headers
      });
    }
    /**
     * 验证手机号和短信验证码是否匹配
     * @param mobile 待验证手机号
     * @param code 收到的短信验证码
     */
    verifyCaptcha(token: string, code: string) {
      return this.http.post<{ validate_token: string }>(
        `${environment.apiBaseUrl}auth/captcha`,
        JSON.stringify({ captcha_token: token, captcha_code: code }),
        {
          headers: this.headers
        }
      );
    }
    /**
     * 检查用户名是否唯一
     * @param username 用户名
     */
    usernameExisted(username: string) {
      const params = new HttpParams().append('username', username);
      return this.http.get<{ existed: boolean }>(
        `${environment.apiBaseUrl}auth/search/username`,
        {
          headers: this.headers,
          params: params
        }
      );
    }
    /**
     * 检查电子邮件是否唯一
     * @param email 电子邮件
     */
    emailExisted(email: string) {
      const params = new HttpParams().append('email', email);
      return this.http.get<{ existed: boolean }>(
        `${environment.apiBaseUrl}auth/search/email`,
        {
          headers: this.headers,
```

```
          params: params
        }
      );
    }
    /**
     * 检查手机号是否唯一
     * @param mobile 手机号
     */
    mobileExisted(mobile: string) {
      const params = new HttpParams().append('mobile', mobile);
      return this.http.get<{ existed: boolean }>(
        `${environment.apiBaseUrl}auth/search/mobile`,
        {
          headers: this.headers,
          params: params
        }
      );
    }
  }
```

HttpClient 都支持那些操作，有哪些方法？最好的方法是看看它的类定义，下面的代码列出了它的主要方法。可以看到对于常见的 Rest API 操作来说，DELETE、GET 方法是没有 body 参数的，而 patch、put、post 是有的，这个 body 就是 Http Request Body，对于 GET 和 DELETE 操作，都不会携带诸如 JSON、XML 等对象，而是直接在 URL 上体现参数。

```
  class HttpClient {
    constructor(handler: HttpHandler)
    request(first: string | HttpRequest<any>, url?: string, options:
{...}): Observable<any>
    delete(url: string, options: {...}): Observable<any>
    get(url: string, options: {...}): Observable<any>
    head(url: string, options: {...}): Observable<any>
    jsonp<T>(url: string, callbackParam: string): Observable<T>
    options(url: string, options: {...}): Observable<any>
    patch(url: string, body: any | null, options: {...}): Observable<any>
    post(url: string, body: any | null, options: {...}): Observable<any>
    put(url: string, body: any | null, options: {...}): Observable<any>
  }
```

有的读者观察到了，很多方法带了一个 options 的参数，这个参数是用来指定一些请求的特殊配置，如果需要写入特定的 Header，那么需要使用 HttpHeaders，如果需要使用参数（也就是 /×××?a=value1&b=value2），就使用 HttpParams。那么这个 options 的对象结构如下所示，这就可以理解我们上面代码中的诸如 { headers: this.headers, params:

params } 的参数形式。

```
{
    body?: any;
    headers?: HttpHeaders | {
        [header: string]: string | string[];
    };
    observe?: HttpObserve;
    params?: HttpParams | {
        [param: string]: string | string[];
    };
    reportProgress?: boolean;
    responseType?: 'arraybuffer' | 'blob' | 'json' | 'text';
    withCredentials?: boolean;
}
```

6.1.4 更改注册表单控件

我们在之前的表单提交函数中直接把 form.value 传到了父组件。这个有点问题，因为这个组件中的 passwords 是一个 FormGroup，里面包含 password 和 repeat。所以这个结构如果传到后端是会出错的，因为后端认为没有 password，我们来改造 submit 函数，构造 User。

```
// 省略导入
export class RegisterFormComponent implements OnInit {
  // 省略
  submit({ valid, value }: FormGroup, ev: Event) {
    if (!valid) {
      return;
    }
    const user: User = {
      login: value.login,
      password: value.passwords.password,
      email: value.email,
      name: value.name,
      mobile: value.mobile,
      avatar: value.avatar
    };
    this.submitEvent.emit(user);
  }
}
```

6.2 RxJS 进阶

6.2.1 改造登录表单

我们这里给登录表单增加一个图形验证码功能，当然后端已经集成了这个功能。这个功能对于“笨组件”来说非常简单，模板上增加一个 <input>和<img>控件，注意这个控件此时并没有作为表单控件出现（没有设置 formControlName），因为现在鉴权接口中没有图形验证码的参数。那可能有的读者会问，这个验证如果和鉴权无关，那么不就可以直接绕过去了？比如使用 Postman 直接访问鉴权接口，就不用验证码了啊。对的，这个地方是有问题的，但是我们放到后面处理，现在先实现验证码的显示和检验。

```
<mat-card>
  <!--省略头部-->
  <mat-card-content>
    <form fxLayout="column" fxLayoutAlign="stretch center"
[formGroup]="form" (ngSubmit)="submit(form, $event)">
      <!--省略其他 field -->
      <mat-form-field class="full-width">
        <input matInput placeholder="图形验证码"
(input)="verifyCaptcha($event.target.value)">
        <img matSuffix [src]="captchaUrl" (click)="processClick()">
      </mat-form-field>
      <!--省略按钮 -->
    </form>
  </mat-card-content>
  <!--省略 Action -->
</mat-card>
```

在组件文件中，我们简单地把图片点击事件和输入框的输入事件发射出去，并提供一个图片地址的输入型属性。

```
// 省略
export class LoginFormComponent implements OnInit {
  // 省略
  @Input() captchaUrl = '';
  @Output() refreshCaptcha = new EventEmitter<void>();
  @Output() codeInput = new EventEmitter<string>();
  // 省略
  processClick() {
    this.refreshCaptcha.emit();
  }
  verifyCaptcha(code: string) {
    this.codeInput.emit(code);
```

```
    }
}
```

在“聪明组件”（LoginComponent）中，模板文件要设置上“笨组件”的输入和输出属性。

```
<div fxLayout="row" fxLayout.xs="column" fxLayoutAlign="center space-around">
  <app-login-form
    (submitEvent)="processLogin($event)"
    (refreshCaptcha)="refreshCaptcha()"
    (codeInput)="verifyCaptcha($event)"
    [captchaUrl]="(captcha$ | async)?.captcha_url">
  </app-login-form>
</div>
```

6.2.2 RxJS 的高阶操作符

LoginComponent 中涉及图片验证码的逻辑中有以下几个流。

- 点击图片的点击事件流 clickSub。
- 输入框的输入事件流 verifySub。
- 请求验证码图片的 API 事件流 this.authService.requestCaptcha()。
- 验证输入的验证码的 API 事件流 this.authService.verifyCaptcha(token, code)。

这几个流并不是孤立存在的，而是有一定的关联关系。

（1）clickSub 触发之后才引发 this.authService.requestCaptcha()。

（2）verifySub 触发之后才引发 this.authService.verifyCaptcha(token, code)。

这里面体现了两层意思，第一个是时间顺序，第二个是流中嵌套了流。先来看组件代码：

```
// 省略元数据和导入声明
export class LoginComponent implements OnDestroy, OnInit {
  // 省略 quote$ 声明
  captcha$: Observable<Captcha>;
  clickSub = new Subject();
  captchaSub = new BehaviorSubject<Captcha>(null);
  verifySub = new Subject<string>();
  sub = new Subscription();
  // 省略构造函数
  ngOnInit(): void {
```

```
    this.captcha$ = this.clickSub.pipe(
      startWith({}),
      switchMap(_ => this.authService.requestCaptcha()),
      tap(captcha => {
        this.captchaSub.next(captcha);
      })
    );

    this.sub = this.verifySub
      .pipe(
        withLatestFrom(this.captchaSub),
        switchMap(([code, captcha]) =>
          this.authService
            .verifyCaptcha(captcha.captcha_token, code)
            .pipe(
              map(res => res.validate_token),
              catchError(err => of(err.error.title))
            )
        )
      )
      .subscribe(t => console.log(t));
  }
  ngOnDestroy() {
    if (this.sub) {
      this.sub.unsubscribe();
    }
  }
  processLogin(auth: Auth) {
    this.authService
      .login(auth)
      .pipe(take(1))
      .subscribe(u => console.log(u));
  }
  refreshCaptcha() {
    this.clickSub.next();
  }
  verifyCaptcha(code: string) {
    this.verifySub.next(code);
  }
}
```

在上面代码中，switchMap 是一个高阶操作符，很多读者看到“高阶”这个词就有点懵，简单类比一下，X+2=4 这是 X 的一次方程，那么 X * X +2=4 这就是 X 的二次方程，也可以说是 X 的高阶方程。所以类似地，Observable<Type> 这是“一次”的 Observable，那么 Observable<Observable<Type>> 就是高阶的 Observable。

拿上面的例子来说，this.clickSub 得到的是 void 的数据流，因为 next()时没有传递数据，这会发送一个 void，也就是我们只需知道事件发生了，不需要关心具体的值。如果我们不采用 switchMap，那么可以写成 this.clickSub.pipe(map(_ => this.authService.requestCaptcha()))，但是这样转换之后我们得到流中的值就由 void 变成了 Observable<Captcha>，看，我们在流中又发现了流。那么设想一下，如果这个时候我们订阅这个流：

```
this.clickSub.pipe(map(_ => this.authService.requestCaptcha()))
.subscribe(val => console.log(val))
```

大家可以自行实验一下，输出的应该是一个 Observable 对象。但如果想要得到的是那个 Captcha 怎么办？由于 val 是 Observable<Captcha>，所以我们就再订阅一下：

```
this.clickSub.pipe(map(_ => this.authService.requestCaptcha()))
.subscribe(val => {
  val.subscribe(captcha => console.log(captcha));
})
```

大家再实验一下，问题解决了，但，这个解决方案看起来总是有点别扭。

首先，需要知道一点，subscribe 意味着流的终结。一旦 subscribe 之后，就无法再利用各种操作符了。

```
his.clickSub.pipe(map(_ => this.authService.requestCaptcha()))
.subscribe(val => {
  val.subscribe(captcha => console.log(captcha))
  .filter(captcha => captcha !== null); // <--- 不正确，无法再应用操作符
})
```

其次，如果采用上面的方式，将无法在希望的时候取消内部订阅，异常处理也是非常麻烦。

最后，你将陷入一个嵌套地狱，这正是 RxJS 要解决的问题，但你这么使用其实就又回到了老路上。

请记住以下几个原则。

- 永远不要在 subscribe 中嵌套 subscribe，高阶操作符就是处理这类需求的。
- 尽可能少使用 subscribe，尽量合并流，或者利用 async 管道。
- 如果使用 subscribe，一定记得要在适当的时候取消订阅 unsubscribe，否则会内存泄漏。

高阶操作符的作用就是将高阶转为低阶，起到流的“拍扁”的作用，很多读者初学时不知

道什么时候该使用高阶操作符，有一个小技巧，就可以先写成低阶，然后看操作符内返回的是流还是你要的数据类型，如果是流就使用高阶操作符。

1. mergeAll/mergeMap/switchMap

（1）mergeAll

还记得我们前面提到的嵌套 subscribe 的例子吗？我们首先使用普通的 map 操作符得到一个高阶流。

```
// 高阶流
this.captcha$ = this.clickSub.pipe(
  map(_ => this.authService.requestCaptcha()),
);
// 用 mergeAll “拍扁”这个高阶流
this.captcha$.mergeAll().subscribe(captcha => console.log(captcha));
```

mergeAll 其实做的事情是订阅内层流（this.authService.requestCaptcha ()），然后把内层的值发送到外层。

（2）mergeMap

这种操作在 RxJS 中太频繁了，所以有了一个快捷方式——mergeMap，你不用分两步处理，可以一步到位。

```
this.captcha$ = this.clickSub.pipe(
  mergeMap(_ => this.authService.requestCaptcha()),
);
```

所以， mergeMap 相当于 Map + mergeAll。

这种方式有个问题，就是对于每个外层 Observable 的值都会产生内层流，用我们具体的例子来看，就是每次点击图片（外层流）都会去发送 API 请求（内层流）。这个如果是你想要的，那么就用 mergeMap。

但通常情况下，如果用户快速点击图片，则会产生多个 HTTP 的 API 请求，但我们其实只关心最新的这次，之前的不需要。这就需要请出 switchMap 了。

（3）switchMap

switchMap 究竟起到什么作用呢？首先，它为什么有个 switch 的前缀，就是因为要切换的意思，它代表如果外层 Observable 有值发出，就立刻取消之前的订阅，重新开始一个新的订阅。我们只维护一个内层流的订阅，就是最新的这个外层 Observable 对应的内层流。

这样解释还是太抽象了，我们来看上面的例子中的一段代码，如果 this.clickSub 有新值，也就是用户又点了一下，那么我们就取消之前的内层流订阅，只关心这次的点击产生的内层流 this.authService.requestCaptcha()。

```
this.captcha$ = this.clickSub.pipe(
  // 忽略其他操作符
  switchMap(_ => this.authService.requestCaptcha()),
  // 忽略其他操作符
);
```

2. Subject 以及事件流的“冷”和“热”

重点看一下组件的改造，我们之前都是 from 或 fromEvent 由现存对象或事件构建一个流，现在尝试一个新的方式，使用 Subject。

我们之前讲过 Subject，它既是 Observer 又是 Observable。它是 Observer，所以有 next(v)、error(e) 和 complete()。它又是 Observable，所以适用于 Observable 的操作符都适用于它，也可以成为 subscribe()。

另一个明显的区别是，Subject 构成的流和普通 Observable 不一样，怎么个不一样法呢？看下面的例子的代码，我们声明了 clickSub，在图片点击事件的处理函数 refreshCaptcha 中，发射了一个事件 this.clickSub.next()。那么无论订阅与否，这个流中都有数据，对吗？因为事件触发了，这个流就有值发射出来。但如果我们使用类似 fromEvent(imageElement, 'click') 这种形式构建 Observable，那这个 Observable 在没有任何订阅的时候是没有值的。

这就引出了一个流的“冷”和“热”的概念，类似 Subject 这种流就是“热”的流，而普通的 Observable 是“冷”的流。举一个更形象的例子来说明“冷”和“热”的区别：“冷”的流就好像我们去视频网站看电影，你和我分别在两台电脑上点进去，我是下午 2:00 进去看的，你是下午 2:30 进去看的，但我们都会从头看起，而不是你点进来一看已经进行了半个小时了。而“热”的流就像我们分别在各自家中看球赛的直播，那么我是 2:00 打开电视的，你是 2:30 打开的，你就只能错过前半个小时了，我们看到的内容是一样的。

再举一个小例子，下面的两个订阅，你认为应该分别打印出什么呢？对的，都是 1,2,3,4。

```
const data$ = Observable.from([1,2,3,4]);
const subscriptionA = data$.subscribe(d => console.log(d));
const subscriptionB = data$.subscribe(d => console.log(d));
```

如果是下面的代码呢？ subscriptionA 和 subscriptionB 这两个订阅会打印什么？

是的，subscriptionA 的输出是 1,2,3,4，而 subscriptionB 的输出是 3，4。subscriptionB 由于订阅的时间点在 data$.next(2); 之后，所以完美地错过了前两个值。

```
const data$ = new Subject<number>();
const subscriptionA = data$.subscribe(d => console.log(d));
data$.next(1);
data$.next(2);
const subscriptionB = data$.subscribe(d => console.log(d));
data$.next(3);
data$.next(4);
```

那么，究竟是使用“冷”的流还是“热”的流呢？答案是依据你的需求和场景，各有各的用途。当然很多时候，其实它们可能都可以解决你的问题，但要注意区别，在遇到结果和你的期望不符时，请仔细想想是否忽略了流的性质。

在 LoginComponent 中，我们还使用了一个不同的 Subject——BehaviorSubject（captchaSub = new BehaviorSubject<Captcha>(null);），它的不同之处在于它可以记住最近的一个值，这个太有用了。我们来看一下 captcha 获取和验证的逻辑。

（1）发送 GET 请求到 /api/auth/captcha。

（2）后端 API 返回 { captcha_token: 'blabalbla',captcha_url: 'http://×××.yyy.zzz/someimage.jpg' }。

（3）用户输入图片中的代码。

（4）发送 POST 请求到 /api/auth/captcha，但需要携带一个 JSON 数据 { captcha_token: 'blabalbla', captcha_code: '用户输入值'}。

如果仔细看一下第（4）步，我们需要得到在第（2）步中得到的服务器返回值，也就是 this.authService.requestCaptcha()这个流的发送的最近的值。但是由于这个流是一个“冷”的流，如果我们使用 withLatestFrom 操作符，这个流就会再执行一遍，我们得到的就不是上次的值，而变成了一个新的图片的 token，这可就麻烦了。

但是没关系，我们请出 BehaviorSubject 定。在得到验证码之后将其值用 BehaviorSubject 发出（this.captchaSub.next(captcha)），然后合并。

```
this.captcha$ = this.clickSub.pipe(
  startWith({}),
  switchMap(_ => this.authService.requestCaptcha()),
  tap(captcha => {
    this.captchaSub.next(captcha); // <--- 这里发射
  })
);
```

```
  this.sub = this.verifySub.pipe(
    withLatestFrom(this.captchaSub), // <--- 这里使用最近的值
    switchMap(([code, captcha]) =>
      this.authService
        .verifyCaptcha(captcha.captcha_token, code)
        .pipe(
          map(res => res.validate_token),
          catchError(err => of(err.error.title))
        )
    )
  )
  .subscribe(t => console.log(t));
```

6.2.3 合并操作符

我们之前其实已经用上合并类操作符了，这里就进行一下集中的介绍。RxJS 提供的操作符太多了，我们不会逐一介绍，而是把最常用的几种做一个讨论。

1. merge

merge 是最简单的一种合并，就是把两个流按各自的顺序合并成一个。

```
const dataA$ = interval(1000).pipe(take(5), mapTo('A'));
const dataB$ = interval(2000).pipe(take(5), mapTo('B'));
const merge$ = merge(dataA$, dataB$);
merge$.subscribe(val => console.log(val));
```

输出结果应该是，注意 AB 不是一起输出的，而是时间间隔非常小，几乎同时，但流是不可能一下给出两个值的，所以有先后顺序，A 在 B 前仅仅作为示意，实际上有可能 B 在 A 前。

```
源流 A: •---A---A---A---A---A
源流 B: -------B-------B-------B-------B-------B
合并流:----A--AB--A---AB--A---B-------B-------B
```

2. concat 和 startWith

concat 是有顺序地把一个流排在另一个流的末尾，也就是严格按顺序输出，即使排在后面的流实际的速度较快，也要等到前面的流完成后才能开始。这和 merge 不同，merge 是不保证顺序的。

```
const dataA$ = of(4, 5, 6);
const dataB$ = of(1, 2, 3);
const concat$ = dataA$.pipe(concat(dataB$));
concat$.subscribe(val => console.log(val));
```

上面代码输出的结果是 4,5,6,1,2,3，对于 concat，一定要注意的是，不要把 Observable 放在一个无尽流之后。什么是无尽流？就是永远不结束，比如 fromEvent、interval(1000) 就是无尽流。如果把一个流 concat 放到一个无尽流后就会导致根本等不到那个流的数据了。

startWith 的作用和 concat 相反，是在流发射前加一个值，也就是先发射一个初始值。这个操作符很有用，因为有时候，我们构造了一个流，但在没有事件触发时，我们也希望有初始值，或者需要一个初始值才能激活这个流。

3．combineLatest/zip/withLatestFrom

为什么这几个操作符要放在一起讲？因为它们很相似，但又有差别，这个差别有时很微妙，所以需要放在一起比较。

（1）combineLatest

combineLatest 的作用是，只要参与合并的任何一个流发射一个值，就把所有参与合并的流的最新值都发出来。但是需要注意的一点是，这个合并流是要等参与合并的每个流都发射过值之后才会有值。

这个操作符的最佳使用场景是几个流彼此依赖去进行某种计算或操作，也就是说每个流的变化都会影响计算结果，比如下面的 BMI 计算器中 BMI 的值是由身高和体重两个值计算而得，任何一个值的改变都会影响计算结果（见图 6-4）。

```
const weight$ = fromEvent(weight, 'input').pluck('target', 'value');
const height$ = fromEvent(height, 'input')
  .pluck('target', 'value');

const bmi$ = combineLatest(weight$, height$, (w, h) => {
  return w/(h/100*h/100);
});

bmi$.subscribe(val => {
  bmi.innerText = Math.round(val * 100) / 100;
});
```

图 6-4

我们来简单模拟一下各个流的情况。

- 首先，身高处输入 170，BMI 值不会显示，因为此时体重数据流还没有值发出。记住 combineLatest 需要每个参与流都至少有一个值发射出来才会有值。
- 然后输入体重 70，这时身高流的最新值是 170，所以（170，70）作为合并后的数据参与计算。
- 我们又把身高改成了 180，这时体重的最新值是 70，所以（180，70）参与计算。
- 体重被改成 75 后，身高的最新值为 180，那么（180，75）参与计算。

```
身高：----170-----------180------------
体重：---------70------------------75--
BMI：------(170,70)---(180,70)---(180,75)
```

（2）zip

zip 和 combineLatest 非常像，但它要求任何一个参与合并的流发出数据后，等待其他流发出和它位置一致的数据。换句话说，如果流 A 和 B 参与合并，那么流 A 发出第一个元素 A1，合并流会等待 B 发出第一个元素 B1。如果发出，合并流就会有（A1，B1）；如果 B 发出第二个元素 B2，合并流会等待 A 发出第一个元素 A2，如果发出，合并流就会有（A2，B2）。

```
const dataA$ = interval(1000);
const dataB$ = interval(1000).pipe(take(2));
const zip$ = zip(dataA$, );
//输出：[0,0]...[1,1]
const subscribe = zip$.subscribe(val => console.log(val));
```

上面代码的输出只有 [0,0] 和 [1,1]，尽管 dataA$ 是一个无尽序列，一直有值输出，但合并流并不会，因为它始终等不到 dataB$ 的第三个元素。

大家如果了解 zip 这个单词有拉链的意思，就会有助于你记住这个操作符的含义，一个齿对应一个齿，有对齐的效果。而且 zip 会等待合并流中最慢的那个，所以有时也会起到延时的作用。

（3）withLatestFrom

withLatestFrom 就是要在一个流发出最新值时，得到另一个流的最新值。这个听上去也有点像 combineLatest，但是 withLatestFrom 有“以我为主”的意思。举个例子，只要输入框的输入事件发生了，那么此时去合并 this.captchaSub；如果 this.captchaSub 发射了新值，则我们并不在意。

```
this.verifySub.pipe(
  withLatestFrom(this.captchaSub), // <--- 这里使用最近的值
  switchMap(([code, captcha]) =>
```

```
      this.authService
        .verifyCaptcha(captcha.captcha_token, code)
        .pipe(
          map(res => res.validate_token),
          catchError(err => of(err.error.title))
        )
    )
  )
```

6.3 HTTP 拦截

这几章来回在 Angular 和 Spring 之间切换，是否也感觉到 Angular 和 Spring 之间有很多共享的概念，比如依赖注入、比如注解。HTTP 拦截这个概念对于 Angular 也是存在的。前端是要发起 HTTP 请求到后端的，那么如果我们需要，比如统一在前端发起的请求中加入某些指定的 Header，这个当然可以在每个服务中实现。但是不是有点烦琐啊，而且也不利于代码的重构，因为这些加 Header 的地方分散在程序的各个地方。当然，HttpInterceptor（HTTP 拦截器）能做的事情不止是这些，我们也可以截获返回的 Response，根据不同的返回状态做特定的处理。

6.3.1 实现一个简单的 HttpInterceptor

Angular 提供了 `HttpInterceptor` 接口，用于拦截 `HttpRequest` 并处理。`HttpInterceptor` 提供了 `intercept()` 方法，它通常在调用下一个拦截器之前拦截传出请求。`intercept()` 方法有两个参数 `HttpRequest` 和 `HttpHandler`。

下面我们就来实现一个拦截，当客户端发送 API 请求到服务器时，如果这个 API 的路径传递有误，那么后端会返回 404，我们这个截断器的作用就是拦截到请求后，如果返回的错误码是 404，就导航到前端的 404 路由。

```
// 省略导入

@Injectable()
export class NotFoundInterceptor implements HttpInterceptor {
  constructor(private _injector: Injector) {}

  intercept(
    req: HttpRequest<any>,
    next: HttpHandler
  ): Observable<HttpEvent<any>> {
    return next.handle(req).pipe(
      tap(
```

```
          event => {},
          err => {
            if (err.status === 404) {
              const router = this._injector.get(Router);
              router.navigate(['/404']);
            }
          }
        )
      );
    }
  }
```

很常见的一个场景是当用户浏览某个不存在的链接时，系统会提供一个 404 页面。传统的 Web 开发中，这一块后端会统一处理。但在前后端分离的场景中，Angular 生成的是一个单页面应用（最终只有一个 index.html），所谓的路由其实只是用来装载不同的组件，浏览器地址栏的那个路由链接其实只是显示而已。所以在我们的场景中，只需要建立一个 404 的组件，在恰当的时候显示它就好了。

6.3.2 鉴权 HttpInterceptor

在实际的开发中，通常后台的 Rest API 需要基于 token 机制的鉴权，这种鉴权在项目中是在 HttpRequest 的头中写鉴权信息，如 Authorization: Bearer <jwtToken>。当然可以利用 Angular 的 HttpHeader 对象进行写入，但如果 API 较多，系统较复杂，这样做的重复性劳动就太多了。

这是一个比较典型的使用 HttpInterceptor 来简化这个流程的场景，我们拦截请求，并将鉴权头设置进去，这样在业务模块中就无须考虑鉴权的问题了，大大简化了业务模块的逻辑。

```
  // 省略导入

  @Injectable()
  export class AuthHeaderInterceptor implements HttpInterceptor {
    constructor(private store: Store<fromAuth.State>, private authService:
AuthService) {}
    intercept(req: HttpRequest<any>, next: HttpHandler) {
      if (!req.url.includes('/api/') || req.url.includes('/api/auth')) {
        return next.handle(req);
      }
      if (isRefreshTokenExpired()) {
        this.store.dispatch(new Logout());
        return next.handle(req);
      }
      if (isIdTokenExpired()) {
```

```
        return this.authService.refreshTokens(getRefreshToken()).pipe(
          tap(pair => this.store.dispatch(new RefreshTokenSuccess(pair))),
          map(pair => pair.id_token),
          switchMap(token => {
            return next.handle(req.clone({ setHeaders: { Authorization:
`Bearer ${token}` } }));
          })
        );
      }
      const id_token = getIdToken();
      const reqClone = req.clone({ setHeaders: { Authorization: `Bearer
${id_token}` } });
      return next.handle(reqClone);
    }
  }
```

创建 HttpInterceptor 之后，我们需要在要使用的模块中（如果是全局的 HttpInterceptor，一般放在 AppModule 或 CoreModule 中）的 `providers` 数组中列出来。这样系统就会自动使用注册的 HttpInterceptor 进行请求的拦截了。

```
  // 省略
  import { AuthFailureInterceptor } from './interceptors/auth-
failure.interceptor';
  // 省略
  @NgModule({
    // 省略
    providers: [
      { provide: HTTP_INTERCEPTORS, useClass: NotFoundInterceptor, multi:
true },
      { provide: HTTP_INTERCEPTORS, useClass: AuthHeaderInterceptor, multi:
true },
      { provide: HTTP_INTERCEPTORS, useClass: AuthFailureInterceptor,
multi: true }
    ]
  })
  export class CoreModule {
    constructor(
      @Optional()
      @SkipSelf()
      parentModule: CoreModule,
      ir: MatIconRegistry,
      ds: DomSanitizer
    ) {
```

```
    // 省略
  }
}
```

6.3.3 一个日志拦截器

在开发中，如果我们可以知道 HTTP 请求的信息、返回的响应是成功还是失败，以及执行了多长时间这些信息，那对于我们在开发中调试问题是很方便的。可能有的读者说 Chrome 的开发者工具已经可以看到这些信息啊，是的，但是这个信息是在你打开开发者工具的情况下才能看到。设想一下，你可能会要求客户或者是测试人员都一直打开开发者工具，遇到问题再把信息发给你吗？而如果我们有一个日志可以记录下发出的请求，就可以在适当时机将日志传回服务器端进行分析了。

```
// 省略导入

@Injectable()
export class LoggingInterceptor implements HttpInterceptor {
  intercept(req: HttpRequest<any>, next: HttpHandler) {
    const startTime = Date.now();
    let status: string;

    return next.handle(req).pipe(
        tap(
          event => {
            status = '';
            if (event instanceof HttpResponse) {
              status = '成功';
            }
          },
          error => status = '失败'
        ),
        finalize(() => {
          const elapsedTime = Date.now() - startTime;
          const message = req.method + " " + req.urlWithParams +" "+
status
          + " : " + elapsedTime + "毫秒";

          this.logDetails(message);
        })
    );
  }
  private logDetails(msg: string) {
    console.log(msg);
```

```
    }
}
```

6.4 Angular 路由

我们前面或多或少地接触了 Angular 的路由，那究竟什么是路由？路由就是用户可以从一个视图“导航”到另一个视图的机制。但是，请注意，这里的导航和浏览器的导航不是一个概念，因为 Angular 中的路由是基于单页应用的，所以不存在真正的链接页面。这个链接的形式其实往往更多意味着以 URL 这种形式定义视图的跳转逻辑。

浏览器一般有以下典型行为。

- 在地址栏中输入 URL，浏览器会导航到指定页面。
- 点击页面链接，浏览器会导航到对应页面。
- 点击浏览器的前进或后退按钮，浏览器会导航前进或后退到历史页面。

Angular 的路由也充分借鉴了这种行为。

- 可以通过 URL 的解析导航到一个前端视图（注意不是页面）。
- 可以传递可选参数给目标视图组件，从而视图组件可以根据参数动态显示某些逻辑。
- 也可以在页面上绑定路由链接，用户点击时就会导航到指定的视图。
- 也支持编程动态跳转，也就是我们除了可以在点击链接的时候跳转，也可以在点击按钮、选择下拉框等情况下，使用代码来去进行视图跳转。这些跳转的行为会记录在浏览器历史中，所以前进和后退按钮也是好用的。

6.4.1 基准锚链接

大多数的路由应用需要在 `<head>` 下添加一个 `<base>` 元素，这个元素指定了一个基准链接。比如如果把 Angular 编译后的文件，也就是 `dist` 中的内容（不包含 `dist` 目录本身），复制到 Web 服务器的发布根目录中（这一点，每个 Web 服务器都不太一样）。也就是说，我们的访问链接应该是 `http://localhost` 这样的类型，那么此时基准链接是 `/`，设定这个基准链接，那可以直接修改 `src/index.html` 为：

```
<!-- 省略其他代码 -->
<head>
  <base href="/">
  <!-- 省略其他代码 -->
</head>
<!-- 省略其他代码 -->
```

也可以通过 Angular CLI 命令的形式自动更新，这种 Angular CLI 的形式更适合于开发环境和生产环境不一样的情况。比如开发环境下 base href 是 /，而生产环境下是放在 Web 服务器的二级目录，比如 myUrl 这个目录下。此时应用--base-href 可以在不改动 src/index.html 的情况下直接改写编译成功后的文件，也就是 dist/index.html。

```
ng build --base-href /myUrl/
```

Angular 会在任何应用中使用的图片、CSS 和 JavaScript 脚本的相对路径之前添加 base href 的值。换句话说，在应用中如果要引用路径，不需要考虑项目的目录层级结构，只需要写相对于基准路径的相对路径即可。比如我们在 assets 中有一个图片 hello.jpg，那么我们如果在模板中需要引用这个图片，则只需要写成 assets/hello.jpg 即可，无论这个模板文件处于项目工程的哪一层目录。

6.4.2 Router 模块的简介

在任何时候，如果需要在模板中使用 routerLink 指令，或者在代码中使用导航，均需要导入 RouterModule。

```
import { RouterModule } from '@angular/router';
```

如果是一个路由模块，还需要在模块的导入数组中区分是根路由还是子路由模块，分别使用 forRoot 或 forChild 进行路由的构造。

```
// 根路由模块
imports: [ RouterModule.forRoot(routes) ],
// 子路由模块
imports: [ RouterModule.forChild(routes) ],
```

1. 路由表

路由表这个名词听起来很“高大上”，但其实就是一个路径的配置数组。下面是一个典型的路由表的例子：

```
const routes: Routes = [
  {
    path: 'projects',
    loadChildren: '../project#ProjectModule',
    pathMatch: 'prefix',
  },
  {
    path: 'admin',
    loadChildren: '../admin#AdminModule',
    pathMatch: 'prefix',
  },
```

```
  {
    path: '',
    redirectTo: '/auth/login',
    pathMatch: 'full'
  },
  {
    path: '**',
    component: PageNotFoundComponent
  }
];
```

path: '' 路径为空指的就是根路径，比如如果本地开发 Web 服务器的根路径是 http://localhost:4200，那么这个 path: '' 就是说当你在浏览器地址栏输入以上地址时，Angular Router 会进行路径的对比，发现匹配的条目，就由匹配到的规则处理。

这个匹配的规则有不同的策略。

- pathMatch: 'full'——当且仅当路径完全匹配时才会应用这个规则。比如上面的例子中的 path: ''，这个路由定义中我们就使用了 full 这种匹配策略，所以它只会匹配 http://localhost:4200 这种路径（端口号不同服务器配置会有不同），但是不会匹配 http://localhost:4200/×××这种路径。
- pathMatch: 'prefix'——当路径是以 path 指定的值开头的时候就会匹配。比如 projects 的定义就是应用了 prefix 策略，那么它既可以匹配 http://localhost:4200/projects，也可以匹配 http://localhost:4200/projects/123456 这种路径。

2. 积极加载和懒加载

在上面的例子中，我们可以看到几种类型的路由定义，类似 path: '**' 这个定义中，我们直接指定了一个组件 component: PageNotFoundComponent，这个属于最常见的一种定义。在一个模块中，如果我们希望某个路由匹配时就显示某个组件，这种定义就派上用场了。

path: '' 对应的是一个重定向，在此情况下，我们将匹配到的路由重定向到另一个路径。这里重定向到了 /auth/login，但是很奇怪的是我们并没有再定义 /auth/login，那么系统怎么处理这个路径呢？

这就需要我们引出一个路径模块的概念。在大型项目中，系统经常分成很多模块，这些模块的路由如果都在根路由中定义，就会导致路由表过于复杂。另外这样一个架构导致任何一个模块的路由更新都会去更新根路由文件，显然这不是一个良好的设计，维护会因此变得十分困难。

一个 Angular 团队推荐的模式是为每一个功能模块建立该模块自己的路由模块。比如 Feature Module 位于 app/admin 目录下，那么该模块的路由模块可以叫作 admin-routing.module.ts，同样位于 app/admin 目录下。Angular CLI 甚至提供了一个选项，可以让你在创建一个模块的同时创建它的路由模块。

```
ng g m auth --routing
```

我们在这个 auth-routing.module.ts 中定义 auth 模块的路由。

```
// 省略导入

const routes: Routes = [
  {
    path: 'auth',
    redirectTo: 'auth/login',
    pathMatch: 'full'
  },
  {
    path: 'auth/login',
    component: LoginComponent
  },
  {
    path: 'auth/register',
    component: RegisterComponent
  },
  {
    path: 'auth/forgot',
    component: ForgotPasswordComponent
  }
];

@NgModule({
  imports: [RouterModule.forChild(routes)],
  exports: [RouterModule]
})
export class AuthRoutingModule {}
```

在 AuthModule 中导入这个路由模块。需要指出的是，这个模块不是懒加载模块，所以仍然需要在根模块或者核心模块中导入。

```
// 省略导入

@NgModule({
  imports: [
    SharedModule,
```

```
    AuthRoutingModule
  ],
  declarations: [
    ...
  ]
})
export class AuthModule {}
```

回到我们前面提到的重定向特性。系统加载路由时，除了 AppRoutingModule 中的路由定义数组，也会加载根模块或核心模块中导入的其他功能模块的路由模块（如果存在的话）。还记得我们是在这些模块文件中导入的它们各自的路由模块吗？这就是为什么只要在根模块或核心模块中导入，系统路由就能加载它们的原因。由于我们在核心模块中导入了 AuthModule，系统路由定义中就包含了 AuthRoutingModule 中的路由定义，所以重定向到/auth/login，就会显示 AuthRoutingModule 中定义的这个路径对应的组件——LoginComponent。

projects 和 admin 则对应着懒加载模块，使用 loadChildren 指定在路由匹配时加载哪个模块。这个模块的路径定义是以 app 为基准路径的懒加载模块文件的相对路径。比如 project 模块文件位于 app/project/index.ts，而这个例子中我们的 AppRoutingModule 位于 app/core/app-routing.module.ts。所以在 AppRoutingModule 的这个路由定义数组中，我们就需要使用 ../project#ProjectModule 找到对应的模块。这里可以写成 ../project/index#ProjectModule，但是一个目录下的 index.ts 可以作为这个目录的索引文件，所以无须显式指定就可以直接使用目录作为路径作为这个文件的路径。也就是说，../project 会直接寻找 ../project/index.ts，这样就少写了一层路径。而 # 用来区隔模块路径和模块名称，ProjectModule 是实际上的这个模块的名称。

接下来我们可以看一下这个懒加载模块的路由定义。

```
// 省略导入

const routes: Routes = [
  {
    path: '',
    canActivate: [AdminGuard], canLoad: [AdminGuard],
    children: [
      {
        path: '',
        component: ProjectListComponent
      },
      {
        path: ':projectId',
```

```
        component: ProjectDetailComponent,
        children: [
          {
            path: 'groups',
            children: [
              {
                path: '',
                component: TaskGroupListComponent,
              },
              {
                path: ':groupId',
                component: TaskGroupDetailComponent,
                children: [
                  {
                    path: 'tasks',
                    component: TaskListComponent,
                  }
                ]
              }
            ]
          }
        ]
      }
    ]
  }
];

@NgModule({
  imports: [RouterModule.forChild(routes)],
  exports: [RouterModule]
})
export class ProjectRoutingModule {}
```

这个路由定义和积极加载模块不一样的是，模块根路径是空 `path: ''`，但在积极加载模块中，比如 `AuthRoutingModule`，如果设定根路径为空就会导致 `AppRoutingModule` 中的 `path: ''` 的定义无效。因为两者的 path 完全一致，而 `AppRoutingModule` 加载在后，这样先匹配的会生效。

但是为什么在懒加载模块中，我们可以有 `path: ''` 这种空字符的定义呢？因为懒加载模块之前是没有加载到系统路由的，而我们激活懒加载的方式是访问某个特定的路由，比如 http://localhost:4200/projects，这个定义是在系统路由中完成的。遇到这个 URL，系统会加载对应的懒加载模块，这里就是 `ProjectModule`，但是这个模块自身的路由怎么显示是由 `ProjectRoutingModule` 定义的。这里的空字符的路径不会和 `AppRoutingModule` 中的冲突的原因在于，我们在根路由中定义了一个前缀去激活这个模块，所以这里定义的空字符路径会有一个前缀，就是 projects。

3. 路径变量

利用路由传递参数有两种方式，其一是通过路径变量，那么什么叫作路径变量呢？比如 http://localhost:4200/tasks/1234 这样一个 URL 中的 1234 一般是这个 Task 的 id，那么这个 1234 也就是一个变量的值。我们可以把类似这种 URL 表示成 http://localhost:4200/task/:id，其中 :id 代表路径变量，变量名是 id，当然不一定起名叫 id，也可以叫 taskId 等你喜欢的名字。

在 Angular 中取得这个变量的值就可以通过注入 ActivatedRoute，通过它的 paramMap 得到这个变量。

```
constructor(private route: ActivatedRoute) {}
ngOnInit() {
  const id$ = this.route.paramMap.pipe(
      filter(params => params.has('id')),
      map(params => params.get('id'))
    );
}
```

需要指出的是，路径变量可以有多个，比如 http://localhost:4200/projects/123/tasks/456。如果定义是 http://localhost:4200/projects/:projectId/tasks/:taskId，那么 projectId 和 taskId 就是两个路径变量。一般来说，表达的意义和 REST API 很像，就是 id 为 123 的项目下的 id 为 456 的任务。那么这个例子的路由可以定义成下面的样子：

```
{
  path: ':projectId',
  component: ProjectDetailComponent,
  children: [
    {
      path: 'tasks/:taskId',
      component: TaskDetailComponent
    }
  ]
}
```

4. 查询参数

另一种参数传递的方式就是查询参数，比如类似 http://localhost:4200/tasks?active=false&projectId=1234 这样的 URL 中以 ?param1=value1¶m2=value2 的方式传递，就是查询参数。同样可以通过 ActivatedRoute 的 queryParamMap 得到参数。

```
constructor(private route: ActivatedRoute) {}
ngOnInit() {
  const id$ = this.route.queryParamMap.pipe(
      filter(params => params.has('param1')),
      map(params => params.get('param1'))
    );
}
```

和路径变量不一样的是，查询参数无须预先定义在路由定义对象中，在 routerLink 指令或 navigate 时直接指定即可。

大多数情况下，这两种传递参数的方式就已经够用了，但是还是有时候，我们需要传递一些更复杂的参数，这些参数不是单纯的值，而是对象，那么这种情况我们怎么处理呢？这就需要引出路由数据的概念了。

5. 路由数据

在路由定义对象中，可以通过 data 添加需要该路由传递的数据，data 是一个字典类型的对象，如果要传递值，就给出一个 key，如下面例子中就是 breadcrumb，这个 key 对应的值可以是任意类型。

```
{
  path: 'users',
  component: UserHomeComponent,
  canActivate: [AuthGuard, AdminGuard],
  canLoad: [AuthGuard, AdminGuard],
  data: {
    breadcrumb: '用户管理'
  }
},
```

这样定义好之后就可以使用 ActivatedRoute 的 data 得到数据。

```
this.route.data.pipe(
  map(data => data['breadcrumb'])
)
```

6. 匹配策略

系统在匹配到路由表中的第一个符合项时就会导航到对应的 URL，所以一定要注意顺序，不仅是在单个路由数组中的顺序，也要考虑存在多个路由模块时对于路由模块的导入顺序。一般来说，越模糊的（比如使用通配符的）要放在越下面，比如 path: '**'，这种一定要放在最下面。上面例子中的 /auth/login 是一个预加载模块，这个模块在根模块或者核心模块中导入时就需要在 AppRouting 中。如果是懒加载模块，比如例子中的 projects 和 admin

则在数组中需要放在模糊的匹配之前。

6.4.3 获取父路由的参数

有时我们在子路由中需要获取父路由的参数，在 Angular 中，这是很容易办到的，因为路由是有父子结构的。下面的例子中，如果我们的父路由是:floor，子路由是:floor/:room，如果在子路由的组件中，可以通过下面的方法获得父路由的参数 floor。

```
const routefloor$ = this.route.parent.paramMap.pipe(
      filter(params => params.has('floor')),
      map(params => Number.parseInt(params.get('floor')))
    );
```

类似地，this.route.parent.parent 就是往上两层的路由了。

6.4.4 获得前一个路由

有时，我们还会想要获取前一个路由，这个在 Angular 中同样比较简单，只不过需要使用 RxJS 的一个操作符 pairwise。这个 pairwise 操作符是发射之前和当前的值组成的数组，比如 [0,1]，[1,2]，[2,3]，[3,4]，[4,5] 就是一个典型的 pairwise 的输出。利用 Router 的 events 属性可以方便地得到前一个路由。

```
constructor(private router: Router) {
  this.router.events.pipe(
    filter(e => e instanceof RoutesRecognized),
    pairwise()
  )
  .subscribe((event: any[]) => {
      console.log(event[0].urlAfterRedirects);
    });
}
```

6.4.5 Activated Route

前面我们用到了多个 ActivatedRoute 中的属性，表 6-1 给出一个更详尽的属性清单。

表 6-1

属　性	描　述
url	得到路由路径的观察者对象，路由路径是由一个组成路径各部分的数组构成的
data	得到路由数据的观察者对象，包含在路由定义中提供的数据对象
paramMap	观察者对象，其值为路由参数键值对

续表

属　性	描　述
queryParamMap	观察者对象，其值为路由查询参数键值对
fragment	观察者对象，其值为 URL 碎片
outlet	路由插座的名字，如果没有给路由插座起名字，默认的名字就是 primary
routeConfig	返回在路由定义中的路由配置
parent	返回当前路由的父 ActivatedRoute
firstChild	第一个子路由
children	返回所有子路由

6.5 安全守卫

一般来说，所有应用都有一定程度上的权限要求，不可以任意地自由导航。

- 特定的视图只对某些用户授权访问。
- 用户需要先鉴权，也就是登录才能访问。
- 你需要先获取某些数据才能显示对应视图。
- 在离开某一视图之前，你希望提醒用户保存数据或确认取消保存。

在 Angular 中引入了一个守卫的概念来帮开发者处理上面的场景，守卫根据场景不同，分为以下几种类型：

- CanActivate 决定是否可以导航到某个路由。
- CanActivateChild 决定是否可以导航到某个子路由。
- CanDeactivate 决定是否可以离开当前路由。
- CanLoad 决定是否可以导航到某个懒加载模块的路由。
- Resolve 在路由激活之前加载数据。

你在路由的每一层级都可以设置多个守卫，Angular 的规则是只要这些守卫有返回 false 的，那么尚未执行的守卫就会取消，整个导航的动作也会取消。从优先级上说，Angular 会先检查 CanDeactivate 和 CanActivateChild，检查的顺序是从里（最深层的子路由）往外。然后会检查 CanActivate，这个检查是从外往里进行的。

一个简单的路由守卫可以是一个函数，只要它返回 Observable<boolean>，Promise<boolean> 或者 boolean。另外，守卫如果要在各个路由模块中被使用，则需要提供出来。

```
@NgModule({
```

```
  ...
  providers: [
    provide: 'AlwaysActivateGuard',
    useValue: () => {
      return true;
    }
  ],
  ...
})
export class AppModule {}
```

上面就是一个最简单的守卫，它总是返回 true，而且可以通过 AlwaysActivateGuard 进行注入，使用这个守卫那么在路由定义中像下面这样即可。

```
export const routes:Routes = [
  {
    path: '',
    component: SomeComponent,
    canActivate: ['AlwaysActivateGuard']
  }
];
```

大部分情况下，我们还是需要建立一个类来构造一个守卫的，因为守卫里面经常需要注入其他的 service 来判断什么样的逻辑下会返回 true 或 false。比如下面的例子中，我们需要调用 AuthService 来判断用户是否已经登录。

```
// 省略导入

@Injectable({
  providedIn: 'root'
})
export class AuthGuard implements CanActivate {

  constructor(private authService: AuthService) {}

  canActivate(
    route: ActivatedRouteSnapshot,
    state: RouterStateSnapshot
  ): boolean | Observable<boolean> | Promise<boolean> {
    return this.authService.isLoggedIn();
  }
}
```

6.5.1 激活守卫

没有权限的应用是很少见的。一般来说，应用都需要根据用户角色限制其访问某些区域，或者根据其账户是否激活而让用户看到或看不到某些内容。CanActivate 守卫就是一个管理这些导航规则的利器，在 Angular 中实现 CanActivate 接口，只需要实现 canActivate 这个方法，这个方法返回的有几种可能 Observable<boolean> | Promise<boolean> | boolean，这个写法就是返回 Observable<boolean> 或者 Promise<boolean> 或者 boolean。

```
export interface CanActivate {
    canActivate(route: ActivatedRouteSnapshot, state: RouterStateSnapshot): Observable<boolean> | Promise<boolean> | boolean;
}
```

在上面例子中的 AuthGuard 其实就是用来判断用户是否已经登录的，以及阻止未登录的用户访问该模块。

6.5.2 激活子路由守卫

6.5.1 节使用 CanActivate 阻止了未授权的访问，但如果该路由存在子路由，那可能需要更细化的权限控制来允许或禁止某些角色访问子路由。此时就是 CanActivateChild 派上用场的时候，CanActivateChild 守卫和 CanActivate 守卫非常相似，区别是 CanActivateChild 是在子路由激活前调用的。同样，如果要实现一个子路由守卫，则只需实现 CanActivateChild 接口。

```
export interface CanActivateChild {
    canActivateChild(childRoute: ActivatedRouteSnapshot, state: RouterStateSnapshot): Observable<boolean> | Promise<boolean> | boolean;
}
```

在下面的例子中，canActivateChild: [ProjectOwnerGuard] 用于只能显示用户参与的项目。

```
// 省略导入

const routes: Routes = [
  {
    path: '',
    canActivate: [AdminGuard],
    canActivateChild: [ProjectOwnerGuard],
    children: [
      {
        path: '',
```

```
          component: ProjectListComponent
        },
        {
          path: ':projectId',
          component: ProjectDetailComponent,
          children: [
            {
              path: 'groups',
              children: [
                {
                  path: '',
                  component: TaskGroupListComponent,
                },
                {
                  path: ':groupId',
                  component: TaskGroupDetailComponent,
                  children: [
                    {
                      path: 'tasks',
                      component: TaskListComponent,
                    }
                  ]
                }
              ]
            }
          ]
        }
      ]
    }
  ];
```

6.5.3 加载守卫

加载守卫和激活守卫很多读者会分不清楚，加载守卫其实是用于懒加载模块——阻止用户异步加载整个模块。从效果上看，如果使用 `CanActivate` 而不使用 `CanLoad`，那虽然无法访问该模块，但在浏览器中却可以看到这个模块的源码（见图 6-5）。

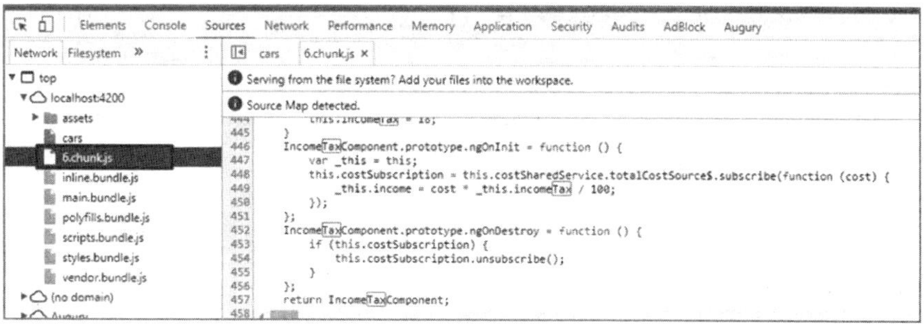

图 6-5

如果使用了 CanLoad，那么是无法看到模块源码的，也就是说整个模块都不会加载（见图 6-6）。

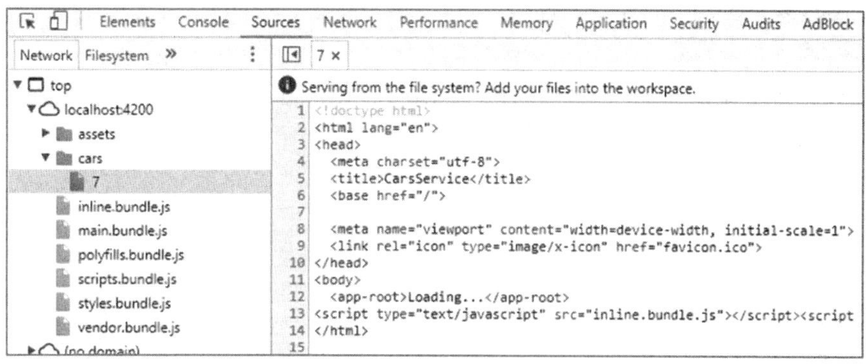

图 6-6

6.5.4 退出守卫

CanDeactivate 守卫经常使用的一个场景就是用户如果要离开该路由时，提醒用户保存或放弃更改。下面的例子中，我们使用了一个小技巧来将 CanDeactivate 守卫变成一个可以在整个应用共享的形式，使用一个 CanComponentDeactivate 让每个需要使用退出提醒功能的组件实现这个接口，从而将退出的具体逻辑交给组件去判断。

```
// 省略导入

export interface CanComponentDeactivate {
  canDeactivate: () => Observable<boolean> | Promise<boolean> | boolean;
}

@Injectable()
export class CanDeactivateGuard implements
```

```
CanDeactivate<CanComponentDeactivate> {

    canDeactivate(component: CanComponentDeactivate) {
      return component.canDeactivate ? component.canDeactivate() : true;
    }

  }
```

6.5.5 数据预获取守卫

这个守卫和其他的有明显不同，这个 `Resolve` 接口返回的不是一个布尔型，而是数据本身，这个守卫的目的就是在导航到路由之前，预先得到某些数据。

```
  interface Resolve<T> {
    resolve(route: ActivatedRouteSnapshot, state: RouterStateSnapshot):
Observable<T> | Promise<T> | T
  }
```

上面的例子中，我们根据路由参数的 `projectId` 取得项目数据，如果没有数据，则导航到项目列表页面。

第 7 章 后端不只是 API

前后端分离的是后端对于视图的渲染和路由功能被剥离给了前端，似乎后端的工作就只剩下了构建 CRUD 这种 API 了，事实真的是这样吗？后端处理工作其实远远不止是 API，除非你想做的仅仅是为了配合前端搭建的一个“脚手架”。

剥离前端后，后端大部分的工作是真的比较“后端”了，因为实现的功能是没有用户界面的，比如性能、安全、搜索、数据的审计历史等。本章就来看看后端的这些功能，需要说明的是，这些功能是非常专业的，需要深入学习的，这里我们只是带大家一起开个头，相关的细节需要我们一起在今后的工作中不断学习和体会。

7.1 缓存

缓存是提升性能的一大利器，但缓存的处理，比如何时失效、如何失效等也是非常复杂的。本节我们一起来学习 Spring 的基于注释的 Cache 配置方法，展现了 Spring Cache 的强大之处，然后介绍了其基本的原理，以及扩展点和使用场景的限制。

Spring 从 3.× 开始引入了基于注解的缓存技术，本质上是一个对缓存使用的抽象，通过在现有代码中添加注解，就可以够达到缓存方法的返回对象的效果。Spring 的缓存具备相当的灵活性，不仅能够使用 SpEL（Spring Expression Language）来定义缓存的 key 和各种条件，也支持各自主流 Cache 框架的集成。其特点总结如下。

- 通过注解即可使得现有代码支持缓存。
- 支持开箱即用，即不用安装和部署额外第三方组件就可以使用缓存。
- 支持 SpEL 表达式，能使用对象的任何属性或者方法来定义缓存的键值和条件。

- 支持 AOP。
- 支持自定义键值和自定义 CacheManager，具有相当的灵活性和扩展性。

7.1.1 配置 Cache

最简化的配置其实只需要添加注解@EnableCaching 即可，也就是我们可以在 Application.java 的类上面加注解就可以让应用支持 Cache 。后面如果要做一些定制化，还是新建一个配置文件，在 config 包下新建 CacheConfig.java 文件。

```
package dev.local.gtm.api.config;

import org.springframework.cache.annotation.EnableCaching;
import org.springframework.context.annotation.Configuration;

@EnableCaching
@Configuration
public class CacheConfig {
}
```

是的，就是这么简单，在我们没有指定其他的 Cache 实现方式之前，Spring Boot 会自动帮我们配置一个简单的 Cache 实现。这个实现方式是使用 ConcurrentHashMap 作为缓存的存储。默认情况下，系统会在需要时创建缓存，但你也可以限定缓存的数量，这可以通过设置 cache-names 来实现。如果使用了这个列表之外的缓存，Spring Boot 应用会无法启动。

```
spring:
  cache:
    cache-names: usersByLogin, usersByMobile, usersByEmail
```

除这种默认的简单缓存外，Spring 支持以下类型的缓存框架。

- Generic
- JCache (JSR-107)
- EhCache 2.x
- Hazelcast
- Infinispan
- Couchbase
- Redis
- Caffeine

为了在后期可以看到 Cache 是否生效了，可以配置 Cache 的日志，更改 application.yml 设置 dev 环境下的 org.springframework.cache=DEBUG。

```
spring:
  application:
    name: api-service
  profiles:
    active: dev
  data:
    mongodb:
      database: gtm-api
  cache:
    type: redis
  redis:
    host: redis
server:
  port: 8080
logging:
  level:
    org.apache.http: ERROR
    org.springframework:
      web: ERROR
      data: ERROR
      security: ERROR
      cache: ERROR
    org.springframework.data.mongodb.core.MongoTemplate: ERROR
    dev.local.gtm.api: ERROR

---

spring:
  profiles: dev
  devtools:
    remote:
      secret: thisismysecret
  data:
    mongodb:
      database: gtm-api-dev
  redis:
    host: localhost
logging:
  level:
    org.apache.http: DEBUG
    org.springframework:
        web:
          client.RestTemplate: DEBUG
        data: DEBUG
```

```
        security: DEBUG
        cache: DEBUG
    org.springframework.data.mongodb.core.MongoTemplate: DEBUG
    dev.local.gtm.api: DEBUG

---

spring:
  profiles: test
  data:
    mongodb:
      database: gtm-api-test
  cache:
    type: none
logging:
  level:
    org.apache.http: DEBUG
    org.springframework:
          web:
            client.RestTemplate: DEBUG
          data: DEBUG
          security: DEBUG
    org.springframework.data.mongodb.core.MongoTemplate: DEBUG
    dev.local.gtm.api: DEBUG
---

spring:
  profiles: prod
```

7.1.2 常用的缓存注解

还记得我们在 UserRepo 中曾经设置过几个注解吗？@Cacheable(cacheNames =×××)。

```
@Repository
public interface UserRepo extends MongoRepository<User, String> {
    String USERS_BY_LOGIN_CACHE = "usersByLogin";

    String USERS_BY_MOBILE_CACHE = "usersByMobile";

    String USERS_BY_EMAIL_CACHE = "usersByEmail";

    @Cacheable(cacheNames = USERS_BY_MOBILE_CACHE)
    Optional<User> findOneByMobile(@Param("mobile") String mobile);

    @Cacheable(cacheNames = USERS_BY_EMAIL_CACHE)
```

```
        Optional<User> findOneByEmailIgnoreCase(@Param("email") String
email);

        @Cacheable(cacheNames = USERS_BY_LOGIN_CACHE)
        Optional<User> findOneByLogin(@Param("login") String login);

        Page<User> findAllByLoginNot(Pageable pageable, @Param("login")
String login);
    }
```

Spring 中提供以下几个注解。

- @Cacheable 触发缓存填充（triggers cache population）
- @CacheEvict 触发缓存的回收
- @CachePut 更新缓存
- @Caching 一组应用到某个方法上的缓存
- @CacheConfig 在类级别上共享一些缓存相关的设置

@Cacheable

顾名思义，@Cacheable 是用来标记可以缓存的方法，也就是说方法的返回值会存储在缓存中，下次如果以同样的参数访问这个方法，会返回缓存中的值而不是又一次执行这个方法。在 UserRepo 中使用的就是 @Cacheable 的最简化形式。

```
@Cacheable(cacheNames = USERS_BY_MOBILE_CACHE)
Optional<User> findOneByMobile(@Param("mobile") String mobile);
```

这个例子中，findOneByMobile 和名字为 usersByMobile（String USERS_BY_MOBILE_CACHE = "usersByMobile";) 的缓存关联了。每次这个方法被调用时，缓存会检查这个方法是否已经调用过，如果调用过，就直接返回缓存值。

默认键值策略

由于缓存很多时候是键值对（key-value），所以每一次方法的调用需要映射成一个 key，以便我们可以通过这个 key 访问缓存中的值。Spring 的默认 key 生成是通过一个简单的 KeyGenerator，遵循以下规则。

- 如果没有参数，则返回 SimpleKey.EMPTY。
- 如果只有一个参数，则返回这个参数的实例。
- 两个及两个以上的参数，返回一个包含所有参数的 SimpleKey，参数需要实现 equals() 和 hashCode() 方法避免 key 值重复。

有时，对于缓存来说，一个方法的参数并不是都有同等重要性。如果参数只是在方法中使用，而不是影响结果，那么它可能不适合作为 Cache 的 key。比如下面的例子中，id 作为唯一识别文件的标识，我们指定其作为 key，而另一个参数不参与（仅仅作为示例，实际的商业逻辑不一定是这样的，请勿照搬）。

```
@Cacheable(cacheNames="files", key="#id")
public String findFile(String id, boolean argNotAffectingResult)
```

当然，这个 key 可以使用 SpEL 表达式，比如 key="T(someparam).hash(#someparam) 等，也可以指定自定义 key 生成器，比如 @Cacheable(cacheNames="files", keyGenerator="customKeyGenerator")，这可以通过实现 org.springframework.cache.interceptor.KeyGenerator 接口来自定义。

7.1.3 测试缓存是否生效

如果我们在 API 文档界面 http://localhost:8080/swagger-ui/index.html 对 /api/auth/search/username 进行测试，则可以观察 Console 里面的日志，如图 7-1 所示。

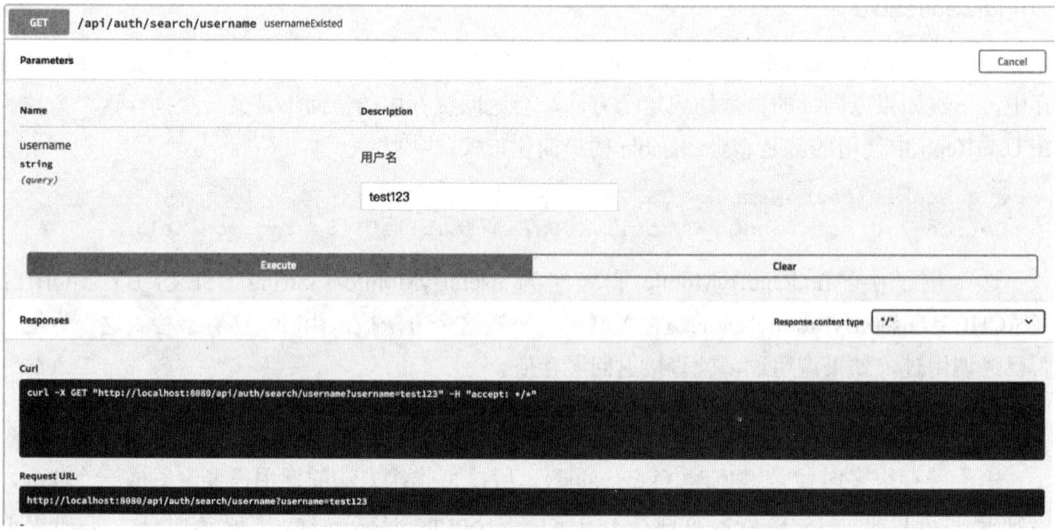

图 7-1

在 Spring Boot 启动的日志中可以看到类似下面这种日志，这是 Cache 配置成功的标志，系统 Cache 框架将我们注解标记的方法缓存起来了。

```
   : Adding cacheable method 'findOneByEmailIgnoreCase' with attribute:
[Builder[public abstract j
  ava.util.Optional
dev.local.gtm.api.repository.UserRepo.findOneByEmailIgnoreCase(java.lang.Str
```

```
ing)] caches=[usersByEmail] | key='' | keyGenerator='' | cacheManager='' |
cacheResolver='' | condition='' | unless='' | sync='false']
   2018-04-27 19:50:20.207 DEBUG 10821 --- [           main]
o.s.c.a.AnnotationCacheOperationSource
     : Adding cacheable method 'emailExisted' with attribute:
[Builder[public dev.local.gtm.api.web.
   rest.AuthResource$ExistCheck
dev.local.gtm.api.web.rest.AuthResource.emailExisted(java.lang.String)]
caches=[usersByEmail] | key='' | keyGenerator='' | cacheManager='' |
cacheResolver='' | condition='' | unless='' | sync='false']
```

我们可以观察到，日志中有 AuthServiceImpl.usernameExisted() 方法进入的条目，也有 MongoTemplate 进行数据库查询的条目，这证明第一次进行查询的时候，没有进行缓存，这是正常情况。

```
   2018-04-27 19:29:41.981 DEBUG 9561 --- [ XNIO-2 task-19]
d.l.gtm.api.aop.logging.LoggingAspect : Enter:
dev.local.gtm.api.web.rest.AuthResource.usernameExisted() with argument[s] =
[test123]
   2018-04-27 19:29:41.981 DEBUG 9561 --- [ XNIO-2 task-19]
dev.local.gtm.api.web.rest.AuthResource : REST 请求 -- 用户名是否存在 test123
   2018-04-27 19:29:41.982 DEBUG 9561 --- [ XNIO-2 task-19]
d.l.gtm.api.aop.logging.LoggingAspect : Enter:
dev.local.gtm.api.service.impl.AuthServiceImpl.usernameExisted() with
argument[s] = [test123]
   2018-04-27 19:29:41.988 DEBUG 9561 --- [ XNIO-2 task-19]
o.s.d.m.r.query.MongoQueryCreator : Created query Query: { "login" :
"test123" }, Fields: { }, Sort: { }
   2018-04-27 19:29:41.990 DEBUG 9561 --- [ XNIO-2 task-19]
o.s.data.mongodb.core.MongoTemplate : find using query: { "login" :
"test123" } fields: Document{{}} for class: class
dev.local.gtm.api.domain.User in collection: api_users
   2018-04-27 19:29:41.996 DEBUG 9561 --- [ XNIO-2 task-19]
d.l.gtm.api.aop.logging.LoggingAspect : Exit:
dev.local.gtm.api.service.impl.AuthServiceImpl.usernameExisted() with result
= false
   2018-04-27 19:29:41.997 DEBUG 9561 --- [ XNIO-2 task-19]
d.l.gtm.api.aop.logging.LoggingAspect : Exit:
dev.local.gtm.api.web.rest.AuthResource.usernameExisted() with result =
dev.local.gtm.api.web.rest.AuthResource$ExistCheck@32ef7921
   2018-04-27 19:29:41.999 DEBUG 9561 --- [ XNIO-2 task-19]
o.s.s.w.header.writers.HstsHeaderWriter : Not injecting HSTS header since it
did not match the requestMatcher
org.springframework.security.web.header.writers.HstsHeaderWriter$SecureReque
stMatcher@e228328
```

```
   2018-04-27 19:29:42.002 DEBUG 9561 --- [ XNIO-2 task-19]
o.s.s.w.a.ExceptionTranslationFilter : Chain processed normally
   2018-04-27 19:29:42.002 DEBUG 9561 --- [ XNIO-2 task-19]
s.s.w.c.SecurityContextPersistenceFilter : SecurityContextHolder now cleared,
as request processing completed
```

接下来，我们再次执行同样的请求，还是观察日志输出，这一次找不到任何 MongoDB 查询的日志，这证明缓存起作用了。

```
   2018-04-27 19:42:22.733 DEBUG 9561 --- [ XNIO-2 task-20]
d.l.gtm.api.aop.logging.LoggingAspect : Enter:
dev.local.gtm.api.web.rest.AuthResource.usernameExisted() with argument[s] =
[test123]
   2018-04-27 19:42:22.733 DEBUG 9561 --- [ XNIO-2 task-20]
dev.local.gtm.api.web.rest.AuthResource : REST 请求 -- 用户名是否存在 test123
   2018-04-27 19:42:22.733 DEBUG 9561 --- [ XNIO-2 task-20]
d.l.gtm.api.aop.logging.LoggingAspect : Enter:
dev.local.gtm.api.service.impl.AuthServiceImpl.usernameExisted() with
argument[s] = [test123]
   2018-04-27 19:42:22.758 DEBUG 9561 --- [ XNIO-2 task-20]
d.l.gtm.api.aop.logging.LoggingAspect
    : Exit: dev.local.gtm.api.service.impl.AuthServiceImpl.usernameExisted()
with result = false
   2018-04-27 19:42:22.758 DEBUG 9561 --- [ XNIO-2 task-20]
d.l.gtm.api.aop.logging.LoggingAspect : Exit:
dev.local.gtm.api.web.rest.AuthResource.usernameExisted() with result =
dev.local.gtm.api.web.rest.AuthResource$ExistCheck@53c05c2
   2018-04-27 19:42:22.766 DEBUG 9561 --- [ XNIO-2 task-20]
o.s.s.w.header.writers.HstsHeaderWriter : Not injecting HSTS header since it
did not match the requestMatcher
org.springframework.security.web.header.writers.HstsHeaderWriter$SecureReque
stMatcher@e228328
   2018-04-27 19:42:22.773 DEBUG 9561 --- [ XNIO-2 task-20]
o.s.s.w.a.ExceptionTranslationFilter : Chain processed normally
   2018-04-27 19:42:22.773 DEBUG 9561 --- [ XNIO-2 task-20]
s.s.w.c.SecurityContextPersistenceFilter : SecurityContextHolder now cleared,
as request processing completed
```

7.2 Redis 作为缓存框架

Redis 是一个开源的、基于内存的基于键值对的数据库，是 NoSQL 的一种。一般用作数据库、缓存服务或消息服务使用。Redis 支持多种数据结构，包括字符串、哈希表、链表、集合、有序集合、位图等。Redis 具备 LRU 淘汰、事务实现，以及不同级别的硬盘持久化等能力，并且支持副本集和通过 Redis Sentinel 实现的高可用方案，同时还支持通过 Redis Cluster 实现的数据自动分片能力。Redis 支持的数据类型包括：字符串、哈希表、链表、集合、有序集合，以及基于这些数据类型的相关操作。Redis 使用 C 语言开发，在大多数 *nix 系统上无须任何外部依赖就可以使用。Redis 支持的客户端语言也非常丰富，常用的计算机语言如 C、C#、C++、Object-C、PHP、Python、Java、Perl、Lua、Erlang 等均有可用的客户端来访问 Redis 服务器。当前 Redis 的应用已经非常广泛，如新浪、淘宝、Github 等均在使用 Redis 的缓存服务。

7.2.1 Redis 的安装配置

我们之前提到过容器的一大优点就是安装配置极为容易，和网上的一些教程不同，我们这里不讲如何原生安装 Redis，直接从使用容器讲起。

第一步，取得 Redis 的镜像，redis-alphine 是一个最小化的镜像版本。

```
docker pulll redis-alpine
```

你也可以不使用 redis-alpine 版本，那就使用 redis，“ : ” 后面跟的是 tag，一般可以是版本号，具体可以去网站中查看[1]。

```
docker pulll redis:4.2.1
```

第二步，在 docker-compose.yml 中配置。

```
version: '3.2'
services:
  redis:
    # 使用 tag 的 image
    image: redis:4-alpine
    # 如果希望启动 Redis 时带有参数，那么可以在此处定义
    command: [ "redis-server", "--protected-mode", "no" ]
    ports:
      - "6379:6379"
   volumes:
     # 映射文件夹
     - redis-data:/data
    hostname: redis
  mongo:
    image: mongo:3.6.4
    ports:
      - "27017:27017"
```

1 https://hub.docker.com/_/redis/

```
      volumes:
        - api_db:/data/db
    api-server:
      image: api
      ports:
        - "8080:8080"
        - "5005:5005"
      links:
        - mongo
        # 连接到api-server，分开写成 redis:redis
        # 前面的是在这个 services 中的名字，后一个是在 api-service 中可见的 hos 名称。
        - "redis:redis"
  volumes:
    api_db: {}
    redis-data: {}
```

第三步，启动服务，我们只需在 docker-compost.yml 所在目录下执行下面的命令，就可以启动 Redis 了，是不是很简单，到此服务就启动了，我们可以节省很多时间专注在开发上。

```
docker-compose up -d redis
```

有的读者说了，那如果想同时启动 Redis 和 mongoDB 怎么办呢？输入下面这个命令。

```
docker-compose up -d redis mongo
```

对的，你猜到了，如果要连 api-service 一起启动，那就是：

```
docker-compose up -d
```

如果要停止服务呢？下面的一句代码就解决问题了。

```
docker-compose down
```

到这里，Redis 的安装和配置就完成了，下面看看如何在 Spring Boot 中集成。

7.2.2 在 Spring Boot 中集成 Redis

Spring Boot 对于 Redis 有着开箱即用的体验，只需添加 spring-boot-starter-data-redis，即可完成 Redis 的集成。

```
// 省略
dependencies {
    // 省略
    implementation("org.springframework.boot:spring-boot-starter-data-
mongodb")
    implementation("org.springframework.boot:spring-boot-starter-data-
redis") // <--- 这里
    testImplementation("org.springframework.security:spring-security-
```

```
test")
    }
```

使用 Redis 作为缓存框架更是简单得不能再简单，只需要在 application.yml 中设置 spring.cache.type 为 redis，随后就可以在 spring.redis.host 中配置主机名或 IP 地址，用 spring.redis.port 配置端口。在下面的配置中，我们为开发环境（dev）配置了 localhost，而为生产环境配置了 redis 作为主机名，我们在生产环境中会使用容器，你还记得我们在 docler-compose.yml 中是怎么样配置的吗？ api-service 连接了两个容器，一个是 mongo，另一个就是 redis。

```
    spring:
      application:
        name: api-service
      profiles:
        active: dev
      data:
        mongodb:
          database: gtm-api
      cache:
        type: redis # <--- 默认使用 redis 作为缓存
      redis:
        host: redis # <---redis 主机名，需要和 docker-compose.yml 中的 `link` 中
定义保持一致
    server:
      port: 8080
    logging:
      level:
        org.apache.http: ERROR
        org.springframework:
          web: ERROR
          data: ERROR
          security: ERROR
          cache: ERROR
        org.springframework.data.mongodb.core.MongoTemplate: ERROR
        dev.local.gtm.api: ERROR

    ---

    spring:
      profiles: dev
      devtools:
        remote:
          secret: thisismysecret
      data:
        mongodb:
```

```
      database: gtm-api-dev
  redis:
    host: localhost # <--- 开发模式下
logging:
  level:
    org.apache.http: DEBUG
    org.springframework:
        web:
          client.RestTemplate: DEBUG
        data: DEBUG
        security: DEBUG
        cache: DEBUG
    org.springframework.data.mongodb.core.MongoTemplate: DEBUG
    dev.local.gtm.api: DEBUG

---

spring:
  profiles: test
  data:
    mongodb:
      database: gtm-api-test
  cache:
    type: none  # 测试模式下，不使用 Cache
logging:
  level:
    org.apache.http: DEBUG
    org.springframework:
          web:
            client.RestTemplate: DEBUG
          data: DEBUG
          security: DEBUG
    org.springframework.data.mongodb.core.MongoTemplate: DEBUG
    dev.local.gtm.api: DEBUG
---

spring:
  profiles: prod
```

现在如果我们先启动 MongoDB 和 Redis：

```
docker-compose up -d redis mongo
```

然后启动自己的应用，你就可以看到缓存还是正常工作的，像之前一样。

在容器中配置 Redis

配置容器时，只需要在 ENTRYPOINT 配置启动的 Java 命令时传入-Dspring.redis.host=redis 即可，这里的 host 设置成 redis，是因为 docker-compose 配置时，我们为 api-server 连接的就是 redis。当然如果端口不是默认的 6379，则还需要传入-Dspring.redis.port=××××。

```
// 省略
task createDockerfile(type:
com.bmuschko.gradle.docker.tasks.image.Dockerfile, dependsOn: ['bootJar']) {
    description = "自动创建 Dockerfile"
    destFile = project.file('src/main/docker/Dockerfile')
    from 'openjdk:8-jdk-alpine'
    volume '/tmp'
    addFile "${project.name}-${project.version}.jar", "app.jar"
    instruction { 'ENTRYPOINT [' +
            '"java", ' +
            '"-
agentlib:jdwp=transport=dt_socket,address=5005,server=y,suspend=n", ' +
            '"-Dspring.data.mongodb.uri=mongodb://mongo/taskmgr", ' +
            '"-Dspring.redis.host=redis", ' + // <--- 这里
            '"-Dspring.profiles.active=prod", ' +
            '"-Djava.security.egd=file:/dev/./urandom", ' +
            '"-jar","/app.jar"]'}
    maintainer 'Peng Wang "wpcfan@gmail.com"'
}
// 省略
```

7.2.3 Redisson

Redisson 是一个在 Redis 的基础上实现的 Java 驻内存数据网格（In-Memory Data Grid）。它底层采用的是 Netty 框架，不仅提供了一系列的分布式的 Java 常用对象，还提供了许多分布式服务。其中包括（BitSet、Set、Multimap、SortedSet、Map、List、Queue、BlockingQueue、Deque、BlockingDeque、Semaphore、Lock、AtomicLong、CountDownLatch、Publish / Subscribe、Bloom filter、Remote service、Spring cache、Executor service、Live Object service、Scheduler service）Redisson 提供了使用 Redis 的最简单和最便捷的方法。

提供一个 Redisson 的配置，需要在 config 包下新建 CacheConfig.java 文件。

```
@RequiredArgsConstructor
@EnableCaching
@Configuration
@AutoConfigureAfter(RedisAutoConfiguration.class)
public class CacheConfig {
```

```
        private final RedisProperties redisProperties;

        @Bean(destroyMethod = "shutdown")
        RedissonClient redissonClient() {
            Config config = new Config();
            // sentinel
            if (redisProperties.getSentinel() != null) {
                SentinelServersConfig sentinelServersConfig =
config.useSentinelServers();

sentinelServersConfig.setMasterName(redisProperties.getSentinel().getMaster(
));
                val nodes = redisProperties.getSentinel().getNodes();
                sentinelServersConfig.addSentinelAddress(nodes.toArray(new
String[0]));

sentinelServersConfig.setDatabase(redisProperties.getDatabase());
                if (redisProperties.getPassword() != null) {

sentinelServersConfig.setPassword(redisProperties.getPassword());
                }
            } else { // 单个 Server
                SingleServerConfig singleServerConfig =
config.useSingleServer();
                String schema = redisProperties.isSsl() ? "rediss://" :
"redis://";
                singleServerConfig.setAddress(schema +
redisProperties.getHost() + ":" + redisProperties.getPort());

singleServerConfig.setDatabase(redisProperties.getDatabase());
                if (redisProperties.getPassword() != null) {

singleServerConfig.setPassword(redisProperties.getPassword());
                }
            }
            return Redisson.create(config);
        }
    }
```

为了避免多处配置同样的信息，我们在这个文件中注入了 Redis 的属性设置类 RedisProperties。根据是否设置了前哨而决定是构造一个分布式的配置还是单机配置。在单机配置中，我们利用 RedisProperties 取得 Redis 的信息，并应用到 Redisson 的配置中。在单机配置中，Redisson 设置的地址的格式是“协议://地址:端口号”，和 MongoDB 类似，redis://localhost:6379 就是采用非 SSL 协议连接本机的 6379 端口，而 rediss://redis:6479 就是以 SSL 协议连接主机 Redis 的 6479 端口。

7.3 使用 ElasticSearch 提升搜索性能

Elasticsearch 也是一个著名的 NoSQL 数据库，它是基于原来社区中大名鼎鼎的 lucence 文本搜索引擎的基础发展而来的，所以它的强项就是搜索。那么问题来了，为什么我们在项目中要使用两个 NoSQL 数据库？既然有了 MongoDB，而且 MongoDB 的查询能力也很强的情况下，为什么又引入了 Elasticsearch？

这个话题其实同样可以应用到 SQL 和 NoSQL，其实没有任何一种技术框架是万能的，都有各自适合的应用场景。在一个现实世界的复杂项目中，可能有些模块或功能适合 SQL 数据库，而有些适合 NoSQL 数据库，没有一个规则说你只能使用一种类型的数据库。相反如果我们可以在不同场景使用不同的数据库才会让我们的工程更加灵活强大。Elasticsearch 在全文搜索方面几乎无人可敌，所以我们需要搜索或全文索引时就应该使用 Elasticsearch，而 MongoDB 是一个更通用的文档数据库，在一般性的增、删、改、查操作上显然更为合适。其实细心的读者会发现前一章我们还用到了一个 NoSQL 数据库——Redis，它擅长以键值对的形式在内存中高速的读写，所以我们的缓存框架采用了 Redis。

7.3.1 配置

1. 安装 ElasticSearch

Elasticsearch 的安装很简单，还是通过 Docker 来进行。Elasticsearch 官方提供的镜像现在改到了 elastic 自己的网站发行，如果想从官网镜像，则需要使用下面的命令。

```
docker pull docker.elastic.co/elasticsearch/elasticsearch:5.5.0
```

这个镜像是含有 x-pack 扩展包的，这个扩展包提供了很多功能，包括一些付费才能使用的商业特性。在我们的例子中，还是希望有一个相对纯净的 Elasticsearch，所以我们自己写一个 Dockerfile 将 x-pack 移除掉。请注意，如果你决定保留 x-pack，则需要自定义配置用来支持 x-pack 的特性。在根目录下建立 docker/elasticsearch 目录，在此目录下建立 Dockerfile。

```
FROM docker.elastic.co/elasticsearch/elasticsearch:5.5.0
RUN elasticsearch-plugin remove x-pack --purge
```

一般情况下，下载速度会很慢，所以最好使用像阿里云之类的云服务建立一个自己的仓库，以便提高下载的速度。

2．集成 Spring Data Elasticsearch

对于一个 Spring Boot 应用，如果可以使用 Spring Data，那么给生产效率带来的提升是非常大的，好在 Spring Data 有针对 Elasticsearch 的子项目，也就是 Spring Data Elasticsearch 了。在 api/build.gradle 中添加 spring-boot-starter-data-elasticsearch 依赖。

```
// 省略
dependencies {
    //省略
    implementation("org.springframework.boot:spring-boot-starter-data-redis")
    implementation("org.springframework.boot:spring-boot-starter-data-elasticsearch") // <--- 这里
    testImplementation("org.springframework.security:spring-security-test")
}
```

需要注意的是，也许是 Elasticsearch 版本管理的问题，也许是 Spring Data Elasticsearch 的成熟度不高的问题，但总之两者配合时需要注意的细节配置问题比使用其他 Spring Data 的子项目时遇到的要多。

其中，最明显的一个需要细心配置的地方是：使用的 Spring Data Elasticsearch 版本号和 Elasticsearch 的版本号有对应关系，从笔者的经验来看，最好和官方推荐的版本一致，否则在使用的过程中会出现很多问题（见表 7-1）。本书定稿时 Spring Data Elasticsearch 3.1 还处于 Milestone 阶段，所以我们采用的 Elasticsearch 版本是 5.5.0。

表 7-1

Spring Boot 版本 (x)	Spring Data Elasticsearch 版本 (y)	Elasticsearch 版本 (z)
x <= 1.3.5	y <= 1.3.4	z <= 1.7.2*
2.x> x >= 1.4.x	2.0.0 <=y < 3.0.0**	2.0.0 <= z < 3.0.0**
x >= 2.x	3.1.0 > y >= 3.0.0	5.5.0
x >= 2.x	y >= 3.1.0	6.2.2

3．添加配置类

对于 Elasticsearch 的 Java 配置类，我们只需简单地提供一个 ElasticsearchTemplate，使用一个自定义的实体映射转换类。这里需要说明的是，对于 Elasticsearch 5.× 来说，TransportClient 是框架推荐的客户端，但到了 Elasticsearch 6.× 之后，官方推荐使用新的 HighLevelRestClient 以 HTTP 协议访问服务器。

```
package dev.local.gtm.api.config;
```

```
// 省略导入

@RequiredArgsConstructor
@Configuration
@EnableConfigurationProperties(ElasticsearchProperties.class)
@ConditionalOnProperty("spring.data.elasticsearch.cluster-nodes")
public class ElasticsearchConfig {

    private final TransportClient transportClient;

    @Bean
    public ElasticsearchTemplate
elasticsearchTemplate(Jackson2ObjectMapperBuilder
jackson2ObjectMapperBuilder) {
        return new ElasticsearchTemplate(transportClient, new
CustomEntityMapper(jackson2ObjectMapperBuilder.createXmlMapper(false).build(
)));
    }

    public class CustomEntityMapper implements EntityMapper {

        private ObjectMapper objectMapper;

        public CustomEntityMapper(ObjectMapper objectMapper) {
            this.objectMapper = objectMapper;

objectMapper.configure(DeserializationFeature.FAIL_ON_UNKNOWN_PROPERTIES,
false);

objectMapper.configure(DeserializationFeature.ACCEPT_SINGLE_VALUE_AS_ARRAY,
true);
        }

        @Override
        public String mapToString(Object object) throws IOException {
            return objectMapper.writeValueAsString(object);
        }

        @Override
        public <T> T mapToObject(String source, Class<T> clazz) throws
IOException {
            return objectMapper.readValue(source, clazz);
        }
```

```
        }
    }
```

一定要注意，上面的 @ConditionalOnProperty("spring.data.elasticsearch.cluster-nodes")注解要求在配置文件中必须指定 cluster-nodes。在 application.yml 中添加 elasticsearch 的属性 cluster-name 和 cluster-nodes。Elasticsearch 是一个分布式的数据库，所以一般情况下有多个节点构成一个集群，cluster-name 是集群的名称，而 cluster-nodes 是集群的节点集合，比如 cluster-nodes: xxx.xxx.xxx.xx:9300, yyy.yy.yyy.yy:9300。但是这里演示的是一个单机环境，所以写一个节点，也就是本机 localhost:9300 即可。

```
spring:
  application:
    name: api-service
  # 省略
  data:
    elasticsearch:
      cluster-name: docker-cluster
      cluster-nodes: localhost:9300
  # 省略
```

这个集群名称以及单节点的设置可以在 Elasticsearch 的配置文件中指定，我们在 docker/elasticsearch 中建立一个子文件夹 config，然后新建一个 elasticsearch.yml 文件。

```
### Elasticsearch 的默认配置
### 摘自 https://github.com/elastic/elasticsearch-docker/blob/master/build/elasticsearch/elasticsearch.yml
# cluster.name: "docker-cluster"
cluster.name: "docker-cluster"
network.host: 0.0.0.0

## 如果配置在一个公网 IP 时，minimum_master_nodes 需要显性配置
## 设置为 1 允许单节点集群
## 相关细节讨论见：https://github.com/elastic/elasticsearch/pull/17288
discovery.zen.minimum_master_nodes: 1

### 使用单节点发现策略，避免启动时的检查
### 见 https://www.elastic.co/guide/en/elasticsearch/reference /current/bootstrap-checks.html
# discovery.type: single-node
discovery.type: single-node
```

在 docker-compose.yml 中指定这个配置文件，在下面的配置中建立 volumes 的映射将配置文件 ./docker/elasticsearch/config/elasticsearch.yml 中映射到容器内 Elasticsearch 的配置文件 /usr/share/elasticsearch/config/elasticsearch.yml，这样就可以让容器启动后使用我们的配置文件了。

```
  version: '3.2'
  services:
    elasticsearch:
      image: docker.elastic.co/elasticsearch/elasticsearch:5.6.8
      volumes:
        - esdata:/usr/share/elasticsearch/data
        -
 ./docker/elasticsearch/config/elasticsearch.yml:/usr/share/elasticsearch/co
nfig/elasticsearch.yml:ro
      environment:
        - discovery.type=single-node
      ports:
        - "9200:9200"
        - "9300:9300"
  ## 省略其他配置
  volumes:
    api_db: {}
    redis-data: {}
    esdata: {}
```

我们这里可以简单地使用 docker-compose up -d elasticsearch 启动 Elasticsearch 服务，如果打开浏览器访问 http://localhost:9200，则可以看到类似如图 7-2 所示的 JSON 输出，这就说明 Elasticsearch 启动正常。9200 是 Elasticsearch 的监控端口，而 9300 是客户端连接 Elasticsearch 服务的端口，这两个不要混淆了，在 Spring Boot 中的 application.yml 配置时一定要使用 9300 端口。

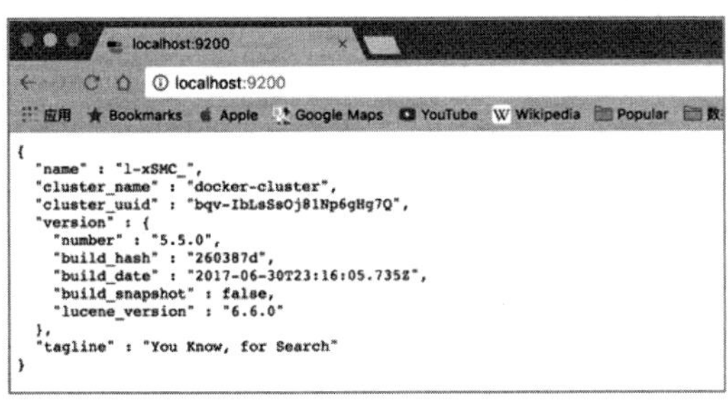

图 7-2

有了这些基础配置之后，可以利用 Elasticsearch 做什么呢？当然是利用它的强项——基于索引的强大数据查询能力。

7.3.2 构建用户查询 API

在 domain 下新建一个 search 包，然后在 domain/search 下新建 UserSearch.java 文件。

```
package dev.local.gtm.api.domain.search;

// 省略导入

@Data
@NoArgsConstructor
@Document(indexName = "users", type = "user")
public class UserSearch implements Serializable {

    private static final long serialVersionUID = 1L;

    @Id
    private String id;

    private String login;

    private String mobile;

    private String name;

    private String email;

    private String avatar;

    private boolean activated;

    private Set<String> authorities;

    public UserSearch(User user) {
        this.id = user.getId();
        this.activated = user.isActivated();
        this.avatar = user.getAvatar();
        this.email = user.getEmail();
        this.login = user.getLogin();
        this.mobile = user.getMobile();
        this.name = user.getName();
        this.authorities = user.getAuthorities().stream()
                .map(Authority::getName)
                .collect(Collectors.toSet());
    }
}
```

这个实体类和 User 很像，为什么要重建一个而不是复用 User 呢？因为无论从数据库角度还是实际的业务角度，这都是两个相对独立的实体，对于搜索来说，我们有些字段是不希望被搜索的，比如密码，有些字段可能对于搜索来说不必以嵌套对象的形式处理，比如权限（authoritg）等。

接下来，我们就需要为这个领域对象建立一个 Respository，用于它的增、删、改、查。在 Repository 下面建立一个子 package 包叫作 search，然后在其中建立一个新的文件 UserSearchRepository.java

```
package dev.local.gtm.api.repository.search;

// 省略导入

@Repository
public interface UserSearchRepository extends
ElasticsearchRepository<UserSearch, String> {
}
```

这时候需要注意的是，我们在同一个项目中，使用了两个 Spring Data 的 Repository。而 Spring Boot 的依赖注入是通过接口类型去匹配和实例化的，尽管我们没有复用领域对象，但最好还是限定一下两种 Repository 的类型的组件扫描范围。

对于 MongoDB 类型的处理在 DatabaseConfig 中增加 @EnableMongoRepositories 注解。

```
// 对于 MongoDB 的 Repository 请限定在 dev.local.gtm.api.repository.mongo 包
中
@RequiredArgsConstructor
@Configuration
@EnableMongoRepositories(basePackages =
"dev.local.gtm.api.repository.mongo")
public class DatabaseConfig {
    // 省略
}
而对于 Elasticsearch 类型的处理在 ElasticConfig 中增加 @EnableElasticsearch
Repositories 注解
// 对于 Elasticsearch 的 Repository 请限定在 dev.local.gtm.api.repository.
search 包中
@RequiredArgsConstructor
@EnableElasticsearchRepositories(basePackages =
"dev.local.gtm.api.repository.search")
@EnableConfigurationProperties(ElasticsearchProperties.class)
@ConditionalOnProperty("spring.data.elasticsearch.cluster-nodes")
@Configuration
public class ElasticConfig {
    // 省略
}
```

1. 改造 Service

我们需要考虑在何时进行 UserSearch 的增、删、改、查。要注意的是，并不是 User 保存时一定会存储 UserSearch，因为 UserSearch 对于有些字段并不关心，比如重置密码的方法就不必涉及 UserSearch。

```
public class AuthServiceImpl implements AuthService {

    private final UserRepository userRepository;

    private final UserSearchRepository userSearchRepository;

    // 省略其他成员变量

    @Override
    public void registerUser(UserDTO userDTO, String password) {

        // 省略

        log.debug("用户 {} 即将创建", newUser);
        userRepository.save(newUser);
        userSearchRepository.save(new UserSearch(newUser)); // <--- 这里
        this.clearUserCaches(newUser);
        log.debug("用户 {} 创建成功", newUser);
    }
    // 省略
}
```

同样地，在 UserServiceImpl 中，我们需要对增加用户（createUser）和更改用户信息（updateUser）中保存 UserSearch，而在删除用户（deleteUser）中之后也需要删除 UserSearch。此外，需要添加一个搜索方法 search，这里采用了 Spring Data Elasticsearch 提供的 queryStringQuery 方法对传入的字符串转换成 Elasticsearch 的查询语句。

```
public class UserServiceImpl implements UserService {

    private final UserRepository userRepository;
    private final UserSearchRepository userSearchRepository;
    // 省略

    @Override
    public User createUser(UserDTO userDTO) {
        // 省略
        userRepository.save(user);
        userSearchRepository.save(new UserSearch(user));
```

```
            this.clearUserCaches(user);
            log.debug("用户: {}", user);
            return user;
        }

        @Override
        public Optional<UserDTO> updateUser(UserDTO userDTO) {
            return
Optional.of(userRepository.findById(userDTO.getId())).filter(Optional::isPre
sent).map(Optional::get)
                    .map(user -> {
                        // 省略
                        userRepository.save(user);
                        userSearchRepository.save(new UserSearch(user));
                        this.clearUserCaches(user);
                        log.debug("用户: {} 的数据已更新", user);
                        return user;
                    }).map(UserDTO::new);
        }

        @Override
        public void deleteUser(String login) {
            userRepository.findOneByLoginIgnoreCase(login).ifPresent(user ->
{
                userRepository.delete(user);
                userSearchRepository.delete(new UserSearch(user));
                this.clearUserCaches(user);
                log.debug("已删除用户: {}", user);
            });
        }

        @Override
        public Page<UserSearch> search(String query, Pageable pageable) {
            return userSearchRepository.search(queryStringQuery(query),
pageable);
        }

        // 省略
    }
```

接下来，就是改造一下 UserResource，把搜索类的路径统一放在 /api/_search/ 下面，把查询的字符串作为路径变量。

```
    @RequiredArgsConstructor
    @RequestMapping("/api")
```

```
    @RestController
    public class UserResource {
        private final UserService userService;

        @GetMapping("/_search/users/{query}")
        public Page<UserSearch> search(@PathVariable String query, final
Pageable pageable) {
            return userService.search(query, pageable);
        }
    }
```

启动应用后，在 http://localhost:8091/swagger-ui.html#/ 中可以看到 /api/_search/users/{query} 的接口，如图 7-3 所示。

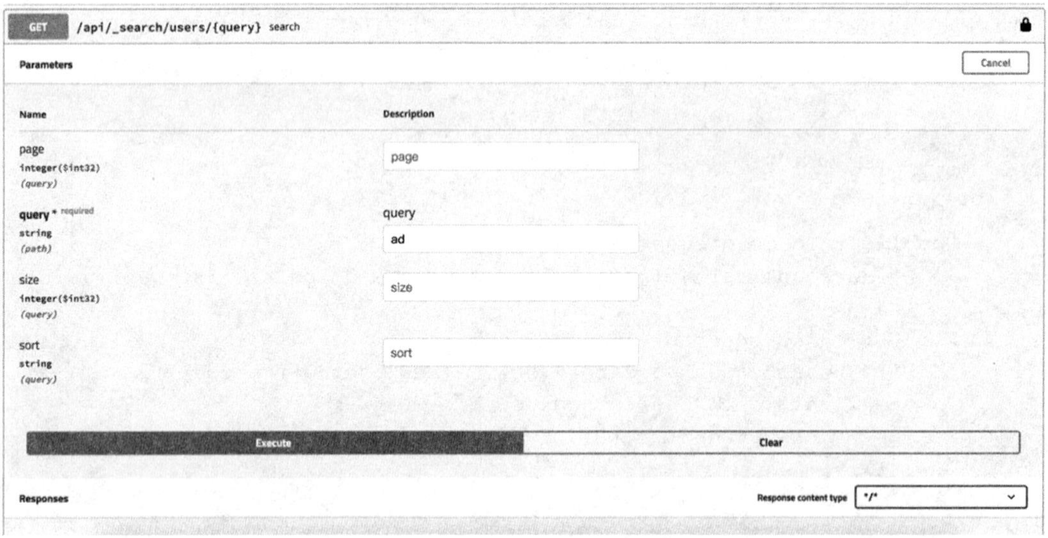

Name	Description
page integer($int32) (query)	page
query * required string (path)	query ad
size integer($int32) (query)	size
sort string (query)	sort

图 7-3

我们可以在 query 中尝试一下 Elasticsearch 的 QueryString Query 的威力，当然要实验之前，请先建立一些必要的数据，新建几个 User，这个可以通过 Swagger 很快地完成。

假设我们的几个用户数据如下：

```
    [
        {
          "id": "5b6919202873f9fe26e24752",
          "login": "lisi",
          "mobile": "13000000002",
          "name": "李四",
          "email": "lisi@local.dev",
```

```
      "avatar": "avatars:svg-2",
      "activated": true,
      "authorities": [
        "ROLE_USER"
      ]
    },
    {
      "id": "5b69193c2873f9fe26e24754",
      "login": "wangwu",
      "mobile": "13000000003",
      "name": "王五",
      "email": "wangwu@local.dev",
      "avatar": "avatars:svg-1",
      "activated": true,
      "authorities": [
        "ROLE_USER",
        "ROLE_ADMIN"
      ]
    },
    {
      "id": "5b6919042873f9fe26e24750",
      "login": "zhangsan",
      "mobile": "13000000001",
      "name": "张三",
      "email": "zhangsan@local.dev",
      "avatar": "avatars:svg-3",
      "activated": true,
      "authorities": [
        "ROLE_USER"
      ]
    }
]
```

如果在 query 中输入“zhangsan”或者“张三”，甚至“张”，都可以看到如下的输出，我们搜索到了这个姓名为“张三”的用户。

```
{
  "content": [
    {
      "id": "5b6919042873f9fe26e24750",
      "login": "zhangsan",
      "mobile": "13000000001",
      "name": "张三",
      "pinyinNameInitials": "zs",
      "email": "zhangsan@local.dev",
```

```
      "avatar": "avatars:svg-3",
      "activated": true,
      "authorities": [
        "ROLE_USER"
      ],
      "createdBy": "admin",
      "createdDate": "2018-08-07T03:59:00.503Z",
      "lastModifiedBy": "admin",
      "lastModifiedDate": "2018-08-07T03:59:00.503Z"
    }
  ],
  "pageable": {
    "sort": {
      "sorted": false,
      "unsorted": true
    },
    "offset": 0,
    "pageSize": 20,
    "pageNumber": 0,
    "paged": true,
    "unpaged": false
  },
  "facets": [],
  "aggregations": null,
  "scrollId": null,
  "totalElements": 1,
  "totalPages": 1,
  "size": 20,
  "number": 0,
  "numberOfElements": 1,
  "sort": {
    "sorted": false,
    "unsorted": true
  },
  "first": true,
  "last": true
}
```

但如果我们搜索“zhang”或者“san”，结果却会返回空，这个主要是用户名的“zhangsan”是一个词，而姓名的“张三”这个中文词其实是两个字（word），所以“张”可以被匹配而“zhang”却没有被匹配到。那如果我们想要匹配“zhang”怎么办呢？可以使用通配符 *，如果在 query 输入框中输入“zhang*”就有可以搜索到同样的结果了。

有的读者可能看到这里会觉得有点奇怪，因为我们没有指定哪个字段，只是给出了要搜索的值，为什么能搜索到结果呢？一般来说在做常规 SQL 查询时应该是 name = "value" 或者模糊查询像 name like %value%，但我们在上面的例子中完全没有指定一个字段。这是由于 Elasticsearch 是一个全文搜索引擎，也就是说它会对所有字段建立索引，从行为上说，非常类似于 Google 这种网络搜索引擎。

那么如果我们希望只对某个字段进行搜索可不可以呢？当然没问题，如果将搜索条件改成“name:张”，那么这个条件就是只应用于 name 这个字段了。你可以试试如果条件改成“name:zhang* ”会返回什么？

另一个问题是如果多个条件怎么办？比如说我们希望找到姓名为“张三”或者“李四”的用户，那么我们可以试试这样的条件“张三 OR 李四”，注意 OR 需要大写，而且之前和之后都有空格。这样的条件会搜索到下面的结果：

```
{
  "content": [
    {
      "id": "5b6919042873f9fe26e24750",
      "login": "zhangsan",
      "mobile": "13000000001",
      "name": "张三",
      "pinyinNameInitials": "zs",
      "email": "zhangsan@local.dev",
      "avatar": "avatars:svg-3",
      "activated": true,
      "authorities": [
        "ROLE_USER"
      ],
      "createdBy": "admin",
      "createdDate": "2018-08-07T03:59:00.503Z",
      "lastModifiedBy": "admin",
      "lastModifiedDate": "2018-08-07T03:59:00.503Z"
    },
    {
      "id": "5b6919202873f9fe26e24752",
      "login": "lisi",
      "mobile": "13000000002",
      "name": "李四",
      "pinyinNameInitials": "ls",
      "email": "lisi@local.dev",
      "avatar": "avatars:svg-2",
      "activated": true,
      "authorities": [
        "ROLE_USER"
```

```
      ],
      "createdBy": "admin",
      "createdDate": "2018-08-07T03:59:28.788Z",
      "lastModifiedBy": "admin",
      "lastModifiedDate": "2018-08-07T03:59:28.788Z"
    }
  ],
  "pageable": {
    "sort": {
      "sorted": false,
      "unsorted": true
    },
    "offset": 0,
    "pageSize": 20,
    "pageNumber": 0,
    "paged": true,
    "unpaged": false
  },
  "facets": [],
  "aggregations": null,
  "scrollId": null,
  "totalElements": 2,
  "totalPages": 1,
  "size": 20,
  "number": 0,
  "numberOfElements": 2,
  "sort": {
    "sorted": false,
    "unsorted": true
  },
  "first": true,
  "last": true
}
```


当然你也可以指定字段“name:张三 OR name:李四”，得到的是同样的结果。

2. 查询语法

QueryString Query 可以使用一套“迷你语言”，QueryString 被解析为一系列术语（term）和运算符（operator）。术语可以是单个单词或短语，也可以用双引号括起来，系统以相同的顺序搜索短语中的所有单词。运算符允许自定义搜索。

（1）字段名称

我们下面会使用一系列的例子来说明。

字段 name 包含 张三

```
name:张三
name 字段包含"张三"或"李四"，如果省略 OR 运算符，将使用默认运算符。
name:(张三 OR 李四)
name:(张三 李四)
name 字段包含完全匹配的 张三 这个词，使用双引号表示要完全匹配。
name:"张三"
字段也可以使用通配符，下面的意思是以 na 开头的字段包含"张三"或"李四"。
na*:(张三 李四)
```

字段 name 包含非空值

```
_exists_:name
```

（2）通配符

可以使用单个术语运行通配符搜索：? 用于替换单个字符，* 用于替换零个或多个字符。

```
zha?g s*
```

请注意，通配符查询可能会占用大量内存并执行性能非常差，只需要想想看我们得需要查询多少个术语才能匹配查询字符串 a * b * c *，就会明白为什么了。尤其是在单词的开头使用通配符（例如 *san），会严重影响性能，因为索引中的所有术语都需要检查是否匹配。

（3）模糊匹配

我们可以使用“模糊”运算符~来搜索与搜索字词类似但不完全相同的字词。

```
zhagnsan~
```

上面的表达式如果你仔细观察，就会发现我们故意把 zhang 写成了 zhagn，其中 g 和 n 颠倒了顺序。但在模糊匹配中这是可以匹配到的，这个功能在有时需要容忍拼写错误时尤其有用。

Elasticsearch 使用 Damerau-Levenshtein distance 来查找最多包含两个更改的所有项，其中更改是单个字符的插入、删除或替换，或是两个相邻字符的转置。默认编辑距离为 2，但编辑距离为 1 应足以捕获所有人为拼写错误的 80%。编辑距离可以在 ~ 后跟一个数字来表示。

```
zhagnsan~1
```

（4）范围

可以为日期、数字或字符串字段指定范围。包含范围边界的使用方括号 [min TO max] 和不含范围边界的使用大括号 {min TO max}。

下面的表达式是 2018 年 8 月。

```
date:[2018-08-01 TO 2018-08-31]
```

下面是 1 ~9 的数字。

```
num:[1 TO 9]
```

大于或等于 10 的数字，下面使用 * 表示开区间，也可以使用 >= 或 <= 表示开区间。

```
num:[10 TO *]
num:>=10
```

大于或等于 1 小于 5 的数字，要注意大括号和中括号可以混合使用。

```
num:[1 TO 5}
```

其实，上面的只是 Elasticsearch 的一小部分功能，更强大的功能可以去官网学习探索。

7.4 Spring Boot Actuator 和数据审计

在企业级开发中，我们经常会碰到需要进行数据审计，以及保留数据变更历史的需求。一般来说，当我们听到审计这个字眼儿的时候，浮现在脑海的应该是构建一个包含每个需要审核的实体的每个版本的日志。这个任务其实比较复杂，但所幸的是，现在我们不需要从头做起了。

Spring Boot Actuator 提供了一个监控和管理生产环境的模块，可以使用 http、jmx、ssh、telnet 等管理和监控应用。审计（auditing）、应用健康（health）、数据采集（metrics gathering）功能会自动加入到应用里面。其中审计功能提供了对于安全事件的发布和订阅，默认的事件是鉴权成功或失败，但是可以自定义事件。这就给了我们一个机制可以处理企业需求中常见的审计功能。

7.4.1 初窥审计事件

如果我们向 build.gradle 中加入 Spring Boot Actuator 的依赖，Spring Boot 就会聪明地、自动地帮我们添加一系列的监控接口。

```
dependencies {
  // 省略
  implementation("org.springframework.boot:spring-boot-starter-actuator")
  // 省略
}
```

1. 监控管理接口

Spring Boot Actuator 内建了一系列的管理接口，这些接口的默认路径前缀是 /actuator，比如 info 的完整路径就是 http://localhost:8080/actuator/info（见表 7-2）。

表 7-2

接口名称	描　述	默认可用	JMX	Web
auditevents	得到当前应用的审计事件	是	是	否
beans	列出当前应用的所有 Bean	是	是	否
conditions	列出用于匹配配置中的类的条件，以及它们是否匹配的原因	是	是	否
configprops	列出所有定义为 @ConfigurationProperties 的属性	是	是	否
env	列出 Spring 中的环境变量	是	是	否
flyway*	列出 flyway 中已经应用的 migration	是	是	否
health	应用的健康状况	是	是	是
httptrace	显示 HTTP 跟踪信息，默认最近的 100 条请求响应信息	是	是	否
info	显示应用信息	是	是	是
loggers	获取或更改应用的日志信息	是	是	否
liquibase**	列出 Liquibase 已经应用的 migration	是	是	否
metrics	列出当前应用的指标信息	是	是	否
mappings	列出所有 @RequestMapping 路径	是	是	否
scheduledtasks	列出应用中的定时任务	是	是	否
sessions	允许获得和删除用户会话，不适用于 Reactive 模式	是	是	否
shutdown	关闭应用	否	是	否
threaddump	执行一个线程的 dump	是	是	否

* flyway 是一个开源的 SQL 数据库版本管理项目，该接口只有在集成了该类库之后才能生效。[1]

** liquibase 是一个开源的 SQL 数据库版本管理项目，该接口只有在集成了该类库之后才能生效。[2]

2. 监控接口的配置

从表 7-2 可以看到，大部分接口默认都是 JMX 可访问，默认 Web 可用的只有 health 和 info 接口，而且默认情况下也禁止了跨域 CORS。但能否设置成 Web 可访问呢？当然可以了，只需要在 application.yml 中做一些设置即可。

```
management:
  endpoints:
    # 默认激活所有监控管理接口
    enabled-by-default: true
    web:
      # 设置路径前缀，默认是 /actuator
```

1 项目地址在：<https://flywaydb.org/>

2 项目地址在：<https://www.liquibase.org/>

```
        base-path: /management
        # 让所有监控接口都接受 Web 访问
        exposure:
          include: "*"
        # 设置跨域
        cors:
          allowed-origins: "*"
          allowed-methods: GET,POST
```

上面，我们通过设置 base-path 也把路径前缀改成了 /management，现在启动应用，然后访问 http://localhost:8080/management，这一次，我们应该得到了更多的接口信息，如图 7-4 所示。

localhost:8080/management

{"_links":{"self":{"href":"http://localhost:8080/management","templated":false},"auditevents":
{"href":"http://localhost:8080/management/auditevents","templated":false},"beans":
{"href":"http://localhost:8080/management/beans","templated":false},"health":
{"href":"http://localhost:8080/management/health","templated":false},"conditions":
{"href":"http://localhost:8080/management/conditions","templated":false},"shutdown":
{"href":"http://localhost:8080/management/shutdown","templated":false},"configprops":
{"href":"http://localhost:8080/management/configprops","templated":false},"env":{"href":"http://localhost:8080/management/env","templated":false},"env-toMatch":{"href":"http://localhost:8080/management/env/{toMatch}","templated":true},"info":
{"href":"http://localhost:8080/management/info","templated":false},"loggers":
{"href":"http://localhost:8080/management/loggers","templated":false},"loggers-name":
{"href":"http://localhost:8080/management/loggers/{name}","templated":true},"heapdump":
{"href":"http://localhost:8080/management/heapdump","templated":false},"threaddump":
{"href":"http://localhost:8080/management/threaddump","templated":false},"metrics-requiredMetricName":
{"href":"http://localhost:8080/management/metrics/{requiredMetricName}","templated":true},"metrics":
{"href":"http://localhost:8080/management/metrics","templated":false},"scheduledtasks":
{"href":"http://localhost:8080/management/scheduledtasks","templated":false},"httptrace":
{"href":"http://localhost:8080/management/httptrace","templated":false},"mappings":
{"href":"http://localhost:8080/management/mappings","templated":false}}}

图 7-4

如果此时启动应用，然后调用 /auth/login 接口，做一次鉴权，成功还是失败是无所谓的，之后访问 http://localhost:8080/management/auditevents 就可以看到鉴权事件了，如图 7-5 所示。

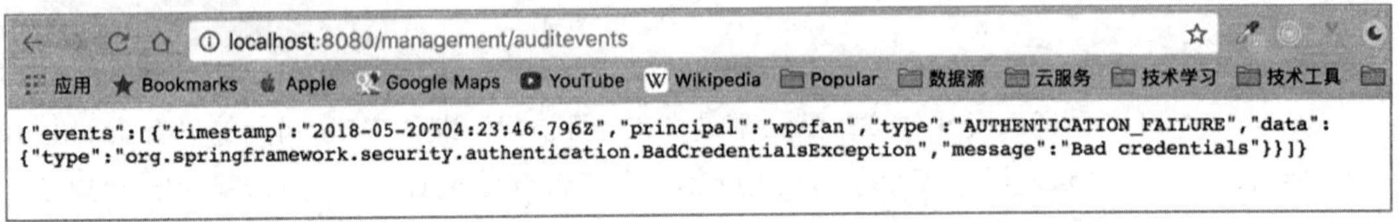

图 7-5

我们把这个返回的 JSON 格式化一下，这样看得更清楚一些，我们得到是一个 events 数组，数组中的每个元素都是一个 org.springframework.boot.actuate.audit.AuditEvent。

```
{
  "events":[
    {
      "timestamp":"2018-05-20T04:23:46.796Z",
      "principal":"wpcfan",
      "type":"AUTHENTICATION_FAILURE",
```

```
        "data":{
"type":"org.springframework.security.authentication.BadCredentialsException",
          "message":"Bad credentials"
        }
      }
    ]
  }
```

3．监控接口的安全

大部分的监控接口我们是不想暴露给普通用户的，否则也太危险了，想想看如果 shutdown 接口被远程的非法用户调用的画面，所以我们需要对接口进行安全保护。所幸的是，由于我们已经集成了 Spring Security，这一点要做到非常容易，所以只需要在 SecurityConfig 中来设置对应接口的权限即可。

```
public class SecurityConfig extends WebSecurityConfigurerAdapter {

    // 省略

    @Override
    protected void configure(HttpSecurity http) throws Exception {
        http.addFilterBefore(corsFilter,
UsernamePasswordAuthenticationFilter.class)
            .exceptionHandling()
            .authenticationEntryPoint(problemSupport)
            .accessDeniedHandler(problemSupport)
            .and()
              .csrf().disable()
              .headers().frameOptions().disable()
            .and()
              .sessionManagement()
              .sessionCreationPolicy(SessionCreationPolicy.STATELESS)
            .and()
              .authorizeRequests()
              .antMatchers("/api/auth/**").permitAll()
              .antMatchers("/api/**").authenticated()
              // 允许健康接口匿名访问
              .antMatchers("/management/health").permitAll()
              // 其他监控接口都需要管理员权限
              .antMatchers("/management/**").hasAuthority(AuthoritiesConst
ants.ADMIN)
              .antMatchers("/v2/api-docs/**").permitAll()
              .antMatchers("/swagger-
```

```
resources/configuration/ui").permitAll()
                .antMatchers("/swagger-ui/index.html").permitAll()
              .and()
                .apply(jwtConfigurer);
        }
    }
```

7.4.2 实现应用的数据审计

通常情况下，审计的需求归结为两个问题。

- 实体的更改或创建发生在什么时间？
- 谁执行的这次变更？

既然这两个问题是属于审计的通用问题，那么所有要求审计的实体就得能够回答这两个问题。因此我们构造一个抽象基类 AbstractAuditingEntity，所有在本应用中需要审计的实体均要继承这个基类。

```
package dev.local.gtm.api.domain;

import com.fasterxml.jackson.annotation.JsonIgnore;
import lombok.Getter;
import lombok.Setter;
import org.springframework.data.annotation.CreatedBy;
import org.springframework.data.annotation.CreatedDate;
import org.springframework.data.annotation.LastModifiedBy;
import org.springframework.data.annotation.LastModifiedDate;
import org.springframework.data.mongodb.core.mapping.Field;

import java.io.Serializable;
import java.time.Instant;


/**
 * 审核特性的领域对象的基础抽象类，包括以下属性：创建人、创建时间、修改人、修改时间
 * 一般情况下如果你希望领域对象有历史操作记录，需要让领域对象继承此基类
 *
 * @author Peng Wang (wpcfan@gmail.com)
 */
@Getter
@Setter
public abstract class AbstractAuditingEntity implements Serializable {

    private static final long serialVersionUID = 1L;
```

```
    @CreatedBy
    @Field("created_by")
    @JsonIgnore
    private String createdBy;

    @CreatedDate
    @Field("created_date")
    @JsonIgnore
    private Instant createdDate = Instant.now();

    @LastModifiedBy
    @Field("last_modified_by")
    @JsonIgnore
    private String lastModifiedBy;

    @LastModifiedDate
    @Field("last_modified_date")
    @JsonIgnore
    private Instant lastModifiedDate = Instant.now();
}
```

我们在这个基类中定义了创建者（createdBy）、创建时间（createdDate）、最近修订者（lastModifiedBy），以及最近修改时间（lastModifiedDate）。而且我们使用了 Spring Data 提供的针对这几个特性的专门注解：@CreatedBy 和 @LastModifiedBy 用于获得谁创建或修改了实体，而 @CreatedDate 和 @LastModifiedDate 用于获取发生的时间点。

当我们使用 @CreatedBy 或 @LastModifiedBy 注解之后，审计框架需要知道当前操作的用户信息，所以要提供一个 AuditorAware<T> 的实现类，让审计框架可以通过 getCurrentAuditor 方法得到当前操作用户。

```
package dev.local.gtm.api.security;

// 省略导入

@Component
public class SpringSecurityAuditorAware implements AuditorAware<String>
{

    @Override
    public Optional<String> getCurrentAuditor() {
        return
Optional.of(SecurityUtils.getCurrentUserLogin().orElse(Constants.SYSTEM_ACCO
UNT));
```

```
    }
}
```

如果要实现比较简单的审计功能，这样就可以了，在实体保存到数据库时，Spring Boot 会自动设置创建/最近修改时间以及创建人和修改人。当然，我们为了让 Spring Boot 激活审计功能，还需要在 DatabaseConfig 中增加一个注解 @EnableMongoAuditing。

```
@EnableMongoAuditing(auditorAwareRef = "springSecurityAuditorAware")
public class DatabaseConfig {
  // 省略
}
```

但是，如果是比较复杂的审计功能，比如要进行对象的版本比较等，自己要实现就比较复杂了。好在有人已经为这种需求提供了强大的解决方案，而且和 Spring Boot 的配合及其良好。下面，我们就有请 JaVers 闪亮登场。

7.4.3 JaVers 和 Spring Boot 集成

JaVer 是一个开源的专注于数据审计和版本控制的软件，它可以和 Spring Boot、Spring Security 进行无缝集成，提供 SQL 数据库和 MongoDB 的良好支持。

集成 JaVers 的步骤非常简单，首先在 build.gradle 中添加依赖。

```
implementation("org.javers:javers-spring-boot-starter-mongo:3.9.7")
```

然后新建一个 AuthorProvider 的实现类，用来提供当前操作用户。

```
package dev.local.gtm.api.config.audit;

// 省略导入

@Component
public class JaversAuthorProvider implements AuthorProvider {

  @Override
  public String provide() {
    return
SecurityUtils.getCurrentUserLogin().orElse(Constants.SYSTEM_ACCOUNT);
  }
}
```

最后，对于要进行数据审计的 Repository，添加一个注解，用 TaskRepository 来举例，我们添加 @JaversSpringDataAuditable 注解之后，JaVers 就会自动处理 Task 的变更。

```
package dev.local.gtm.api.repository.mongo;
```

```
// 省略其他导入
import org.javers.spring.annotation.JaversSpringDataAuditable;

/**
 * 任务存储接口
 */
@JaversSpringDataAuditable
@Repository
public interface TaskRepository extends MongoRepository<Task, String> {
    // 省略
}
```

JaVers 会在 MongoDB 中创建一个叫 jv_snapshots 的 collection，所有的数据变化都会存储在这个 collection 中。

现在需要建立一个 Service，用于获得审计对象的变化历史列表，以及之前一个版本的变更。需要注意的是，我们需要一个带完整包名的类名称，以便得到待审计对象的类型（Class.forName），然后通过 JaVers 提供的 QueryBuilder 构造对应的查询。

```
package dev.local.gtm.api.service;

// 省略导入

@Slf4j
@RequiredArgsConstructor
@Service
public class AuditEventService {

    private static final String basePackageName =
Constants.BASE_PACKAGE_NAME + ".domain.";
    private final Javers javers;

    public Optional<String> getChangesByClassName(String className, Long
commitVersion) {
        try {
            val clazz = Class.forName(basePackageName + className);
            val jqlQuery =
QueryBuilder.byClass(clazz).withVersion(commitVersion - 1);
            val changes = javers.findChanges(jqlQuery.build());
            return
Optional.of(javers.getJsonConverter().toJson(changes));
        } catch (Exception e) {
            return Optional.empty();
        }
```

```
    }

    public Page<EntityAuditEvent> getSnapshots(String entityType,
LocalDate from, LocalDate to, Pageable pageable) throws
ClassNotFoundException {
        log.debug("获得一页指定实体对象类型的审计事件");
        val entityTypeToFetch = Class.forName(basePackageName +
entityType);
        val jqlQuery = QueryBuilder.byClass(entityTypeToFetch)
            .limit(pageable.getPageSize())
            .skip(pageable.getPageNumber() * pageable.getPageSize())
            .withNewObjectChanges(true);
        if (from != null) {
            jqlQuery.from(from);
        }
        if (to != null) {
            jqlQuery.to(to);
        }

        val auditEvents = javers.findSnapshots(jqlQuery.build()).stream()
            .map(snapshot -> {
                EntityAuditEvent event =
EntityAuditEvent.fromJaversSnapshot(snapshot);
                event.setEntityType(entityType);
                return event;
            })
            .collect(Collectors.toList());

        return new PageImpl<>(auditEvents);
    }

    public EntityAuditEvent getPrevVersion(String entityType, String
entityId, Long commitVersion)
        throws ClassNotFoundException {
        val entityTypeToFetch = Class.forName(basePackageName +
entityType);

        val jqlQuery = QueryBuilder.byInstanceId(entityId,
entityTypeToFetch)
            .limit(1)
            .withVersion(commitVersion - 1)
            .withNewObjectChanges(true);

        return
```

```
EntityAuditEvent.fromJaversSnapshot(javers.findSnapshots(jqlQuery.build()).g
et(0));
        }
    }
```

有了 Service 之后，对应的 Controller 就非常简单了，简单地调用上面建好的 AuditEventService 即可。

```
    package dev.local.gtm.api.web.rest;

    // 省略导入

    @RequiredArgsConstructor
    @RestController
    @RequestMapping("/api")
    public class AuditResource {

      private final AuditEventService auditEventService;

      @GetMapping("/management/audits/entities")
      @Secured(AuthoritiesConstants.ADMIN)
      public List<String> getAuditedEntities() {
        return Arrays.asList("User", "Authority", "Task");
      }

      @GetMapping("/management/audits/changes")
      public Page<EntityAuditEvent> getChanges(@RequestParam("entityType")
String entityType, Pageable pageable) {
        try {
          return auditEventService.getChanges(entityType, pageable);
        } catch(ClassNotFoundException e) {
          throw new ResourceNotFoundException("指定的类型没有找到");
        }
      }

      @GetMapping("/management/audits/previous")
      public EntityAuditEvent getPreviousVersion(
        @RequestParam String entityType,
        @RequestParam String entityId,
        @RequestParam Long commitVersion) {
        try {
          return auditEventService.getPrevVersion(entityType, entityId,
commitVersion);
        } catch (ClassNotFoundException e) {
          throw new ResourceNotFoundException("指定的类型没有找到");
```

```
        }
      }
    }
```

现在，如果新建两个用户，然后使用 Swagger 文档界面测试一下，就可以得到类似下面的结果。客户端可以利用这样的结果做出更可视化的历史审计界面。


```
    {
      "content": [
        {
          "id": "3.0",
          "entityId": "5b8140c924a2d075547230a8",
          "entityType": "User",
          "action": "CREATE",
          "entityValue": "{\"resetDate\": \"2018-08-
25T11:43:05.577Z\",\"lastModifiedDate\": \"2018-08-
25T11:43:05.585Z\",\"lastModifiedBy\": \"admin\",\"mobile\":
\"13000000002\",\"avatar\": \"avatars:svg-2\",\"login\":
\"lisi\",\"authorities\":
\"[...Authority/5b813d8524a2d0755472309d]\",\"password\":
\"$2a$10$fWnojM8Xj81/G6WbAa0BwetvS0ZrPacbWhpvjywWQ72.6.OHRG5ny\",\"createdDa
te\": \"2018-08-25T11:43:05.585Z\",\"pinyinNameInitials\":
\"ls\",\"createdBy\": \"admin\",\"name\": \"李四\",\"id\":
\"5b8140c924a2d075547230a8\",\"email\": \"lisi@local.dev\",\"activated\":
\"true\"}",
          "commitVersion": 1,
          "modifiedBy": "admin",
          "modifiedDate": "2018-08-25T19:43:05.591Z"
        },
        {
          "id": "2.0",
          "entityId": "5b81409524a2d075547230a6",
          "entityType": "User",
          "action": "CREATE",
          "entityValue": "{\"resetDate\": \"2018-08-
25T11:42:13.515Z\",\"lastModifiedDate\": \"2018-08-
25T11:42:13.525Z\",\"lastModifiedBy\": \"admin\",\"mobile\":
\"13000000001\",\"avatar\": \"avatars:svg-1\",\"login\":
\"zhangsan\",\"authorities\":
\"[...Authority/5b813d8524a2d0755472309d]\",\"password\":
\"$2a$10$B0cL9.GNwYdxo9qAFGUzF.pJuVWtyNtsomw6AuPjdWXIN6ADLtF3q\",\"createdDa
te\": \"2018-08-25T11:42:13.525Z\",\"pinyinNameInitials\":
\"zs\",\"createdBy\": \"admin\",\"name\": \"张三\",\"id\":
\"5b81409524a2d075547230a6\",\"email\": \"zhangsan@local.dev\",\"activated\":
\"true\"}",
```

```
        "commitVersion": 1,
        "modifiedBy": "admin",
        "modifiedDate": "2018-08-25T19:42:13.538Z"
      },
      {
        "id": "1.0",
        "entityId": "5b813d8524a2d075547230a0",
        "entityType": "User",
        "action": "CREATE",
        "entityValue": "{\"lastModifiedDate\": \"2018-08-
25T11:29:09.376Z\",\"lastModifiedBy\": \"system\",\"mobile\":
\"13999999999\",\"avatar\": \"avatar-0\",\"login\":
\"admin\",\"authorities\":
\"[...Authority/5b813d8524a2d0755472309e, ...Authority/5b813d8524a2d07554723
09d]\",\"password\":
\"$2a$10$NitMS5jTz6lA.C/HHvzPPu30xdqqYh0mvtDzTWiupQzjO7ko0U6pK\",\"createdDa
te\": \"2018-08-25T11:29:09.376Z\",\"pinyinNameInitials\":
\"cjgly\",\"createdBy\": \"system\",\"name\": \"超级管理员\",\"id\":
\"5b813d8524a2d075547230a0\",\"email\": \"admin@local.dev\",\"activated\":
\"true\"}",
        "commitVersion": 1,
        "modifiedBy": "system",
        "modifiedDate": "2018-08-25T19:29:09.452Z"
      }
    ],
    "pageable": "INSTANCE",
    "totalElements": 3,
    "last": true,
    "totalPages": 1,
    "size": 0,
    "number": 0,
    "numberOfElements": 3,
    "first": true,
    "sort": {
      "sorted": false,
      "unsorted": true
    }
  }
```

7.5 WebSocket 实时通信服务

WebSocket 协议（RFC 6455）提供了一种标准化的方式来进行服务器和客户端之间进行双

工、双向的基于 TCP 连接的通信方式。它既不同于 HTTP，又构建于 HTTP 之上，它可以使用 HTTP 的协议端口，比如 80 或 443。

一个 WebSocket 交互是由一个 HTTP 请求开始的，这个 HTTP 请求使用 "Upgrade" 头去转换到 WebSocket 协议。

```
GET ws://localhost:8080/websocket/app HTTP/1.1
Host: localhost:8080
Connection: Upgrade
Pragma: no-cache
Cache-Control: no-cache
Upgrade: websocket
Origin: chrome-extension://heeinifidncnmblmddfajlikkiihfgai
Sec-WebSocket-Version: 13
User-Agent: Mozilla/5.0 (Macintosh; Intel Mac OS X 10_13_4)
AppleWebKit/537.36 (KHTML, like Gecko) Chrome/66.0.3359.181 Safari/537.36
Accept-Encoding: gzip, deflate, br
Accept-Language: zh-CN,zh;q=0.9,en-US;q=0.8,en;q=0.7
Cookie: _ga=GA1.1.13036277.1526997915; m=
Sec-WebSocket-Key: I88buTTIKcJE1QKxHAMpnA==
Sec-WebSocket-Extensions: permessage-deflate; client_max_window_bits
Sec-WebSocket-Protocol: v10.stomp, v11.stomp, v12.stomp
支持 WebSocket 的服务器端收到这个请求后，会返回
HTTP/1.1 101 Switching Protocols
Expires: 0
Cache-Control: no-cache, no-store, max-age=0, must-revalidate
X-XSS-Protection: 1; mode=block
Origin: chrome-extension://heeinifidncnmblmddfajlikkiihfgai
Upgrade: WebSocket
Pragma: no-cache
Sec-WebSocket-Accept: G+NHta8H3Mjiol8xeT3T3Tilabw=
Date: Thu, 07 Jun 2018 04:45:14 GMT
Connection: Upgrade
Sec-WebSocket-Location: ws://localhost:8080/websocket/app
X-Content-Type-Options: nosniff
Sec-WebSocket-Protocol: v10.stomp
```

在成功的“握手”之后，TCP Socket 会保持对服务器和客户端打开以便发送和接收消息。注意，如果 WebSocket 服务器在 Web 服务器后面，比如有时采用的 Web 服务器是 Nginx，通过反向代理把某些请求指向我们的 API 服务，这时就需要配置 Nginx 将 WebSocket 的 Upgrade 请求发送到 WebSocket 服务器。

7.5.1 HTTP 和 WebSocket 的区别和联系

我们在考虑使用 WebSocket 时要注意，通常的 Rest API 编程模型和 WebSocket 编程模型的区别是巨大的，这一点无论对服务端还是客户端来说都是如此。在 Rest API 模型中，我们把资源划分成若干 URL，客户端需要使用 Request/Response 的机制去访问这些 URL，服务端也会根据这些 Request 的 URL、方法和请求头来决定对应的处理方式。

而对于 WebSocket 来说，通常是建立一个连接后就通过这个连接发送后继的消息，这个是一个一直保持连接的模型，而不是像 REST 方式那样每次建立新的连接。一般来说，这种模式要求的编程模型是事件驱动类型的，和 Request/Response 的模型区别较大。

另外，WebSocket 是一个较基础的传输协议，没有像 HTTP 那样规定内容的格式、语义，我们一般需要在其之上再封装一层协议来处理内容、格式等，比如我们下面要谈到的 STOMP。

7.5.2 何时使用 WebSocket

初学者往往觉得既然可以实时，那就所有业务都实时化好了，但实时是有成本的，而且大多数场景不需要实时，对这些场景进行实时化处理代价高昂，却收益甚微。

WebSocket 主要用于处理实时性较高的需求，但它并不是这类问题的唯一解决方案，在很多时候，利用 Ajax 或者 HTTP 流、HTTP 长连接等方式可以获得类似的效果和体验，甚至很多场景下，用其他方案会比 WebSocket 更好。

并非所有应用都要求实时性，多人在线游戏和股票类应用需要更高的实时性，而新闻类，邮件类或者社交类应用往往只需要阶段性的更新即可。除了延时，数据的传输量也是一个考虑因素，如果数据量不大的情况下，使用 HTTP 长连接（long pooling）应该是更好的选择。此外网络环境也是一个考虑因素，如果我们的 Web 代理并不支持，或者使用的云服务不支持 Upgrade 到一个需要常驻的连接，那么也就无法使用 WebSocket 了。

7.5.3 STOMP

WebSocket 协议定义了两种类型的消息，文本和二进制，但它们的内容格式是未定义的。这些通过客户端和服务器协商子协议的机制，即更高级别的消息传递协议来完成。在 WebSocket 之上使用子协议来定义每个消息可以发送什么类型的消息，每个消息的格式和内容是什么等。子协议的使用是可选的，但无论是客户端还是服务器都需要就定义消息内容的某些协议达成一致。

和 Node.js 生态中的 Socket.js 类似，Spring 采用了一个叫作 STOMP 的子协议。STOMP 是由 Google 开发的一套基于 WebSocket 之上的消息通作协议，和传统的 AMQP 协议以及 JMS 的作用类似，但不像传统的协议那么复杂，只包含最常见的一些操作。

STOMP 是一种简单的，面向文本的消息传递协议，最初是为 Ruby、Python 和 Perl 等脚本语言创建的，用于连接企业消息代理。它旨在解决常用消息传递模式的最小子集。STOMP 可用于任何可靠的双向流网络协议，如 TCP 和 WebSocket。虽然 STOMP 是面向文本的协议，但消息是可以携带文本或二进制内容的。

STOMP 客户端可以使用 SEND 或 SUBSCRIBE 命令发送或订阅消息以及 Destination Header，这个 Destination Header 描述消息的内容以及应由谁接收消息。这启用了一个简单的“发布—订阅”机制，可用于通过代理将消息发送到其他连接的客户端，或者向服务器发送消息以请求执行某些工作。

使用 Spring 的 STOMP 支持时，Spring WebSocket 应用充当客户端的 STOMP 代理。消息被路由到 @Controller 注解的消息处理方法或者路由到一个简单的内存代理，该代理跟踪订阅并向订阅用户广播消息。

使用 STOMP 作为子协议能够提供比 WebSocket 更丰富的编程模型。

- 无自己定义自定义消息传递协议和消息格式。
- 可以使用支持 STOMP 的客户端。
- 可以使用像 RabbitMQ、ActiveMQ 等消息代理软件来管理订阅和广播消息。
- 应用程序逻辑组织成若干由 @Controller 注解的方法中，可以根据 STOMP Destination Header 将对应的消息路由给这些方法。
- 使用 Spring Security 进行安全保障。

7.5.4 WebSocket 配置

在 Spring Boot 中加入 WebSocket 的支持非常简单，只需要创建一个配置类，这个类需要实现 WebSocketMessageBrokerConfigurer 这个接口。

```
package dev.local.smartoffice.api.config;

// 省略导入

@Order(Ordered.HIGHEST_PRECEDENCE + 99)
@RequiredArgsConstructor
@Configuration
@EnableWebSocketMessageBroker
public class WebSocketConfig implements WebSocketMessageBrokerConfigurer
{
```

```
    private final AppProperties appProperties;
    private final ChannelRegistrationInterceptor
channelRegistationInterceptor;

    @Override
    public void configureClientInboundChannel(ChannelRegistration
registration) {
        registration.interceptors(channelRegistationInterceptor);
    }

    @Override
    public void configureMessageBroker(MessageBrokerRegistry config) {
        config.enableSimpleBroker("/topic");
        config.setApplicationDestinationPrefixes("/app");
    }

    @Override
    public void registerStompEndpoints(StompEndpointRegistry registry) {
        String[] allowedOrigins =
Optional.ofNullable(appProperties.getCors().getAllowedOrigins())
            .map(origins -> origins.toArray(new String[0]))
            .orElse(new String[0]);
        registry.addEndpoint("/websocket/tracker")
            .setAllowedOrigins(allowedOrigins);
        registry.addEndpoint("/websocket/tracker")
            .setAllowedOrigins(allowedOrigins)
            .withSockJS();
    }
}
```

WebSocketConfig 使用 @Configuration 注解，表明它是 Spring 配置类。它还使用了注解 @EnableWebSocketMessageBroker，@EnableWebSocketMessageBroker 支持 WebSocket 消息处理，而这个处理是交由消息的代理进行的。

configureMessageBroker() 方法在 WebSocketMessageBrokerConfigurer 中配置消息代理。它首先调用 enableSimpleBroker() 启用一个简单的基于内存的消息代理，以在前缀为 /topic 的目标 URL 上将消息传回客户端。它还为使用 @MessageMapping 注解的方法中的消息指定了 /app 前缀，这个前缀会用于定义所有消息映射。

registerStompEndpoints() 方法注册 /websocket/tracker 这个 Endpoint，也就是建立连接的地址。我们通过 withSockJS() 启用 SockJS 作为后备选项，以便在 WebSocket 不可用时可以使用备用传输协议。SockJS 客户端将尝试链接到 ws://localhost:8091/websocket/tracker 并使用可用的

最佳的、可用的传输协议（websocket、xhr-streaming、xhr-polling 等）。

请注意，我们注册了两个 Endpoints，但它们的 URL 是一样的，区别仅仅是是否配置成启用 SockJS。这么做的原因是一般在前端可以使用 SockJS，但在 App 客户端，比如 Android 或 iOS 上可能会采用其他支持 STOMP 协议的类库，这种情况下是无法访问启用了 SockJS 的 URL 的。所以这里面注册了两个 Endpoints，分别对应不同的客户端。

configureClientInboundChannel() 方法中我们配置了拦截器，这个拦截器主要是为 WebSocket 的安全连接服务的，下面的章节会讲到。

7.5.5 WebScoket 安全

WebSocket 消息传递会话中的每个 STOMP 都以 HTTP 请求开始——可以是升级到 WebSockets 的请求（即 WebSocket 握手），也可以是在 SockJS 回退使用 HTTP 传输请求的情况。

Web 应用程序已经具有用于保护 HTTP 请求的身份验证和授权。通常，用户通过 Spring Security 使用某种机制（例如登录页面、HTTP 基本身份验证等）进行身份验证。经过身份验证的用户的安全上下文保存在 HTTP 会话中。Spring Security 提供 WebSocket 子协议授权，该授权使用 ChannelInterceptor 根据其中的用户头来授权消息。

请注意，STOMP 协议在 CONNECT 帧上确实有“登录”和“密码”标头。这些最初设计用于并且仍然需要例如用于 TCP 上的 STOMP。但是，对于基于 WebSocket 的 STOMP，Spring 默认忽略 STOMP 协议级别的授权标头，并假定用户已在 HTTP 传输级别进行了身份验证，期望 WebSocket 或 SockJS 会话包含经过身份验证的用户。

但是我们的应用使用并不是基于 Session 的，而是基于令牌的，因此使用 token 的应用无法在 HTTP 协议级别进行身份验证。它们可能更喜欢在 STOMP 消息传递协议级别使用标头进行身份验证：

- 使用 STOMP 客户端在连接时传递身份验证标头。
- 使用 ChannelInterceptor 处理身份验证标头。

如果我们想和 REST API 中一样采用 token 进行鉴权处理，就需要使用截断器对 STOMP 的 CONNECT 命令对 Header 进行处理。这里我们构建一个 ChannelRegistrationInterceptor。

```
package dev.local.smartoffice.api.config.websocket;

// 省略导入

/**
 * 使用截断器对 WebSocket 连接进行鉴权检验
```

```
   */
  @RequiredArgsConstructor
  @Component
  public class ChannelRegistrationInterceptor implements
ChannelInterceptor {

      private final AppProperties appProperties;
      private final TokenProvider tokenProvider;

      @Override
      public Message<?> preSend(Message<?> message, MessageChannel channel)
{
          StompHeaderAccessor accessor =
MessageHeaderAccessor.getAccessor(message, StompHeaderAccessor.class);
          assert accessor != null;
          if (StompCommand.CONNECT.equals(accessor.getCommand())) {
              val bearerTokenHeader =
accessor.getNativeHeader(appProperties.getSecurity().getAuthorization().getH
eader());
              if(bearerTokenHeader == null) {
                  throw new AuthenticationCredentialsNotFoundException("没
有找到鉴权头信息");
              }
              val bearerToken = bearerTokenHeader.get(0);
              String prefix =
appProperties.getSecurity().getJwt().getTokenPrefix();
              if (!StringUtils.hasText(bearerToken)
|| !bearerToken.startsWith(prefix)) {
                  throw new AuthorizationServiceException("鉴权信息不存在或格
式错误");
              }
              String token = bearerToken.substring(prefix.length());
              Authentication user = tokenProvider.getAuthentication(token);
// access authentication header(s)
              accessor.setUser(user);
          }
          return message;
      }
  }
```

在 WebSocketConfig 中配置该截断器，Spring 将记录并保存经过身份验证的用户，并将其与同一会话中的后续 STOMP 消息相关联。

```
  public class WebSocketConfig implements WebSocketMessageBrokerConfigurer
{
```

```
        private final ChannelRegistrationInterceptor
channelRegistationInterceptor;

        @Override
        public void configureClientInboundChannel(ChannelRegistration
registration) {
            registration.interceptors(channelRegistationInterceptor);
        }
        // 省略其他
    }
```

此外，我们还可以对 STOMP 中的端点和代理进行权限控制，非常类似我们在 SecurityConfig 中做的配置，建立一个 WebsocketSecurityConfiguration。

```
    package dev.local.smartoffice.api.config;

    // 省略导入

    @RequiredArgsConstructor
    @Configuration
    public class WebsocketSecurityConfiguration extends
AbstractSecurityWebSocketMessageBrokerConfigurer {

        @Override
        protected void
configureInbound(MessageSecurityMetadataSourceRegistry messages) {
            messages
                .nullDestMatcher().authenticated()
                .simpDestMatchers("/websocket/tracker").hasAuthority(Authori
tiesConstants.USER)
                // 对于以/topic/开头的所有目标位置要求鉴权
                .simpDestMatchers("/topic/**").authenticated()
                .simpDestMatchers("/app/**").authenticated()
                // 对于消息类型为 MESSAGE and SUBSCRIBE 之外的消息禁止访问
                .simpTypeMatchers(SimpMessageType.MESSAGE,
SimpMessageType.SUBSCRIBE).denyAll()
                // 其他情况全部禁止访问
                .anyMessage().denyAll();
        }

        /**
         * 对于 WebSocket，禁用 CSRF
         */
        @Override
        protected boolean sameOriginDisabled() {
```

```
            return true;
        }
    }
```

7.5.6 建立一个实时消息 Controller

为简单起见，我们建立一个简单的回声服务，在收到客户端发出消息后将该消息广播到 /topic/echo，此时所有订阅了 /topic/echo 的人也会收到这个消息。

```
package dev.local.smartoffice.api.web.websocket;

// 省略导入

/**
 * JdProfileController
 */
@RequiredArgsConstructor
@Controller
public class EchoController {

    private final SimpMessagingTemplate template;

    @MessageMapping("/echo")
    @SendTo("/topic/echo")
    public String echoMessage(@Payload String message) {
        return "reply: " + message;
    }
}
```

7.5.7 测试 WebSocket

为了测试 WebSocket，我们需要有一个可以支持 STOMP 的客户端，这里采用一个 Chrome 扩展插件叫作 Websocket STOMP Client，这个插件需要到 Chrome 商店去下载。安装好插件后，我们就可以在 URL 中输入 ws://localhost:8080/websocket/tracker，然后在 Connection Headers 中填写一个鉴权头，和之前讲过的 Web 鉴权方式一样，是一个 Authorization: Bearer <token> 这样的格式。此时我们点击 Connect 按钮就会发现 State 变成了 CONNECTED，表示连接成功了，如图 7-6 所示。

连接上之后，在 Subscribe 文本框中输入 /topic/echo 以订阅这个主题。然后在 Send to topic 的 Topic 中输入 /app/echo（还记得我们定义的前缀 /app 吗？），如图 7-7 所示，在 Message 中填写 hello。

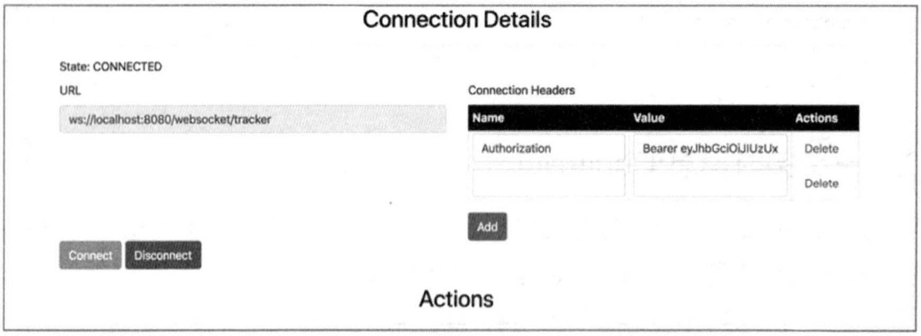

图 7-6

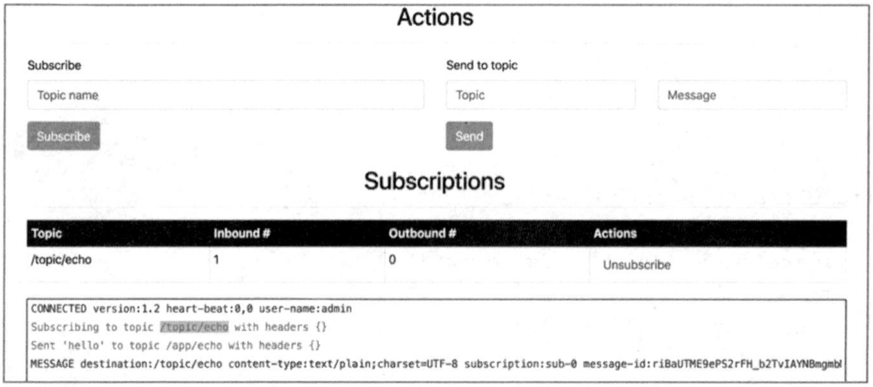

图 7-7

我们可以从下面的文本中看到连接、订阅、发送、接收消息的全过程。

```
// 连接消息
CONNECTED version:1.2 heart-beat:0,0 user-name:admin
// 订阅 /topic/echo 主题
Subscribing to topic /topic/echo with headers {}
// 向 app/echo 发送 hello 这个消息
Sent 'hello' to topic /app/echo with headers {}
// 订阅的 /topic/echo 收到了服务器返回的消息
MESSAGE destination:/topic/echo
content-type:text/plain;charset=UTF-8
subscription:sub-0
message-id:riBaUTME9ePS2rFH_b2TvIAYNBmgmbMvPRa9M4U7-0
content-length:12
reply: hello
```

还可以同时打开多个 Websocket STOMP Client，订阅同一个主题，看看是否所有 Client 都会收到实时消息。

7.6 Spring Boot 的自动化测试

一个大型项目如果没有自动化测试，那交付质量可就岌岌可危了。因为一个大型项目，会有较多的开发团队参与分工，代码的复杂加上协同人员的增多，导致产生的 bug 往往需要多个项目组配合才能发现根本原因，这一方面增加了项目成本，另一方面严重影响生产效率。因为现代软件的迭代速度很快，如果完全使用人工测试，则会严重拖慢项目进度。自动化测试在大工程中会确保每次提交的代码保证一定程度的质量。

但自动化测试也不是万能的，并不能替代人工测试，比如有些需要较复杂操作才能呈现的 bug，比如由于软件架构的限制，无法或者难以实现测试自动化的地方。而且自动化测试也并不适合一个项目在原型开发的阶段大规模推进，因为此时的代码修改会比较频繁。此外笔者也并不赞成追求测试的覆盖度，因为有一些类和方法完全没有测试的必要。比如单纯的数据对象，只有 getter 和 setter 这样的方法，我们对它们进行自动化测试的意义就仅仅在于测试覆盖度提升的百分点，这种工作既耗时又无聊，不做也罢。但对于对外部的接口，以及内部重要的业务逻辑来说，这个自动化测试就很有必要。因为我们希望不会因为某个改动导致公开给其他团队的接口不好用了，或者影响到内部以前好用的业务逻辑，这时自动化测试就是一个提升生产效率的利器。

如何开始构建一个测试

按照标准的 Java 项目结构，src/main 用于存储源代码，而 src/test 用于测试的代码。我们在 src/test 中使用项目主工程同样的包结构。只不过对于测试来说，我们一般对于类的命名后面都会有一个 Test 以标识这是一个测试类。

在开始写测试类之前，首先确定你在 build.gradle 中是否已经引入测试所需要的依赖，我们对测试的大部分依赖通过 org.springframework.boot:spring-boot-starter-test 即可提供。

```
dependencies {
    testImplementation("org.springframework.boot:spring-boot-starter-
test")
}
```

Spring 会自动导入：

- JUnit——Java 单元测试框架，已经基本是业界标配了。
- Spring Test & Spring Boot Test——提供 Spring Boot 集成测试的支持类库和工具类库。
- AssertJ——断言类库。
- Hamcrest——用于条件约束和匹配的类库。
- Mockito——用于模拟对象的框架。

- JSONassert——JSON 的断言类库。
- JsonPath——JSON 路径表达语言类库。

在 Spring 中进行测试是非常简单的，这是因为 Spring 提供良好的依赖注入特性，而这使得我们的代码更容易进行单元测试。在进行单元测试的时候，经常使用的一个技巧是使用 Mock Object 而不是真正的依赖项。除去单元测试外，很多时候还需要进行更真实的测试——集成测试，这在 Spring 中意味着我们需要使用 Spring ApplicationContext。在集成测试中，我们经常碰到的挑战是怎样才能在不需要部署应用程序或连接到其他服务的情况下执行测试。

Spring Boot 提供 @SpringBootTest 注解，当需要 Spring Boot 特性时，使用它就可以了，这个注解配合 @RunWith(SpringRunner.class) 注解即可标识一个测试类。

默认情况下，@SpringBootTest 不会启动服务器，但可以使用 webEnvironment 属性来进一步指定测试的运行方式。

- MOCK——这是默认值，提供模拟的 Web 环境，不会启动服务器。它可以与@AutoConfigureMockMvc 或@AutoConfigureWebTestClient 配合使用进行基于模拟的 Web 环境的测试。
- RANDOM_PORT——随机端口，加载 WebServerApplicationContext 并提供真实的 Web 环境。嵌入式服务器的端口是随机分配的。
- DEFINED_PORT——指定端口，和 RANDOM_PORT 类似，只不过端口是通过 application.yml 指定的。
- NONE——不提供任何 Web 环境。

1. Web 层单元测试

下面我们就来看看如何写一个测试，这个测试中，我们将会使用 @WebMvcTest 这个注解，和 @SpringBootTest 不同，@WebMvcTest 注解是单独为 Web 层测试准备的。@WebMvcTest 只扫描 @Controller、@ControllerAdvice、@JsonComponent、Converter、GenericConverter、Filter、WebMvcConfigurer 和 HandlerMethodArgumentResolver。其他的比如 @Component 之类就不会被扫描到了。如果需要使用其他类型，则可以使用 @Import 导入。

```
package dev.local.gtm.api.rest;

// 省略导入

@RunWith(SpringRunner.class)
@WebMvcTest(controllers = {AuthResource.class}, secure = false)
public class AuthResourceTest {
```

```
    @MockBean
    private AppProperties appProperties;

    @Autowired
    private MockMvc mockMvc;

    @Autowired
    private ObjectMapper objectMapper;

    @MockBean
    private AuthService authService;

    @Autowired
    private HttpMessageConverter[] httpMessageConverters;

    @Autowired
    private ExceptionTranslator exceptionTranslator;

    @Before
    public void setup() {
        val authResource = new AuthResource(authService, appProperties);
        mockMvc = MockMvcBuilders.standaloneSetup(authResource)
            .setMessageConverters(httpMessageConverters)
            .setControllerAdvice(exceptionTranslator)
            .setViewResolvers((ViewResolver) (viewName, locale) -> new
MappingJackson2JsonView()).build();
    }

    @Test
    public void testCaptchaRequestSuccessfully() throws Exception {
        val captcha = new Captcha();
        captcha.setUrl("http://someplace/somepic.jpg");
        captcha.setToken("someToken");
        given(this.authService.requestCaptcha()).willReturn(captcha);
        mockMvc.perform(get("/api/auth/captcha"))
            .andExpect(status().isOk())
            .andExpect(jsonPath("$.captcha_url").isNotEmpty())
            .andExpect(jsonPath("$.captcha_token").isNotEmpty());
    }

    @Test
    public void testCaptchaVerificationSuccessfully() throws Exception {
        val code = "testCode";
        val token = "testToken";
```

```
        val verification = new AuthResource.CaptchaVerification();
        verification.setCode(code);
        verification.setToken(token);
        given(this.authService.verifyCaptcha(code,
token)).willReturn("testValidateToken");
        mockMvc.perform(post("/api/auth/captcha")
            .contentType(MediaType.APPLICATION_JSON_UTF8)
            .content(objectMapper.writeValueAsString(verification)))
            .andDo(print())
            .andExpect(status().isOk())
            .andExpect(jsonPath("$.validate_token").isNotEmpty());
    }

    @Test
    public void testRegisterSuccess() throws Exception {

        val validateToken = "testValidateToken";
        val user = UserVM.builder()
            .login("test1")
            .mobile("13000000000")
            .email("test1@local.dev")
            .name("test 1")
            .password("12345")
            .validateToken(validateToken)
            .build();
        val security = new AppProperties.Security();
        doNothing()
            .when(this.authService)
            .verifyCaptchaToken(validateToken);
        doNothing()
            .when(this.authService)
            .registerUser(user.toUserDTO(), "12345");
        given(this.authService.login(user.getLogin(),
user.getPassword()))
            .willReturn(new JWTToken("idToken", "refreshToken"));
        given(this.appProperties.getSecurity())
            .willReturn(security);

        mockMvc.perform(post("/api/auth/register")
            .contentType(MediaType.APPLICATION_JSON_UTF8)
            .content(objectMapper.writeValueAsString(user)))
            .andDo(print())
            .andExpect(status().isOk())
            .andExpect(jsonPath("$.id_token").isNotEmpty());
```

```
        }

    }
```

上面的例子中由于我们只想测试 Web 层，所以对于 Web 层所依赖的 AuthService 和 AppProperties 使用了 @MockBean 注解。

运行测试时，有时需要在上下文中模拟某些组件或服务。例如，刚刚的例子中我们只想单独测试 Web 层，对于服务层可能并不关心，再比如在应用依赖的某些开发期间不可用的某些远程服务，或者模拟在真实环境中可能难以触发的故障时等，我们就会发现需要模拟某些对象的行为。

Spring Boot 包含一个 @MockBean 注解，这个注解将会为修饰的类型创建一个基于 Mockito 的模拟对象。对于每个 @Test 方法，该模拟对象都会自动重置。

那么我们该如何使用这些模拟对象呢？我们通过一个简单的例子来说明，上面代码的 testCaptchaRequestSuccessfully 测试方法中，我们是要测试 /api/auth/captcha 这个 Rest API。那么先来看看要测试的目标方法是怎么定义的？

```
@GetMapping("/auth/captcha")
public Captcha requestCaptcha() {
    log.debug("REST 请求 -- 请求发送图形验证码 Captcha");
    return authService.requestCaptcha();
}
```

这个方法很简单，就是直接调用了服务层的对应方法而已，从实践角度来说，这种就是典型的不需要测试的那种方法，但这里为了更好地说明一些测试技巧，就把它当作例子了。

这个 requestCaptcha 方法中有一个依赖，就是 AuthService，也就是说如果访问 /api/auth/captcha，那么就会进入 requestCaptcha 方法，但是这个方法中又调用了服务层的方法：authService.requestCaptcha()。如果要单独测试 requestCaptcha，就需要提供一个 AuthService 的模拟对象。这个模拟对象我们在测试类中已经使用 @MockBean 提供了，而且在 setup() 中通过 val authResource = new AuthResource(authService, appProperties); 将模拟对象使用构造注入了 AuthResource 中（请回过头再想想依赖注入的妙处）。也就是说，在我们的测试中，AuthResource 使用的是这个模拟的 AuthService，而非真正的 AuthService。mockMvc.perform (get("/api/auth/captcha")) 说的是向 /api/auth/captcha 发送一个 GET 请求。然后使用 andExpect 来指定一个期望结果，比如第一个期望结果就是 status().isOk()，返回码是 OK，一般来说就是 200。而 jsonPath 是可以使用类似 XML 中 xpath 的形式来表达 JSON 中的节点路径的，就是说在返回的 Http Response 中的 JSON 对象中的第一级节点中的 captcha_url 节点。我们期待这个节点不为空。

```
@Test
public void testCaptchaRequestSuccessfully() throws Exception {
    val captcha = new Captcha();
    captcha.setUrl("http://someplace/somepic.jpg");
    captcha.setToken("someToken");
    given(this.authService.requestCaptcha()).willReturn(captcha);
    mockMvc.perform(get("/api/auth/captcha"))
      .andExpect(status().isOk())
      .andExpect(jsonPath("$.captcha_url").isNotEmpty())
      .andExpect(jsonPath("$.captcha_token").isNotEmpty());
}
```

细心的读者此时会发现一个问题，如果方法有返回值，使用上面的 given(...).willReturn(...) 这种方法去模拟当然没有问题，但没有返回值的怎么办？这种情况下我们可以使用 doNothing().when(someService).someVoidMethod() 这种模拟方式：就是说在 someService 的 someVoidMethod() 被调用时什么都不做。下面的代码就很好地体现了这个范式。

```
@Test
public void testRegisterSuccess() throws Exception {

    val validateToken = "testValidateToken";
    val user = UserVM.builder()
        .login("test1")
        .mobile("13000000000")
        .email("test1@local.dev")
        .name("test 1")
        .password("12345")
        .validateToken(validateToken)
        .build();
    val security = new AppProperties.Security();
    doNothing()
        .when(this.authService)
        .verifyCaptchaToken(validateToken);
    doNothing()
        .when(this.authService)
        .registerUser(user.toUserDTO(), "12345");
    given(this.authService.login(user.getLogin(), user.getPassword()))
        .willReturn(new JWTToken("idToken", "refreshToken"));
    given(this.appProperties.getSecurity())
        .willReturn(security);

    mockMvc.perform(post("/api/auth/register")
        .contentType(MediaType.APPLICATION_JSON_UTF8)
        .content(objectMapper.writeValueAsString(user)))
```

```
            .andDo(print())
            .andExpect(status().isOk())
            .andExpect(jsonPath("$.id_token").isNotEmpty());
    }
```

接下来的事情就比较有趣了，上面代码中，我们使用了一个类似自然语言的表达 given(this.authService.requestCaptcha()).willReturn(captcha)。这个表达的意思就是如果遇到 this.authService.requestCaptcha() 方法调用时，请返回 captcha 这个对象。这样就摆脱了对于真实 AuthService 的依赖，如果我负责 Rest API，而你负责服务层，在你真正的服务开发好之前，我已经可以测试了。

讲完了模拟对象，接下来看一下 MockMVC，Spring 中提供的这个对象让我们可以非常快速方便地测试 Controller，而无须启动一个真实的 HTTP 服务器。

2. 集成测试

通常情况下，一个单元测试只关注自己这个逻辑单元，而集成测试则需要整个系统启动起来，其实不只是系统本身需要启动，可能还涉及系统依赖的外围系统，比如数据库、缓存服务器等。但如果每次测试都得把所有系统启动起来，这测试的准备过程也过于烦琐，而且还容易出错，比如某一个依赖的系统出问题就导致测试无法运行。而且使用真实的周边系统带来的另一个问题就是测试环境的清理问题，每次在数据库中可能会产生大量测试数据，这些数据都需要清理以便进行下一轮测试。

所以，在自动化测试中，就像我们使用 Mock 对象一样，一般采用一个模拟的周边系统嵌入待测试系统中。在工程中使用了 Redis、ElasticSearch 和 MongoDB。所以我们也引入 3 个嵌入式的服务：embedded-redis、embedded-elasticsearch 和 de.flapdoodle.embed.mongo，其中嵌入式 MongoDB 没有指定版本号的原因是 Spring Boot 内建了这个依赖。

```
    testImplementation("com.github.kstyrc:embedded-
redis:${embeddedRedisVersion}")
    testImplementation("pl.allegro.tech:embedded-
elasticsearch:${embeddedElasticsearchVersion}")
    testImplementation("de.flapdoodle.embed:de.flapdoodle.embed.mongo")
```

引入依赖之后，我们还需要建立两个配置类，分别在测试启动时启动 Redis 和 ElasticSearch 服务，在测试关闭时关闭对应服务。如果使用 MongoDB，Spring Boot 则会自动配置，所以就不需要我们手动写配置类了。

```
    package dev.local.gtm.api.config;

    // 省略导入
```

```
@Configuration
@Profile("test")
public class EmbeddedRedisTestConfig {

    private final redis.embedded.RedisServer redisServer;

    public EmbeddedRedisTestConfig(@Value("${spring.redis.port}") final
int redisPort) throws IOException {
        this.redisServer = new redis.embedded.RedisServer(redisPort);
    }

    @PostConstruct
    public void startRedis() {
        this.redisServer.start();
    }

    @PreDestroy
    public void stopRedis() {
        this.redisServer.stop();
    }
}
```

在 Redis 配置中，我们通过 @Value("${spring.redis.port}") 读取配置文件 application.yml 中定义的 Redis 端口，然后使用 @PostConstruct 注解在构造函数之后启动该服务，使用 @PreDestroy 注解在该对象销毁之前停止服务。需要指出的是，测试配置文件，一般会使用 @Profile("test") 注解规定只在测试环境下加载该配置类。

类似地，对于 ElasticSearch 我们创建一个配置类，ElasticSearch 的模拟服务甚至可以设置版本号，这一点非常方便。

```
package dev.local.gtm.api.config;

// 省略导入

@Configuration
@Profile("test")
public class EmbeddedElasticsearchTestConfig {
    private final EmbeddedElastic embeddedElastic;

    public EmbeddedElasticsearchTestConfig() throws IOException {
        this.embeddedElastic = EmbeddedElastic.builder()
            .withElasticVersion("5.5.0")
            .withSetting(PopularProperties.TRANSPORT_TCP_PORT, 9300)
            .withSetting(PopularProperties.CLUSTER_NAME, "docker-
```

```
cluster")
                .withPlugin("analysis-stempel")
                .build();
        }

        @PostConstruct
        public void startRedis() throws IOException, InterruptedException {
            this.embeddedElastic.start();
        }

        @PreDestroy
        public void stopRedis() {
            this.embeddedElastic.stop();
        }
    }
```

对集成测试来说，就不需要模拟（Mock）各种对象了，直接进行各种依赖对象的注入，并执行测试即可。需要特别指出的一点是，在默认情况下，Spring Boot 中的分页对象 Pageable 在测试中会引发异常，需要在构造 MockMvc 时设置一下 .setCustomArgumentResolvers(new PageableHandlerMethodArgumentResolver())。

```
    package dev.local.gtm.api.rest;

    // 省略导入

    @RunWith(SpringRunner.class)
    @WebAppConfiguration
    @SpringBootTest(classes = Application.class)
    @ActiveProfiles("test")
    public class TaskResourceTest {

        private MockMvc mockMvc;

        @Autowired
        private UserRepository userRepository;

        @Autowired
        private TaskRepository taskRepository;

        @Autowired
        private ObjectMapper objectMapper;

        @Autowired
        private HttpMessageConverter[] httpMessageConverters;
```

```
    @Autowired
    private ExceptionTranslator exceptionTranslator;

    @Autowired
    private AppProperties appProperties;

    @Before
    public void setup() {
        taskRepository.deleteAll();
        userRepository.deleteAll();
        val taskResource = new TaskResource(taskRepository,
userRepository);

        mockMvc = MockMvcBuilders.standaloneSetup(taskResource)
            .setMessageConverters(httpMessageConverters)
            .setControllerAdvice(exceptionTranslator)
            .setCustomArgumentResolvers(new
PageableHandlerMethodArgumentResolver())
            .setViewResolvers((ViewResolver) (viewName, locale) -> new
MappingJackson2JsonView()).build();
    }

    @Test
    @WithMockUser
    public void testGetTasks() throws Exception {

mockMvc.perform(get("/api/tasks").accept(MediaType.APPLICATION_JSON)).andExp
ect(status().isOk());
    }

}
```

现在，我们可以运行一下测试，既可以从 IDEA 中点击类或方法左边的图标启动测试，进行整个类的测试，也可以单独测试类中的某个测试方法，如图 7-8 所示。

当然，也可以使用命令行进行测试，如果使用持续集成或持续发布系统，就可以在对应的脚本中使用命令进行测试了。

```
./gradlew test
```

执行命令后，如果没有错误就可以看到下面这样的输出了。

```
> Task :gtm-api:test
// 省略 log 输出
BUILD SUCCESSFUL in 1m 5s
9 actionable tasks: 9 executed
```

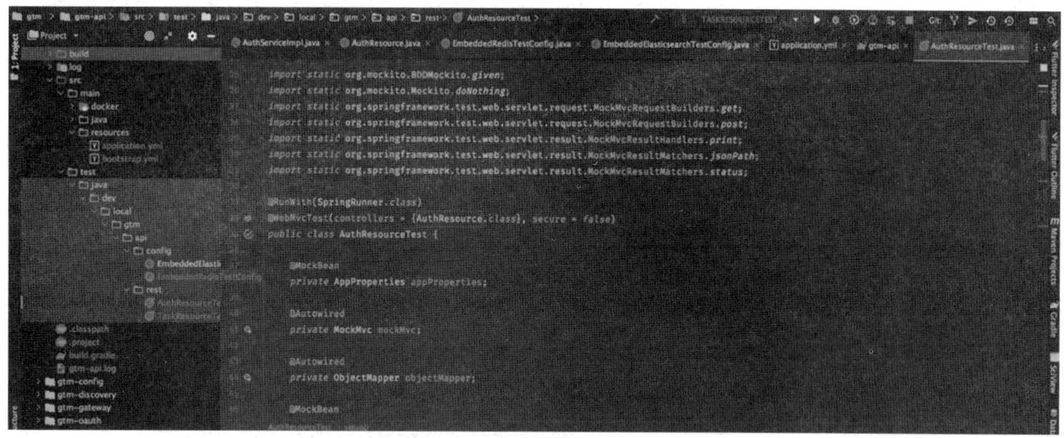

图 7-8

如果测试失败，在 IDEA 中，我们可以看到如下输出，如图 7-9 所示。

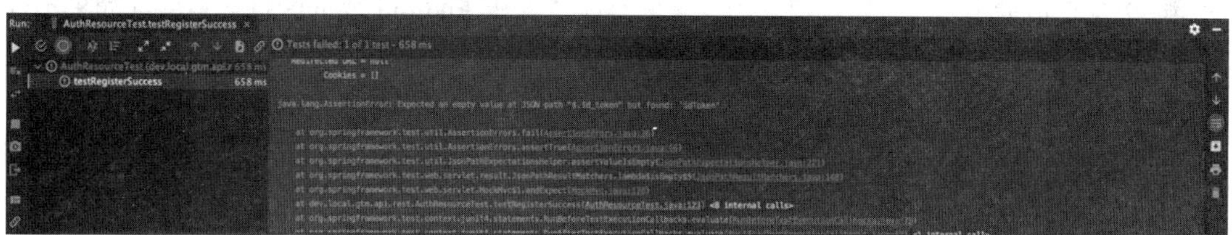

图 7-9

其中，java.lang.AssertionError: Expected an empty value at JSON path "$.id_token" but found: 'idToken' 这句代码说明失败的原因是，我们期待第一层节点 id_token 是空 jsonPath ("$.id_token")，但实际返回的是 idToken 这个字符串。如果我们将断言改成 jsonPath ("$.id_token").isNotEmpty()，那么测试就会通过了。

第 8 章 前端的工程化

今天的前端已经远远不是用一些简单 HTML、CSS 和 JavaScript 可以应对的了，越来越多的功能要求在前端实现，可以说前端的复杂度已经要求和 Android 或 iOS 同等量级（如果不是更强）的生态支持了。所以我们看到了前端从 JQuery 这种类库级别的支持发展到类似 Knockout.js 的双向绑定机制，直到今天以 Angular、React 和 Vue 为代表的各种框架级别的生态。

总体来看，目前的三大主流框架均具备以下特点。

- 在开发模型上更接近传统的客户端编程，很多有客户端、移动端或后端开发经验的读者会感到使用这些框架时和自己以往的开发习惯非常类似。
- 大量的函数式编程的应用使得前端的开发又区别于传统开发，即可以快速而漂亮地实现业务逻辑，但这确实也提高了门槛。当然这个函数式编程并不是必须要做的，但如果掌握了，就可以更漂亮、简捷地写出“健壮”的代码逻辑。
- 内建或社区提供的 UI、路由、状态管理等生态支持。

其中，Angular 在几大框架中属于“大而全”的风格，好处就是官方提供了包括 CLI、路由、动画、服务端渲染、UI 等支持，是选择恐惧症患者的福音。在风格上非常类似于 Java、.Net，如果你有相关语言背景，那么入门是极快的。但有利就有弊，这也使得没有面向对象经验的读者会觉得 Angular 比其他框架要难。从工程支持上来说，Angular 更适合大型团队做大型工程，而 Vue 适合更轻量级的快速开发。

8.1 使用 Redux 管理状态

从前面的使用情况来看，我们的状态管理由于采用了 RxJS，效果还是不错的。但是这种状态管理方案由于没有统一的标准，在大型项目的具体实施中，很可能会走了样。这里介绍一下 Redux。Redux 自从在 React.js 社区火爆之后，其思想很大地影响了其他框架，包括 Vue 和 Angular，现在可以认为 Redux 是业界比较认可的一个状态管理方案，虽然它也不是完美的。

Redux 是为了解决应用状态（State）管理而提出的一种解决方案。那么什么是状态呢？简单来说，对应用开发来讲，UI 上显示的数据、控件状态、登录状态等全部可以看作状态。

我们在开发中经常会碰到，这个界面的按钮需要在某种情况下变灰、那个界面上需要根据不同情况显示不同数量的 Tab、这个界面的某个值的设定会影响另一个界面的某种展现，等等。这些的背后就是状态，状态分为 UI 可视部分和数据部分两大类，数据有的可以体现在界面上，比如增加一个新的 item 到列表，但也有无法直接体现在界面上的，比如某种设置，是否存储到本地 IndexDB 等。应该说在应用开发中最复杂的部分就在于这些状态的管理。很多项目随着需求的迭代，代码规模逐渐扩大、团队人员水平参差不齐就会遇到各种状态管理极其混乱的情况，导致代码的可维护性和扩展性降低。

8.1.1 何时需要使用 Redux

前面说了一堆问题，好像不用 Redux 天就会塌下来了。其实不是的，Redux 就像其他任何一种设计模式一样，有它的适用范围，千万不要认为它是万能的。不分场合地使用 Redux 有的时候会增加业务上不必要的复杂度，Redux 的开发者曾经写过一篇著名的帖子《你也许不需要 Redux》，建议大家可以读一下。在文章中，作者已经非常清晰地列出了什么时候适合采用 Redux，何时不需要使用 Redux。这里我简单摘要并补充一些自己的理解。

Redux 要求开发者改变编程的范式：

- 使用 POJO（Plain Old Javascript Object）描述应用状态。
- 使用 POJO 描述变化。
- 使用纯函数处理变化的逻辑。

这几个要求是使用 Redux 需要做的妥协，在你考虑使用 Redux 时，一定要认真评估，因为这几个要求还是侵入式很强的约束。一般来说，在你可以使用本地状态很清晰地完成业务时，你是不需要使用 Redux 的。当然这样说你可能还是不清楚什么时候适合使用 Redux，这里笔者尝试给出一些具体的场景。

- 你有一些数据，这些数据需要在多个界面展示。换句话说，如果仅仅需要在组件内部使用的状态是不需要使用 Redux 的。
- 当你发现需要在组件中维护多个成员变量，而这些变量已经构成了非常复杂的条件判断，导致代码的可读性和可维护性出现了严重问题时，Redux 往往可以帮你解耦状态逻辑和组件本身，而 Redux 中的逻辑由于是纯函数，也非常易于测试。
- 当你需要将本地状态保存到 local storage 或者在服务端渲染，希望预先组装状态时，因为我们在 Redux 中使用 POJO 描述状态和变化，这个特性让我们可以较容易地实现类似的功能。

Angular 中有哪些现有的状态管理机制？

如前面所说，Redux 是脱胎于 React 社区的，那么问题来了，在 Angular 中是否适用呢？Angular 中现有的状态管理机制是什么样子的？从笔者的经验来看，Redux 在 Angular 生态的地位要低于它在 React 生态中的地位，这是由于 Angular 内建的依赖注入特性其实在很大程度上解决了状态的管理问题。我们可以通过向组件内注入 service 的方式让多个组件共享数据。所以单纯的状态管理并不是在 Angular 中使用 Redux 的根本原因，Redux 之所以非常流行的主要因素之一是它的工具支持简直是太完美了。对开发者来说，如果可以看到每时每刻状态的值的变化以及这些变化产生的来源，这大大地提升了开发人员的生产效率要太简单啊，如图 8-1 所示。

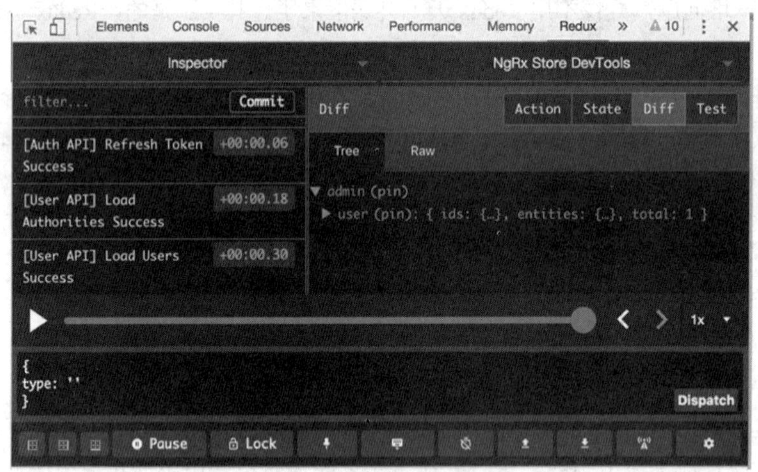

图 8-1

8.1.2 Redux 的核心概念

Redux 的核心概念非常简单，只有 3 个基础元素：Action、State、Reducer。

1. Action——信号

Action 是什么呢？可以将其理解成一个信号或者事件：这个信号不仅存在于页面的交互中，同样存在于应用和后台 API 的交互中。比如点击一个增加项目的按钮，也就是发射了一个信号。那么预期的结果是项目列表页面上会多一个项目，也就是说项目的状态在应用收到这个信号后改变了。

在 Redux 中的 Action 是一个非常简单的 POJO，有两个属性：type 和 payload。type 用于区分信号的类型，而 payload 是这个信号携带的数据，这个是可选的。用 TypeScript 来定义一下就是下面的样子：

```
export interface Action {
    type: string;
    payload?: any
}
```

比如，上述增加项目的信号可以定义成下面的样子，type: 'Add Project' 就是我们用于区分 Action，以便后面在 Reducer 中采取不同的处理。也就是说系统的信号很多，那么在处理时总要知道这个信号是干什么的，然后根据其携带的数据去处理。这个例子中的 payload 是一个项目对象，添加一个项目当然要把这个项目的数据发出来，否则就算我们收到这个信号也没有办法处理数据。

```
{
  type: 'Add Project',
  payload: {
    id: '1234',
    name: '测试项目',
    owner: 'admin'
  }
}
```

前面我们提过，Action 不只存在于页面之上。具体来说，比如增加项目如果需要先访问后台 API，后台 API 添加之后才在前端应用添加项目并展示。那么这个调用后台 API 的动作也应该是由一个信号触发的。这个触发动作可以是刚刚我们定义的那个 Action 吗？当然可以，但这样做下去就会发现问题，HTTP 请求并不是永远都能成功的，如果失败了，怎么办呢？其实失败的情况还是挺多的：比如由于传递的参数没有满足服务端要求，比如访问的 API 路径错误，比如断网或访问超时了，等等。

所以，最好的方式是重新规划 Action：当点击页面上的按钮时发射一个信号，接收到信号之后访问后端 API，如果成功则发射一个添加项目成功的 Action，否则发射一个添加项目失败的 Action。当应用接收到添加项目成功的 Action 后改变应用状态，在列表中添加这个新的项目，如果 Action 是添加项目失败的类型的，那么应用项目列表的状态不变。值得指出的一点是，Add Project 这个信号的发射源是页面，而 Add Project Success 和 Add Project Fail 的发射源是调用后台 API 的逻辑。

```
// 由于一般后台 API 会为项目生成一个 ID，所以在前端并不在 payload 中包含 ID，因为此时 ID 尚未产生
{
  type: 'Add Project',
  payload: {
    name: '测试项目',
    owner: 'admin'
  }
}
// 服务端添加成功后，会返回这个已经在后台数据库保存的项目，这个里面就有 ID 了
{
  type: 'Add Project Success',
  payload: {
    id: '1234',
    name: '测试项目',
    owner: 'admin'
  }
}
// 如果服务端添加失败，则返回一个添加项目失败的 Action，携带错误信息
{
  type: 'Add Project Fail',
  payload: {
    code: '400',
    message: '非法请求'
  }
}
```

从上面的 Action 来看，虽然是足够简单了，但是有一个问题就是数据类型都是字符，这样在大项目中很容易出现拼写错误或者命名重复。在 Angular 中利用 TypeScript 的特性可以将其强类型化，就像下面的例子这样，这个后面我们会详细介绍。

```
export enum AuthActionTypes {
  Login = '[Login Page] Login',
  LoginSuccess = '[Auth API] Login Success',
  LoginFailure = '[Auth API] Login Failure'
}

export class Login implements Action {
  readonly type = AuthActionTypes.Login;
```

```
  constructor(public payload: Auth) {}
}

export class LoginSuccess implements Action {
  readonly type = AuthActionTypes.LoginSuccess;

  constructor(public payload: TokenPair) {}
}

export class LoginFailure implements Action {
  readonly type = AuthActionTypes.LoginFailure;

  constructor(public payload: AppError) {}
}

export type AuthActions =
  | Login
  | LoginSuccess
  | LoginFailure
```

在设计系统时，笔者的习惯是先从 Action 开始设计，先把页面上和 API 返回的 Action 都列出来，在 Action 的 type 中一般要标识一下来源，比如是页面产生的还是 API 产生的。先规划 Action 可以让我们对系统交互有一个比较完整的梳理。

2. State——应用状态

State 这个概念说起来非常简单，你在页面上展现的数据就是 State，在不同条件下按钮的颜色、开启/关闭进行中的动画都是 State。如果你有后端的开发经验或者移动端的 MVVM 开发经验，甚至可以把 State 类比为 View Model。这个 State 和领域对象是不一样的，领域对象体现的是应用的业务逻辑关系，但 State 一般要和具体的数据展现逻辑有关系。一个 State 的定义其实和普通对象也别无二致，就是一个 POJO。

```
export interface State {
  pending: boolean;
  error?: string;
}
```

Redux 的开发者曾经把 State 和数据库做过类比，如果把整个应用的状态类比成一个数据库，则应用中的每个界面或者功能模块对应的 State 就是数据库中的表。

和数据库中的表不一样的是，State 是可以层层嵌套的，比如在下面的定义中可以看到在应用的根 State 中有一个叫 auth 的子 State，而这个叫 auth 的 State 又包含了 3 个子 State: token、loginPage 和 registerPage。

```
// 省略导入

export interface AuthState {
  token: fromToken.State;
  loginPage: fromLoginPage.State;
  registerPage: fromRegisterPage.State;
}

export interface State extends fromRoot.State {
  auth: AuthState;
}
```

3. Reducer——纯函数的状态处理

讲完了前两个概念，其实 Reducer 就再简单不过了——Reducer 就是接收当前的 State 和 Action 作为参数，返回**新的** State 的一个纯函数。注意这是一个新的 State，不是原来的 State，在 Redux 中我们永远不会修改 State，而是返回一个新的 State，这一点非常重要。这个函数一般情况下就是一个 switch...case 的结构，针对不同的 Action 返回不同的 State。

从下面的例子可以看出，Reducer 一般情况下就是根据 Action 类型做 switch，然后返回不同的 State。

```
export function reducer(state = initialState, action:
authActions.AuthActions): State {
    switch (action.type) {
      case authActions.AuthActionTypes.Register: {
        return { pending: true, error: undefined };
      }
      case authActions.AuthActionTypes.ClearRegisterErrors:
      case authActions.AuthActionTypes.RegisterSuccess: {
        return { pending: false, error: undefined };
      }
      case authActions.AuthActionTypes.RegisterFailure: {
        return { pending: false, error: action.payload.title };
      }
      default:
        return state;
    }
}
```

和 State 类似，Reducer 也是可以分成多级的，从应用的根级别到每个界面或功能的级别，同样是采用 key value 字典这样的形式构造，下面的这个例子描述了一个典型的父级 reducer 长成什么样子。

```
export const reducers: ActionReducerMap<AdminState> = {
```

```
    audit: fromAudit.reducer,
    auditPage: fromAuditPage.reducer,
    user: fromUser.reducer,
    authority: fromAuthority.reducer,
    building: fromBuilding.reducer,
    product: fromProduct.reducer,
    room: fromRoom.reducer,
    workspace: fromWorkspace.reducer
  };
```

8.1.3 在 Angular 中使用 Redux

Angular 团队中的一些成员开发了一个开源项目 ngrx[1]，这个项目差不多算作半个官方性质的 Angular 版的 Redux 。对应的子项目有以下几个。

- @ngrx/store——使用 RxJS 实现的 Redux 状态管理框架。
- @ngrx/effects——对于 Action 产生的副作用管理（副作用我们在后面章节有介绍）。
- @ngrx/router-store——将 Angular 路由连接到 store，让你可以像管理其他状态一样管理路由。
- @ngrx/store-devtools——让 ngrx 应用也能使用强大的 Redux Dev Tool（你一定会喜欢时光机器这么强大的特性）。
- @ngrx/entity——一个帮助类库，用来简化 Reducer 的写法，简化常见的增、删、改、查操作。
- @ngrx/schematics——用于使用 Angular CLI 生成 Reducer、Effects 或 Action 模板文件。

1. 安装 ngrx

使用下面的命令安装，为了方便，我们一次性安装所有的软件包。

```
  cnpm install --save @ngrx/store @ngrx/effects @ngrx/router-store
@ngrx/entity
  cnpm install --save-dev @ngrx/store-devtools @ngrx/schematics @angular-
devkit/core @angular-devkit/schematics
```

或者使如下命令。

```
  yarn add @ngrx/store @ngrx/effects @ngrx/router-store @ngrx/entity
  yarn add @ngrx/store-devtools @ngrx/schematics @angular-devkit/core
@angular-devkit/schematics --dev
```

2. 配置 Angular CLI 命令

在项目工程大了之后，很多读者会觉得 ngrx 的写法很烦琐，因为要建立 Action、Reducrer、

---

1 https://github.com/ngrx/platform

Effects 等。ngrx 团队也意识到了这一点，因此为开发者提供了 @ngrx/schematics，让我们可以使用 Angular CLI 来快速建立这些文件。为了以后自动生成文件比较方便，可以使用下面的命令，将其 @ngrx/schematics 设置为 Angular CLI 的默认状态。

```
ng config cli.defaultCollection @ngrx/schematics
```

@ngrx/schematics 继承了 Angular CLI 的 @schematics/angular 定义的命令集合，但如果想要改变一些默认值，比如要生成使用 scss 的组件文件，我们就需要在 angular.json 加入下面的键值对。

```
"schematics": {
  "@ngrx/schematics:component": {
    "styleext": "scss"
  }
}
```

配置好 @ngrx/schematics 之后，就可以通过 Angular CLI 进行 Reducer、Action、Effects 等模板的快速构建。比如利用 store 子命令可以为应用添加 store 的支持，其中 --root 说明我们要创建应用的根 store。

```
ng generate store State --root --module app.module.ts
```

当然，也可以利用简略的缩写命令进行，其中 g 就是 generate 的缩写，而 st 就是 store 的缩写。

```
ng g st State --root --module app.module.ts
```

这个命令会在 app 目录下新建一个 reducers 目录，并在其中新建一个 index.ts 文件，这个文件就是项目根级别的 Reducer，在这个文件中，我们会定义整个应用的状态类型和 Reducer，可以看到 Angular CLI 为我们生成了下面的模板。

```
// 省略导入

export interface State {

}

export const reducers: ActionReducerMap<State> = {

};


export const metaReducers: MetaReducer<State>[]
= !environment.production ? [] : [];
```

有了这个模板，就可以直接定义状态和 Reducer 了，而且 Angular CLI 聪明地更新了 app.module.ts 文件为我们导入了必要的 ngrx 模块。

```
// 省略导入

@NgModule({
  declarations: [AppComponent],
  imports: [
    BrowserModule,
    StoreModule.forRoot(reducers, { metaReducers }),
    !environment.production ? StoreDevtoolsModule.instrument() : []
  ],
  providers: [],
  bootstrap: [AppComponent]
})
export class AppModule {}
```

3. 构建应用的根

有了这个模板，构建应用的根就方便多了，我们在应用根 State 和 Reducer 中添加 router，这里导入了 @ngrx/router-store。这个类库提供了对于系统路由的 Redux 封装支持，使用这个类库的目的是要让 store 变成系统唯一信任的路由变化源。如果不采用这个类库，就会发现路由的变化是无法体现在 store 中的。这个类库的目的不是替换 Angular Router，而是要监听路由的变化，将其以 Reducer State 的形式存储起来。此外你也可以封装一些路由 Action 来替换系统提供的路由导航方式，这样做的目的也还是为了统一 store 的行为，让 store 成为唯一的信任源。但请注意，传统的 Router Link 和 navigate 一样可以使用，@ngrx/router-store 会监听系统路由的变化。

```
// 省略导入

/**
 * 定义应用的根 State 结构
 */
export interface State {
  router: fromRouter.RouterReducerState<RouterStateUrl>;
}

/**
 * 定义应用的根 Reducer 结构
 */
export const reducers: ActionReducerMap<State> = {
  router: fromRouter.routerReducer
};
```

```
  /**
   * 对于退出登录这个特殊的 Action，我们需要将所有的 State 清空：return
reducer(undefined, action);
   * 当然我们还需要清除 local storage 中的信息，这个通过函数 logout() 来完成
   */
  export function storeStateGuard(reducer: ActionReducer<State>):
ActionReducer<State> {
    return function(state, action) {
      if (action.type !== fromAuth.AuthActionTypes.Logout) {
        return reducer(state, action);
      }
      logout();
      return reducer(undefined, action);
    };
  }

  /**
   * 以日志形式输出状态（state）和动作（action）
   */
  export function logger(reducer: ActionReducer<State>):
ActionReducer<State> {
    return function(state: State, action: any): State {
      console.log('state', state);
      console.log('action', action);

      return reducer(state, action);
    };
  }

  /**
   * 除应用本身的 Reducer 外，@ngrx/store 还可以加载一系列的 meta reducer
   * 你可以把它的作用想象成插件，下面的例子中在生产环境提供 storeStateGuard,而在开发
   * 环境提供 logger、storeFreeze 和 storeStateGuard
   */
  export const metaReducers: MetaReducer<State>[]
= !environment.production
    ? [logger, storeFreeze, storeStateGuard]
    : [storeStateGuard];
```

在上述代码中可能有的读者会注意到有 import * as xxx from yyy 这种写法，这种写法对于我们要从某一个文件中导入多个类或者函数时非常有用，as 后面跟的是一个别名，使用这个别名就可以引用这个文件中的元素了。

除 router 外，我们还根据系统环境的不同构建了不同的 Meta Reducer 。这个 Meta Reducer 是什么？

4．Meta Reducer

Meta Reducer 其实本质上就是一个函数，一个高阶函数。那么什么是高阶函数呢？一个接收函数作为参数的函数就是高阶函数。那么 Meta Reducer 就是：

接收一个 reducer 作为参数，并返回一个新的 reducer 的函数。

下面是一个非常简单的输出 state 和 action 的日志 Meta Reducer，可以看到我们定义了一个函数，这个函数接受 reducer 作为参数，然后进行了日志输出，返回了一个新的 reduer 函数。

```
  export function logger(reducer: ActionReducer<State>):
ActionReducer<State> {
    // 返回一个 reducer 函数
    return function(state: State, action: Action): State {
      console.group(action.type);
      console.log('state', state);
      console.log('action', action);

      return reducer(state, action);
    };
  }
```

ngrx 团队为什么要引入这样一个概念呢？在 React 中使用过 Redux 的读者应该知道 Redux 在 React 生态中有很多的中间件（middleware）。这些中间件的效果就是要提供一些 Redux 本身不提供的特性，比如上面例子中就为 Redux 添加了在控制台输出日志的功能。而 ngrx 团队感觉通过高阶函数可以达到中间件的目的，没有必要照搬 React 生态中的中间件概念。

5．构建 Feature

Redux 主要有两个问题。

一个是在大型项目中，由于整个项目使用的是同一个根级 Reducer。虽然可以通过文件切割的形式分成各个功能模块的 Reducer 文件，但是根 Reducer 中仍然需要维护各个子 Reducer，导致在大型项目中，团队频繁更改同一个文件，这会给项目协作带来麻烦。

另一个是，传统的 Reducer 是一棵全局状态树，大型项目的 Reducer 树会非常庞大，而且这棵树上大部分的状态对于当前页面是没有用的，这不仅造成了资源的浪费和性能的降低，而且复杂度升高，导致维护成本上升。

@ngrx 在 4.X 以上版本提供了一个解决方案——Feature。简单来说。Feature 构成了全局 store 的一部分，和模块（module）在 Angular 中的地位类似，你也可以把它理解成 store 的模块。这样的安排在 Angular 中实在太方便了，可以按照 Angular 模块的划分去进行 Reducer 的设计。

比如，我们新建一个 Admin 模块。

```
ng g m Admin --flat false
# 2 下面的是命令输出
CREATE src/app/admin/admin.module.spec.ts (267 bytes)
CREATE src/app/admin/admin.module.ts (189 bytes)
```

通过下面命令，建立一个 Admin 的 Feature，下面的命令会在 AdminModule 所在的 admin 目录中帮我们构建好完整的 actions、reducers 和 effects 目录，以及对应的模板文件。这样一个文件结构也是我们推荐的——在模块中建立对应的 Redux 文件。

```
ng g f admin/Admin -m admin/admin.module --group
# 8 下面的是命令输出
CREATE src/app/admin/actions/admin.actions.ts (238 bytes)
CREATE src/app/admin/reducers/admin.reducer.ts (387 bytes)
CREATE src/app/admin/reducers/admin.reducer.spec.ts (324 bytes)
CREATE src/app/admin/effects/admin.effects.ts (334 bytes)
CREATE src/app/admin/effects/admin.effects.spec.ts (583 bytes)
UPDATE src/app/admin/admin.module.ts (492 bytes)
```

此外，这个命令还会更新 admin.module.ts 文件，我们可以看到，和根模块中 store 的导入方式不同，在这个模块中使用 StoreModule.forFeature 和 EffectsModule.forFeature 来导入 reducer 和 effects。

```
// 省略导入

@NgModule({
  imports: [
    CommonModule,
    StoreModule.forFeature('admin', fromAdmin.reducer),
    EffectsModule.forFeature([AdminEffects])
  ],
  declarations: []
})
export class AdminModule { }
```

有了这样的工具，构建 Redux 时就会少了很多麻烦的文件结构的维护工作，而专注于业务逻辑本身。

8.1.4 Selector——状态选择器

其实，有了 Reducer 之后，我们就可以很方便地调出状态，只需要在组件内声明一个成员变量，在 store 里面选择 auth 状态即可。

```
import * as fromAuth from '../../reducers';
@Component({
  selector: 'app-register',
  templateUrl: './register.component.html',
  styleUrls: ['./register.component.scss']
})
export class RegisterComponent implements OnInit {
  auth$: Observable<Auth>;
  constructor(private store: Store<fromAuth.State>) {
    this.auth$ =  this.store.select('auth');
  }
}
```

但在实际开发中我们在不同组件中需要的可能会有很多变化，比如有的时候只需要 State 中的某个属性，有的时候还需要计算一下某些属性。这些当然可以放在组件中处理，因为 store 其实是一个 Observable，而 RxJS 有很多好用的操作符可以处理这些数据。

这些操作如果放在组件中处理，则有几个明显的缺点。

- 如果这些操作是在不同组件中都用到的，那么这种做法就缺乏可复用性。
- RxJS 的操作符对很多人来说要想掌握得熟练还是有一定门槛的。
- 使用 store 得到状态，却使用 RxJS 处理数据，用两种技术处理状态不利于形成规范统一的代码。

所以更好的处理方式是用 Reducer 处理，那么 @ngrx/store 就给出了选择器这个方案。

什么是选择器呢？选择器就是一个函数，接受状态作为参数，对状态进行处理，返回需要的对象，比如下面的这两个就是最简单的选择器，它们选择了 task 状态的两个属性。

```
export interface State {
  ids: string[];
  entities: {[id: string]: Task};
  selectedProjectId?: string;
}
// 省略 Reducer 部分
export const getSelectedProjectId = (state: State) =>
state.selectedProjectId;
export const getTasks = (state: State) => state.ids.map(id =>
entities[id]);
```

上面的选择器虽然很简单，但已经体现了一些 Reducer 设计上的技巧。首先是在 State 的设计上，一般来说，如果一个界面上要展现一组数据，比如多个任务的列表，那么很直接的

State 设计应该是下面这个样子：

```
export interface State {
  tasks: Task[];
  selectedProjectId?: string;
}
```

那么，为什么我们不采用这样的设计呢？这是因为数组这种数据结构的优点是插入快，如果知道数组下标，则查找也很快。但是在实际开发中，我们一般是通过 id 去查找的，而不是数组下标，这时通过数组查找就比较麻烦了。在这种情况下字典的优势就显现出来了，字典可以快速地由键值定位元素。可不可以结合这两种数据结构呢？我们的 State 设计正是这样的结合形式：将所有元素的 id 形成一个数组（ids）、以字典形式存储列表数据（ entities ），这样一个结构就既可以利用数组的优势也可以利用字典的优势了。

```
export interface State {
  ids: string[];
  entities: {[id: string]: Task};
  selectedProjectId?: string;
}
```

比如，如果要取得所有的任务，那么 ids.map(id => entities[id]) 这样就得到了，如果想要取得某一个 id 对应的任务就可以输入 entities[id]。事实上，在实际开发中，这种模式太普遍了，ngrx 团队还给出了一个效率工具 @ngrx/entity，可以让我们不必每次都手动创建这样的结构，并提供很多可以简化 State 更新操作的方法。

更复杂一些的状态选择就需要组合这些基本的选择器，@ngrx/store 提供了 createSelector 函数，这个函数是一个高阶函数，也就是说它接受函数作为参数，返回处理后的状态。比如在下面例子中，我们就是把 selectTasks 和 selectProjectId 两个基础 selector 作为参数传入，然后在最后一个参数（也是一个函数）中，对状态进行处理并返回，这个返回的值同时也是 createSelector 的返回值。

```
export const selectTasksByProject = createSelector(
  getTasks,
  getSelectedProjectId,
  (tasks, projectId) => {
    return tasks
      ? tasks.filter(task => task.project.id === projectId)
      : [];
  }
);
```

createSelector 可以添加 2~9 个参数，这些参数需要是 State 或者 Selector，而最后一个参数也是一个函数，但这个函数是一个投影函数，将前面的状态作为参数传入，返回一个处理后的数据。在组件中就可以像下面这样去调用了。

```
import * as fromAdmin from '../../reducers';
// 省略组件其他部分
ngOnInit() {
  this.tasks$ = this.store.select(fromAdmin.selectTasksByProject);
}
```

刚刚这个 Selector 其实要求在 Reducer 的 State 中存有 seletedProjectId 这个值。

还有一个刚接触 Selector 经常问的一个问题是，如何可以让 Selector 接收一个外部传入的参数，比如路由的路径参数？这个就要利用一个高阶函数来处理了，像下面的例子中我们使用一个高阶函数接收参数，返回一个 createSelector。

```
export const selectTasksByProject = (projectId: string) =>
  createSelector(selectTaskss, tasks => tasks.filter(task =>
task.projectId === projectId));
```

那么，这样一个带参数的函数就相当于一个工厂方法，我们拿到参数后，构造一个 Selector 并返回这个 Selector。在组件中调用时，可以按下面的方法进行，从路由中获得 projectId，传入我们刚才的工厂方法中 fromAdmin.selectTasksByProject(projectId)。

```
import * as fromAdmin from '../../reducers';
// 省略组件其他部分
ngOnInit() {
  this.tasks$ = this.route.paramMap.pipe(
    filter(params => params.has('projectId')),
    map(params => params.get('projectId')),
    switchMap(projectId =>
this.store.select(fromAdmin.selectTasksByProject(projectId)))
  );
}
```

8.2 使用 Effects 管理副作用

前面我们在利用 @ngrx/schematics 生成文件时，其实不只是 Reducer 和 Action，还生成了 Effects 目录和文件。那么这个 Effects 是什么呢？

要解释 Effects，我们需要回到 8.1.2-1 节中关于 Action 和 Reducer 的例子讨论：

在这个例子中，我们来看一下 Add Project 这个 Action，这个信号其实是要调用后端 API 的，而这个信号并不会和 State 有什么直接的联系，只有在后端添加数据成功后发出 Add

Project Success 时才会对 State 产生影响。这种不对 State 产生影响，但是在 State 之外的其他方面产生了影响的现象，我们给它起个名字叫 Effects，译成中文就是副作用。为什么叫副作用？因为这个副作用的“副”是参照 Redux 来的，Redux 的主要作用是维护管理 State，那么对于非 State 的其他方面的影响就是副作用了。

除了 HTTP 请求这种常见的 Effects，其他较常见的还有如对于 local storage 的读写、对 indexDB 的操作等。任何非 State 的操作都可以看作 Effects。

下面我们用一个例子来说明 Effects 怎么写。

```
import { Injectable } from '@angular/core';
import { Observable, of } from 'rxjs';
import { Action, Store } from '@ngrx/store';
import { Actions, Effect } from '@ngrx/effects';
import { map, switchMap, catchError } from 'rxjs/operators';

import { ProjectService } from '../services/project.service';

import * as fromAdmin from '../reducers';
import * as fromProject from '../actions/project.action';

@Injectable()
export class ProjectEffects {
  @Effect()
  loadUsers$: Observable<Action> =
this.actions$.ofType<fromProject.Actions>(fromUser.ActionTypes.LoadUsers).pipe(
      switchMap(_ =>
        this.service.getAll().pipe(
          map((projects: Projects) => new
fromProject.LoadProjectsSuccess(projects)),
          catchError(error => of(new
fromProject.LoadProjectsFailure(error)))
        )
      )
  );

  constructor(private actions$: Actions, private service: ProjectService,
private store: Store<fromProject.State>) {}
}
```

首先，在 Effects 中的构造函数中注入 Actions，这个 Actions 就是一个 Action 的事件流，是一个 Obseravble。

它会监听应用发射出来的所有 Action , Ngrx 提供了一个操作符 ofType，这个操作符起到的作用就是过滤。比如上面代码中的 this.actions$.ofType<fromProject.Actions>(fromUser.ActionTypes.LoadUsers)，就意味着我们只关心 LoadUsers，也就是项目中的加载用户列表的 Action。

既然是一个 Observable，我们后面就采用了 switchMap，表示收到这个 Action 信号后要执行另一个流的操作。后续要操作的这个流就是调用服务层中的加载用户的方法 getAll()，这个方法中调用了 HttpClient 访问后端定义的 API，可能的结果有两种情况：成功或者失败。

当请求成功的时候，我们发射 LoadProjectsSuccess 这个 Action，而当请求失败的时候，我们利用 catchError 捕获到异常，然后发射 LoadProjectsFailure 这个 Action。

发射这两个 Action 的意义在哪里呢？因为这个 Effects 只负责发送加载用户列表这个 HTTP 请求，而这个处理已经结束了，其他的事情不是这个 Effects 要处理的，所以我们把对应的信号发射出去。

那么谁负载处理呢？谁关心谁处理，还记得我们在 Reducer 中有这两个 Action 的对应处理吗？所以若 Reducer 关心，就是 Reducer 来处理这两种状态。同样地，如果有其他 Effects 关心这两个状态，那么它们也会处理。

说到这里，大家应该理解了，Action 是一个一直存在的信号流，而 Reducer 和 Effects 都在监听，选择自己关心的处理。其区别是，接到信号后，Reducer 只改变状态，而 Effects 只关心副作用。

不继续发射信号的 Effects

是的，总有一些特殊情况，当有这个副作用之后，你不想再去做什么，也没什么好做的。下面例子中的 navigate$ 就是这样一个 Effects，它首先监听 routerActions.GO 信号，然后就导航到对应路由，然后就没有然后了。所以为了说明这种 Effects 不产生信号，我们需要在 @Effect 注解中指定其 dispatch 属性为 false。

```
// 省略导入

@Injectable()
export class AuthEffects {

  @Effect()
  login$: Observable<Action> = this.actions$.pipe(
    ofType<actions.LoginAction>(actions.LOGIN),
    map((action: actions.LoginAction) => action.payload),
    switchMap((val: { email: string; password: string }) =>
      this.authService.login(val.email, val.password).pipe(
        map(auth => new actions.LoginSuccessAction(auth)),
```

```
        catchError(err =>
          of(new actions.LoginFailAction(err))
        )
      )
    )
  );

  @Effect()
  register$: Observable<Action> = this.actions$.pipe(
    ofType<actions.RegisterAction>(actions.REGISTER),
    map(action => action.payload),
    switchMap(val =>
      this.authService.register(val).pipe(
        map(auth => new actions.RegisterSuccessAction(auth)),
        catchError(err => of(new actions.RegisterFailAction(err)))
      )
    )
  );

  @Effect()
  navigateHome$: Observable<Action> = this.actions$.pipe(
    ofType<actions.LoginSuccessAction>(actions.LOGIN_SUCCESS),
    map(() => new routerActions.Go({ path: ['/projects'] }))
  );

  @Effect()
  registerAndHome$: Observable<Action> = this.actions$.pipe(
    ofType<actions.RegisterSuccessAction>(actions.REGISTER_SUCCESS),
    map(() => new routerActions.Go({ path: ['/projects'] }))
  );

  @Effect()
  logout$: Observable<Action> = this.actions$.pipe(
    ofType<actions.LogoutAction>(actions.LOGOUT),
    map(() => new routerActions.Go({ path: ['/'] }))
  );

  @Effect({ dispatch: false })
  navigate$ = this.actions$.pipe(
    ofType(routerActions.GO),
    map((action: routerActions.Go) => action.payload),
    tap(({ path, query: queryParams, extras }) =>
      this.router.navigate(path, { queryParams, ...extras })
    )
```

```
  );

  /**
   *
   * @param actions$
   * @param authService
   */
  constructor(
    private actions$: Actions,
    private router: Router,
    private authService: AuthService
  ) {}
}
```

在上面的代码中，除了有不继续发射的 Effects，还有一些有趣的技巧，我们在 LoginSuccessAction 和 RegisterSuccessAction 信号发出后，使用 navigateHome$和 registerAndHome$分别监听这两个信号，并且发出 routerActions.Go 的信号，然后由 navigate$监听这个信号，使用 Angular Router 导航到对应的路由。

这个链路充分说明要想使用好一个信号流，我们可以非常清晰地将程序逻辑既做到松耦合，又做到以信号为驱动的业务逻辑触发。而且在这样做之后，页面组件就只需和 store 打交道，不用依赖其他服务，比如路由服务。

8.3 使用 @ngrx/entity 提升生产效率

ngrx 很酷，但是我们经常发现自己为不同类型的数据写几乎完全相同的 Reducer 逻辑和选择器，这很容易出错。ngrx 团队估计也收到了很多这种抱怨，于是他们开发了 @ngrx/entity 这个类库，帮我们简化开发。

在 ngrx 中，我们在 store 中存储不同类型的状态，这通常包括：

- 商业数据，例如任务、项目等；
- 某些 UI 状态，例如 UI 的某些设置项、加载进度等。

Entity 是什么呢？有面向对象编程经验的读者知道这是实体的意思，那什么又是实体呢？一般来说，实体是一个可持久化的领域对象，在后端，实体通常表示关系数据库中的表，并且每个实体实例对应于该表中的行。在 Redux 中，我们可以把 store 看成数据库，所以实体就是代表业务数据，比如如下项目：

```
export interface Project {
  id: string | undefined;
```

```
  name: string;
  desc?: string;
  coverImg?: string;
  enabled?: boolean;
  taskFilterId?: string;
  taskLists?: string[];
  members?: string[];
}
```

一般来说，实体都有一个名为 ID 的唯一标识符字段，可以是字符串或数字。我们存储在 store 中的大多数数据都是实体，这些实体在 store 中以数组的形式存储是非常自然的，但这种方法可能有几个潜在的问题：

- 如果我们想根据已知 ID 查找实体，那么将不得不遍历整个集合，这对于非常大的集合来说可能是低效的；
- 如果使用数组，则可能会不小心在数组中存储相同实体的不同版本（具有相同的 ID）。

由于 store 充当内存客户端数据库，所以将业务实体存储在它们自己的内存数据库“表”中是有意义的，并可以为它们提供类似于主键的唯一标识符。然后可以将数据扁平化，并使用实体唯一标识符链接在一起，较好的建模方法是将实体集合存储在字典中，实体的键是唯一 ID，值是整个对象。

```
{
    projects: {
        0: {
            id: 0,
            name: '测试项目 1',
            desc: '这是一个测试项目 1'
        },
        1: {
            id: 1,
            name: '测试项目 2',
            desc: '这是一个测试项目 2'
        },
        ...
    }
}
```

这种存储方式使得通过 ID 查找实体非常简单，例如，为了查找 ID 为 1 的项目，我们只需写成下面的样子，注意不要和数组弄混了，下面是取对象的 key 为 1 的值：

```
projects[1]
```

这种结构会带来一个排序的问题，因为字典是没有顺序的，那么我们是否可以结合两种数据结构的优点呢？可以，那就是既使用字典，又使用数组就好。但是为了减少数据的重复，我们的数组是一个只有 ID 的数组：

```
{
    projects: {
        ids: [0, 1, ...]
        entities: {
            0: {
                id: 0,
                name: '测试项目 1',
                desc: '这是一个测试项目 1'
            },
            1: {
                id: 1,
                name: '测试项目 2',
                desc: '这是一个测试项目 2'
            },
            ...
        }
    }
}
```

这个 ID 构成的数组在实际开发中起到的作用是索引，如果需要数组形式的实体，那么使用 ids.map(id => entities[id]) 即可。那么这种由 ID 构成数组，由字典构成实体集合的类型我们可以定义成下面的样子：

```
export interface ProjectState {
    ids: number[];
    entities: {[key:number]: Project};
}
```

当然，这只是项目的状态，整个应用的状态可以这样定义：

```
export interface State {
    projects: ProjectState:
    tasks: TaskState;
    ...
}
```

在这样的结构下面，而且大部分实体的 Reducer 是非常类似的，大致看起来是下面的样子，只是实体不同而已：

```
const initialProjectState: ProjectState = {
    ids: [],
```

```
        entities: {}
    }

    export function sortByName(a: Project, b: Project): number {
      return a.name.localeCompare(b.name);
    }

    export function reducer(
        state = initialProjectState,
        action: ProjectActions): ProjectState {
        switch (action.type) {
            case ProjectActionTypes.AddProjectSuccess: {
                const project = action.payload;
                if(state.entities[project.id]) return state;
                const ids = [...state.ids, project.id];
                const entities = {...state.entities, {[project.id]:
project}};
                return {...state, {ids: ids, entities: entities}};
            }
            default:
                return state;
        }
    }
```

而且，不只是 Reducer，其实在 Selectors 中的代码重复度也很高，比如：

```
// 得到索引数组
export const ids = (state) => state.ids;
// 得到实体字典
export const entities = (state) => state.entities;
// 得到实体数量
export const total = (state) => state.ids.length;
// 得到数组形态的实体
export const allProjects = (state) => state.ids.map(id => entity[id]);
```

这就引出了我们开头说的，在大项目中写这种重复度较高的代码是很烦的，所以 @ngrx/entity 就来解救我们了。要使用 @ngrx/entity 提供的便利，实体必须有 ID 字段，数字或字符串都行，只要满足这一点，我们就可以像下面代码一样定义状态，看到了吗？ ids 和 entities 哪去了？它们定义在 EntityState 中，所以只需要定义这两个之外的属性，比如 selectedId：

```
export interface State extends EntityState<Project> {
  selectedId: string | null;
}
```

为了能够使用 ngrx/entity 的其他功能，需要首先创建一个 EntityAdapter。它提供了一系列实用程序函数，使操作实体状态变得非常简单。EntityAdapter 允许我们以更简单的方式编写所有初始实体状态，Reducer 和选择器。

在构建 EntityAdapter 时，还需要提供一个排序函数，这个用于集合中实体的排序，下面的代码中给出一个依照项目名称排序的函数。

```
export function sortByName(a: Project, b: Project): number {
  return a.name.localeCompare(b.name);
}

export const adapter: EntityAdapter<Project> =
createEntityAdapter<Project>({
  selectId: (project: Project) => <string>project.id,
  sortComparer: sortByName,
});
```

有了这样一个 adapter 之后，对于状态的增、删、改、查就比较方便了，依据不同的情况可以使用 adapter 提供的 addOne、removeOne、updateOne、addMany、removeMany、updateMany、addAll、removeAll。这些方法的作用从名字上一看就知道了，所以就不展开说了，可以参考下面代码中的用法。

```
// 省略导入

export interface State extends EntityState<Project> {
  selectedId: string | null;
}

export function sortByName(a: Project, b: Project): number {
  return a.name.localeCompare(b.name);
}

export const adapter: EntityAdapter<Project> =
createEntityAdapter<Project>({
  selectId: (project: Project) => <string>project.id,
  sortComparer: sortByName,
});

export const initialState: State = adapter.getInitialState({
  // additional entity state properties
  selectedId: null
});

export function reducer(state = initialState, action: actions.Actions):
```

```
State {
      switch (action.type) {
        case actions.ADD_SUCCESS:
          return { ...adapter.addOne(action.payload, state), selectedId:
null };
        case actions.DELETE_SUCCESS:
          return { ...adapter.removeOne(<string>action.payload.id, state),
selectedId: null };
        case actions.INVITE_SUCCESS:
        case actions.UPDATE_LISTS_SUCCESS:
        case actions.UPDATE_SUCCESS:
        case actions.INSERT_FILTER_SUCCESS:
          return { ...adapter.updateOne({ id: <string>action.payload.id,
changes: action.payload }, state), selectedId: null };
        case actions.LOADS_SUCCESS:
          return { ...adapter.addMany(action.payload, state), selectedId:
null };
        case actions.SELECT:
          return { ...state, selectedId: <string>action.payload.id };
        default:
          return state;
      }
    }

    export const getSelectedId = (state: State) => state.selectedId;
```

结合 @ngrx/schematics 和 @ngrx/entity，我们就可以简化大部分的重复性工作了，大家可以在工作中体会一下。

8.4 服务端渲染

一般的 Angular 应用是运行在浏览器中的，页面在 DOM 中渲染。Angular Universal 通过服务端渲染（SSR - Server Side Rendering）在服务器上生成静态的应用页面。

那么问题来了，说好的前后端分离呢？我们首先来解释一下服务端渲染的好处有哪些。

- 优化 SEO，对搜索引擎爬虫友好。
  - 因为 Angular 本身的控件都是使用 JavaScript 操作 DOM 动态生成的，很多搜索引擎爬虫是无法直接解析出页面内容的。
  - 但是在服务端渲染后，页面的渲染在服务端处理后以静态 HTML 输出，这样爬虫就会抓取到内容了。这对网站的推广是非常有用的，如果你希望在搜索引擎中能够搜到你的网站，那么服务端渲染技术就是比较好的选择了。

- 提高性能。
  - 在 IE，以及不支持 JavaScript 或执行 JavaScript 性能很低的环境，类似 Angular/React/Vue 以 JavaScript 驱动的页面的用户体验会比较差。对于这些情况，服务端渲染由于输出的是静态 HTML，从而消除了 JavaScript 支持程度的影响。
- 快速显示首页，也就是用户看到的第一个页面。
  - 快速显示第一个页面对于网站来说非常重要。调查显示，如果页面加载时间超过 2s，很多用户就会放弃访问了。
  - 使用服务端渲染可以让页面加载得更快，因为一个纯粹的静态 HTML 页面加载是比先下载 JavaScript，然后使用 JavaScript 动态渲染页面要快很多。

8.4.1 Angular Universal 的工作机理

Angular Universal 通过一个 @angular/platform-server 的软件包来支持 DOM 的服务器实现、XMLHttpRequest，以及不依赖于浏览器的其他功能。

因此我们需要使用 platform-server 模块编译客户端应用，并在 Web 服务器上运行生成的应用。服务器端将客户端的页面请求传递给 platform-server 中提供的 renderModuleFactory 函数。

renderModuleFactory 函数接收几个输入参数。

- 模板 HTML（通常是 index.html）。
- 包含组件的 Angular 模块。
- 显示组件的路由 URL。

也就是说，和客户端路由不同，客户端路由只是浏览器地址栏的显示，而在服务器端渲染时，这个路由请求都会生成相应视图的 HTML。

renderModuleFactory 在模板的 <app> 标签内嵌入生成的视图 HTML，为客户端创建完成的 HTML 页面。

最后，服务器端将最终生成的页面返回给浏览器。

8.4.2 安装依赖

Angular CLI 提供了较为方便的步骤，可以帮我们将现有的应用添加服务器端渲染的特性。但在开始之前，我们需要安装以下依赖。

- @angular/platform-server——Angular Universal 服务器端组件。
- @nguniversal/module-map-ngfactory-loader——用于在服务器端渲染中处理懒加载。

- @nguniversal/express-engine——一个基于 Node.js Express 框架的处理引擎。
- ts-loader——使用 Webpack 进行构建时用于 Typescript 文件处理的扩展。
- webpack-cli——一个 Webpack 命令行软件包。

```
yarn add @angular/platform-server @nguniversal/module-map-ngfactory-loader express
yarn add -D ts-loader webpack-cli
```

8.4.3 添加服务器端渲染模块

首先，我们需要修改 AppModule，导入 BrowserModule 的方式改动一下，需要通过 withServerTransition() 方法指定 appId。

```
// 省略导入

@NgModule({
  imports: [
    BrowserModule.withServerTransition({ appId: 'taskmgr' }),
    SharedModule,
    LoginModule,
    CoreModule
  ],
  bootstrap: [AppComponent]
})
export class AppModule {}
```

然后，我们创建一个单独的 AppServerModule，文件位于 src/app/app.server.module.ts。

```
// 省略导入

/**
 * 用于服务端渲染
 */
@NgModule({
  imports: [
    AppModule,
    ServerModule,
    ModuleMapLoaderModule // 处理懒加载
  ],
  bootstrap: [AppComponent]
})
export class AppServerModule {}
```

类似地，我们需要创建一个 src/main.server.ts 为服务端渲染添加一个入口。

```
import { environment } from './environments/environment';
import { enableProdMode } from '@angular/core';

if (environment.production) {
 enableProdMode();
}

export {AppServerModule} from './app/app.server.module';
```

由于 Node.js 服务器端的 JavaScript 导入方式和客户端不太一样，我们需要编译打包成 commonjs 的形式，所以需要创建一个单独的 src/tsconfig.server.json 文件。

```
{
  "extends": "../tsconfig.json",
  "compilerOptions": {
    "outDir": "../out-tsc/app",
    // 注意这里不是 es2015 而是 commonjs
    "module": "commonjs",
    "baseUrl": "./",
    "types": []
  },
  "exclude": [
    "test.ts",
    "**/*.spec.ts"
  ],
  // 添加编译的入口模块
  "angularCompilerOptions": {
    "entryModule": "app/app.server.module#AppServerModule"
  }
}
```

接下来，需要在 angular.json 文件中添加一个 server 段落。

```
"architect": {
  "build": { ... }
  "server": {
    "builder": "@angular-devkit/build-angular:server",
    "options": {
      "outputPath": "dist/server",
      "main": "src/main.server.ts",
      "tsConfig": "src/tsconfig.server.json"
    },
    "configurations": {
      "production": {
        "fileReplacements": [
          {
```

```
                "replace": "src/environments/environment.ts",
                "with": "src/environments/environment.prod.ts"
              }
            ]
          }
        }
      }
    }
```

8.4.4 使用 Node.js Express 构建服务器

目前，Angular Universal 的官方支持两个框架：Node.js 和 .Net。对于 Java 并没有官方支持，但笔者认为影响并不大，因为在微服务架构下，完全可以使用 Node.js 作为 Angular 的服务器端，而 Java 作为 API 提供方，这样的结合丝毫没有违和感。

接下来就开始构建这个服务器端，代码比较简单。

```
import 'zone.js/dist/zone-node';
import 'reflect-metadata';

import { renderModuleFactory } from '@angular/platform-server';
import { enableProdMode } from '@angular/core';

import * as express from 'express';
import { join } from 'path';
import { readFileSync } from 'fs';

// 处理 Lazy Loading 需要导入的
import { provideModuleMap } from '@nguniversal/module-map-ngfactory-loader';

// 激活生产模式可以提供更快的渲染速度
enableProdMode();

// 构建 Express 服务器
const app = express();

const PORT = process.env.PORT || 4000;
const DIST_FOLDER = join(process.cwd(), 'dist');

// index.html 作为模板
const template = readFileSync(
  join(DIST_FOLDER, 'browser', 'index.html')
).toString();
```

```
const {
  AppServerModuleNgFactory,
  LAZY_MODULE_MAP
} = require('./dist/server/main');

app.engine('html', (_, options, callback) => {
  renderModuleFactory(AppServerModuleNgFactory, {
    // index.html
    document: template,
    url: options.req.url,
    // 以 DI 形式提供懒加载处理，在服务端渲染中，我们需要立即渲染
    extraProviders: [provideModuleMap(LAZY_MODULE_MAP)]
  }).then(html => {
    callback(null, html);
  });
});

app.set('view engine', 'html');
app.set('views', join(DIST_FOLDER, 'browser'));

// 对于浏览器访问静态文件提供支持
app.get('*.*', express.static(join(DIST_FOLDER, 'browser')));

// 路由访问支持
app.get('*', (req, res) => {
  res.render(join(DIST_FOLDER, 'browser', 'index.html'), { req });
});

// 启动服务，监听端口
app.listen(PORT, () => {
  console.log(`Node server listening on http://localhost:${PORT}`);
});
```

现在，我们在前端项目根目录建立一个 webpack.server.config.js 文件。这个文件的作用是编译 server.ts 和它的依赖文件，然后生成 dist/server.js。

```
const path = require('path');
const webpack = require('webpack');

module.exports = {
  mode: 'none',
  entry: {
    server: './server.ts',
  },
```

```
  target: 'node',
  resolve: { extensions: ['.ts', '.js'] },
  optimization: {
    minimize: false
  },
  output: {
    // 输出到dist 文件夹
    path: path.join(__dirname, 'dist'),
    filename: '[name].js'
  },
  module: {
    rules: [
      { test: /\.ts$/, loader: 'ts-loader' },
      {
        test: /(\\|\/)@angular(\\|\/)core(\\|\/).+\.js$/,
        parser: { system: true },
      },
    ]
  },
  plugins: [
    new webpack.ContextReplacementPlugin(

      /(.+)?angular(\\|\/)core(.+)?/,
      path.join(__dirname, 'src'), // src 的位置
      {} // 路由表
    ),
    new webpack.ContextReplacementPlugin(
      /(.+)?express(\\|\/)(.+)?/,
      path.join(__dirname, 'src'),
      {}
    )
  ]
}
```

接下来，给 package.json 文件添加几个脚本命令。

```
"scripts": {
  "ng": "ng",
  "start": "ng serve --port=8000",
  "prod": "ng serve --prod --port=8000",
  "build:ssr": "npm run build:client-and-server-bundles && npm run
webpack:server",
  "serve:ssr": "node dist/server.js",
  "build:client-and-server-bundles": "ng build --prod && ng run
taskmgr:server",
```

```
    "webpack:server": "webpack --config webpack.server.config.js --
progress --colors",
    "start:server": "npm run build:ssr && npm run serve:ssr",
    "build": "ng build --prod",
    "test": "ng test",
    "lint": "ng lint",
    "e2e": "ng e2e"
  }
```

现在，使用 npm run start:server 就可以了。此时打开浏览器查看源文件，就会看的 <app-root> 中是有完整的 HTML 的，而在没有使用服务器端渲染的时候，这里面是没有内容的。

8.4.5 服务器端渲染中出现重复请求的处理

在服务器端渲染后，很多读者会发现较为奇怪的现象，有的时候 HTTP 请求会发送两次，服务器端一次，客户端一次。

发生这种现象的原因是，服务器端渲染页面时，会先将 HTTP 请求的内容得到，然后将得到的内容填进页面中，发送给客户端。但客户端并不知道内容已经渲染好了，于是就又发送了一次。

那么解决这个问题在 Angular 6 中已经有非常简单的方案，就是在 AppModule 中导入 TransferHttpCacheModule。

```
import { BrowserModule } from '@angular/platform-browser';
import { TransferHttpCacheModule } from '@nguniversal/common';
import { NgModule } from '@angular/core';
import { CoreModule } from './core';
import { SharedModule } from './shared';
import { LoginModule } from './login';
import { AppComponent } from './core/containers/app';

@NgModule({
  imports: [
    BrowserModule.withServerTransition({ appId: 'taskmgr' }),
    TransferHttpCacheModule,
    SharedModule,
    LoginModule,
    CoreModule
  ],
  bootstrap: [AppComponent]
})
export class AppModule {}
```

在 AppServerModule 中导入 ServerTransferStateModule。

```
// 省略导入

@NgModule({
  imports: [
    AppModule,
    ServerModule,
    ServerTransferStateModule,
    ModuleMapLoaderModule // <-- *Important* to have lazy-loaded routes
work
  ],
  bootstrap: [AppComponent]
})
export class AppServerModule {}
```

这样就可以了？是的，就这么简单。再介绍一下背后机理，解决方案其实是对于每个请求都标记上一个 key，然后，通过 TransferHttpCacheModule 提供的 interceptor 进行请求的拦截，如果发现服务器端已经处理过这个请求就直接返回结果。

第 9 章 Spring Cloud 打造微服务

Spring Cloud 是一个基于 Spring Boot 实现的云应用开发环境，它为基于 Java 的云应用开发中涉及的配置管理、服务发现、断路器、智能路由、微代理、控制总线、全局锁、决策竞选、分布式会话和集群状态管理等操作提供了一种简单的开发环境和模式。

Spring Cloud 包含了多个子项目（针对分布式系统中涉及的多个不同开源产品），比如：Spring Cloud Config、Spring Cloud Netflix（包括 Eureka、Hystrix、Zuul 等 Netflix 全家桶）、Spring Cloud Gateway、Spring Cloud AWS、Spring Cloud Security、Spring Cloud Commons、Spring Cloud Zookeeper、Spring Cloud CLI 等项目。

9.1 微服务的体系架构

什么是微服务呢？笔者理解是微服务允许通过一系列可相互协作的组件来构建一个大型系统，这些组件就是微服务。一个完整的应用可以垂直拆分成多个不同的服务，每个服务都能独立部署、独立维护、独立扩展，服务与服务间通过诸如 RESTful API 或者消息的发布/订阅机制互相调用。

在开始介绍微服务之前，回顾一下单一应用（monolithic application）是什么样子的。一个完整的企业级应用通常被构建成 3 个主要部分。

- 前端用户界面：由运行在 PC 浏览器中的 HTML 页面、JavaScript 和 CSS 组成。
- 数据库：由许多的表构成一个通用的、相互关联的数据管理系统。
- 服务器端应用：服务器端应用处理 HTTP 请求，执行业务逻辑，增、删、改、查数据库中的数据，将结果渲染成适当的 HTML 视图并发送给浏览器。

这样一个应用是一个一体化的、完整的、大而全的应用，任何对系统的改变都需要重新构建和部署一个新版本的服务器端应用程序。

这样的单一应用是一种很自然的构建系统的方式。虽然开发人员可以把应用程序模块化的组织成类、函数、命名空间等，但所有处理请求的逻辑都运行在一个单独的进程中。从拓展性上来说，这种架构可以使用横向扩展，通过负载均衡将多个应用部署到多台服务器上，也可以采用数据库的横向或纵向切分，实现数据库的拓展。这样的开发模型曾经在相当长一段时间内非常成功，但是随着云服务的流行，在云中部署时，我们只是变更应用程序的一小部分，却需要进行整个重新构建和部署。随着功能的增加更新，时间一长，就很难再保持一个好的模块化结构，使得一个模块的变更很难不影响到其他模块。扩展就需要整个应用程序的扩展，而不能进行部分扩展。

这种问题的持续出现自然而然地催生了微服务架构的流行：把应用程序构建为一套服务。服务可以独立部署和扩展，每个服务都有各自的边界，不同的服务可以用不同的编程语言编写，使用不同的数据库，可以让不同的团队编写维护。

9.1.1 服务即组件

组件（component）这个概念在客户端和前端比较普遍，但其实整个软件行业一直希望由组件构建系统，就像在其他工业领域那样，从简单的家具到复杂精密的飞机，从由几十个组件构成到由上百万组件构成，我们一直希望一个复杂的软件系统可以由多个设计精良的组件构成。

微服务架构中的一个重要理念是把服务当成组件，而不是作为库（library）。因为如果应用程序是由若干库构成的，那么对任何一个库的改变都需要重新部署整个应用程序。但是如果把应用程序拆分成很多组件，则只需要重新部署那个改变的组件。而且服务组件化之后会有更清晰的接口和更明确、单一的职责。

需要指出的是，微服务框架不是只有优点而没有缺点的，远程调用会比进程内调用消耗更多资源。也就是说，微服务架构比单一应用会消耗更多资源，在组织上也会更复杂。但是随着应用复杂度的提升，微服务架构的威力也会逐渐显现（见图 9-1）。

9.1.2 微服务架构下的组织机构变化

一个技术团队的传统组织架构按照前端、客户端、服务器端、数据库、UI/UE 和测试来组织团队的构成。这样一个架构在过去几十年也确实体现出强大的威力。但在今天这个节奏越来越快的时代里，这样的组织架构却显得臃肿和效率低下。一个大型团队，按业务线来分割自己会导致很多的依赖。大型应用中每个模块的业务可能都很复杂，在单一应用中，如果问题或需求跨越很多模块边界，则需要协调多个大团队，这种协调和沟通的成本是高昂的，也就意味着短期内修复它们是很困难的。

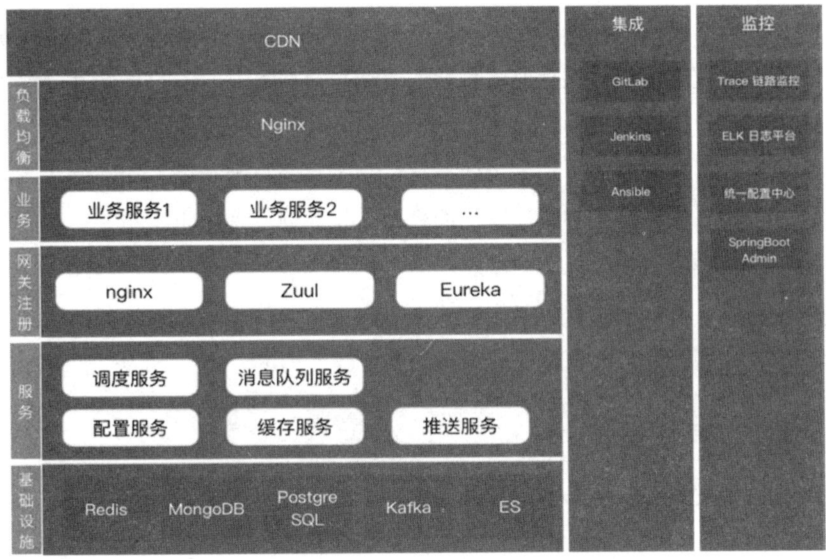

图 9-1

而微服务架构下，组织机构会变成一个个麻雀虽小五脏俱全的小团队，由于微服务有自己的边界，遇到问题的解决基本在小团队内部即可快速对应（见图 9-2）。

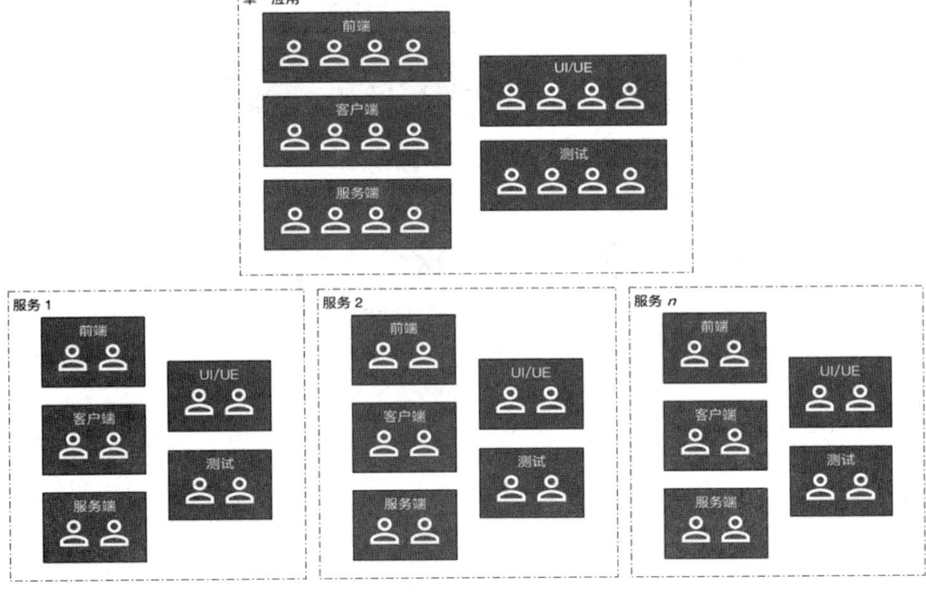

图 9-2

9.1.3 产品化服务

很多公司，尤其是国内的软件公司，大部分都是接到一个项目需求，从需求分析开始逐步地进行设计、开发、测试、交付这样的流程。而且交付完成之后，一般团队都会解散了，大家各自被分配到新的项目组中。

但微服务提倡以产品的角度来去做这些事情，而不是项目的角度。团队应该负责产品的整个生命周期，这要求开发者每天都关注他们的软件运行如何，增加更用户的联系，同时承担一些售后支持。

产品的理念要跟业务能力联系起来，而不是着眼于完成一套功能的软件，这样一个持续产品生命周期的团队，是能够帮助软件及其用户提升业务能力的。

9.1.4 持续集成和持续发布

基础设施自动化技术在近年来得到了长足的发展：云计算减少了构建、发布、运维微服务的复杂性。

使用微服务架构的产品或者系统，一般都采用持续部署（Continuous Delivery，CD）和持续集成（Continuous Integration，CI）。团队使用这种方式构建软件致使更广泛的依赖基础设施自动化技术。图 9-3 说明这种构建的流程：

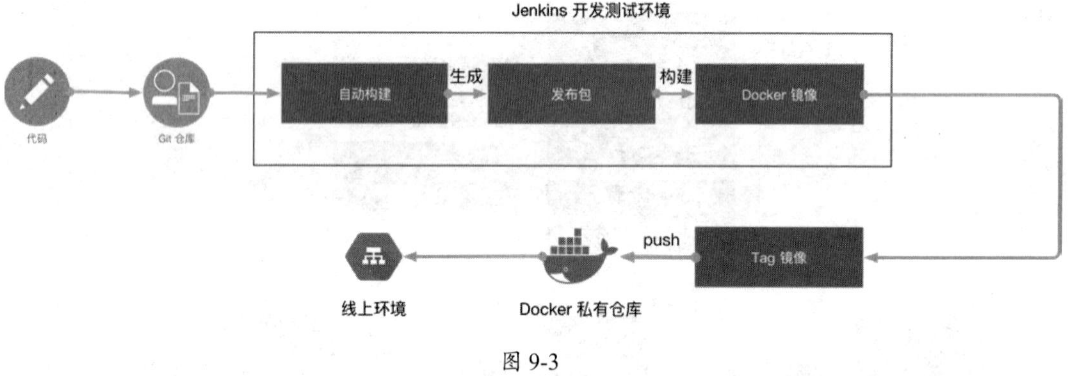

图 9-3

9.1.5 监控和报警

微服务带来的不仅仅是优点，有一些天生的缺点会一并存在。这些缺点带来的后果之一就是，使用服务作为组件的话，组件就可能出错，但整个系统不能因为某一个组件的出错就整体停摆。任务服务都可能发生故障，有可能是第三方服务的原因，也有可能是我们某个微服务的原因，那么应用需要尽可能的优化这种场景的响应。跟单一应用相比，这是一个明显缺点，因为它带来的额外的复杂度。这将让微服务团队时刻想到在服务故障的情况下的用户体验问题。

这就要求我们可以对微服务进行快速故障检测，更进一步，如果可以做到自动恢复变更就非常重要了。微服务应用把实时的监控放在应用的各个阶段中，检测构架元素和业务相关的指标。监控系统可以提供一种早期故障告警系统，让开发团队跟进并调查。

所以，在微服务构架中，服务的监控和报警是必需的，一般我们会监控和记录每个服务的配置，上/下线状态、各种运维和业务相关的指标等。

9.1.6 Spring Cloud 项目依赖

```
/*
 * 这个 build 文件是由 Gradle 的 'init' 任务生成的
 *
 * 更多关于在 Gradle 中构建 Java 项目的信息可以查看 Gradle 用户文档中的
 * Java 项目快速启动章节
 * https://docs.gradle.org/4.0/userguide/tutorial_java_projects.html
 */
// 在这个区块中你可以声明 build 脚本需要的依赖和解析下载该依赖所使用的仓储位置
buildscript {
    ext {
        springBootVersion = '2.0.3.RELEASE'
        springCloudVersion = 'Finchley.RELEASE'
        propDepsVersion = '0.0.9.RELEASE'
        springFoxVersion = '2.9.0'
        redissonVersion = '3.6.5'
        javerVersion = '3.10.0'
        jwtVersion = '0.9.0'
        mongobeeVersion = '0.13'
        problemVersion = '0.23.0'
        javaTuplesVersion = '1.2'
        jacksonDataTypeJsr310Version = '2.9.5'
        jpushVersion = '3.3.6'
        awsVersion = '1.11.136'
        javassistVersion = '3.18.2-GA'
        springbootAdminVersion = '2.0.2'
        jPinyinVersion = '1.1.8'
    }
    ext['spring.data.elasticsearch.version'] = '3.0.9.RELEASE'
    repositories {
        maven
{ setUrl('http://maven.aliyun.com/nexus/content/groups/public/') }
        maven
{ setUrl('http://maven.aliyun.com/nexus/content/repositories/jcenter') }
```

```
            maven { setUrl('http://repo.spring.io/plugins-release') }
        }
        dependencies {
            classpath("org.springframework.boot:spring-boot-gradle-
plugin:${springBootVersion}")
            classpath("io.spring.gradle:propdeps-plugin:${propDepsVersion}")
            classpath("org.springframework:springloaded:1.2.5.RELEASE")
        }
    }

    allprojects {
        group = 'dev.local.gtm'
        apply plugin: 'idea'
    }

    subprojects {
        version = "0.0.1"
        tasks.withType(Jar) {
            baseName = "$project.name"
        }
        apply plugin: 'java'
        apply from: '../gradle/docker.gradle'
        apply plugin: 'propdeps'
        apply plugin: 'propdeps-idea'
        apply plugin: 'io.spring.dependency-management'
        sourceCompatibility = 1.8
        targetCompatibility = 1.8
        repositories {
            maven
{ setUrl('http://maven.aliyun.com/nexus/content/groups/public/') }
            maven
{ setUrl('http://maven.aliyun.com/nexus/content/repositories/jcenter') }
            maven { setUrl('https://repo.spring.io/libs-milestone') }
            jcenter()
        }
        dependencies {
            runtime('org.springframework.boot:spring-boot-devtools')
            compileOnly("org.projectlombok:lombok")
            compileOnly("org.springframework.boot:spring-boot-configuration-
processor")
            testImplementation("org.springframework.boot:spring-boot-
starter-test")
            testImplementation("org.projectlombok:lombok")
        }
```

```
        dependencyManagement {
            imports {
                mavenBom "org.springframework.cloud:spring-cloud-
dependencies:${springCloudVersion}"
            }
        }
    }
```

9.2 配置服务和发现服务

9.2.1 配置中心是什么

一个应用运行起来依赖的不仅仅是代码，还需要连接资源，以及对于某些资源或业务进行参数化的配置。这些配置一般来说经常会采用外部设置的配置文件去调整，如切换不同的数据库、设置功能开关等。

随着系统微服务的不断增加，每个微服务都有自己的配置文件，这种各自管各自的配置管理模式，在开发时没什么问题，部署到生产环境之后管理就会很头疼，到了要大规模更新就更烦了。在这种情况下，如果我们可以统一配置中心就是一个比较好的解决方案，图 9-4 就是一个配置中心的解决方案。

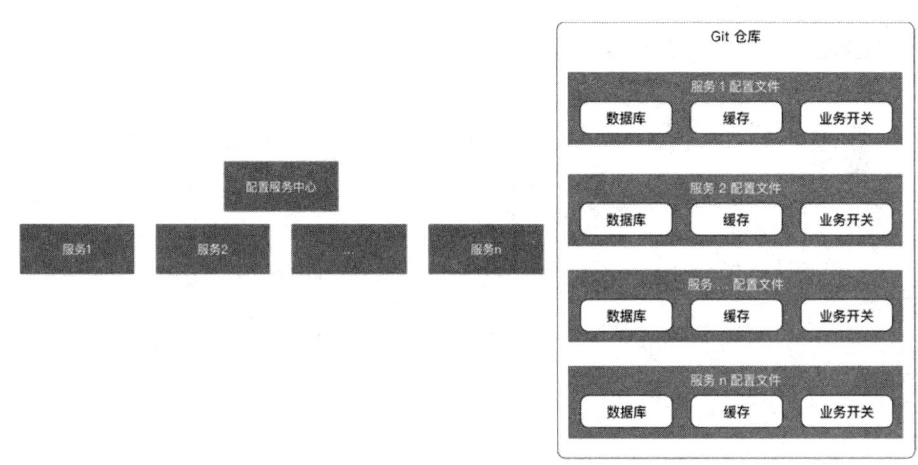

图 9-4

Spring Cloud Config 就是这样一个中心配置服务。

- 提供服务器端和客户端支持。
- 集中式管理分布式环境下的应用配置。

- 基于 Spring 环境，无缝与 Spring 应用集成。
- 可用于任何语言开发的程序。
- 默认实现基于 git 仓库，可以进行版本管理。
- 可自定义实现。

Spring Cloud Config 的服务器端叫作 Spring Cloud Config Server，提供以下功能。

- 拉取配置时更新 git 仓库副本，或指定分支、标签等。
- 支持数据结构丰富，包括 yml、 json、 properties 等。
- 配合 Eureka 可实现服务发现，配合 Spring Cloud Bus 可实现配置推送更新。
- 配置存储基于 git 仓库，可进行版本管理。
- 简单可靠，有丰富的配套方案。

1. 配置中心服务器

在基于 Spring Cloud 的项目中使用建立一个 Spring Cloud Config Server 非常简单，首先建立一个子工程 gtm-config，然后配置 settings.gradle，在项目中添加子工程。

```
include 'gtm-api'
include 'gtm-config'
rootProject.name = 'gtm-backend'
```

接下来需要给子工程配置依赖，新建一个 gtm-config/build.gradle 文件。

```
apply plugin: 'org.springframework.boot'
configurations {
    springLoaded
    // 如果使用 undertow 或 jetty 需要把默认包含的 tomcat 排除在外
    compile.exclude module: 'spring-boot-starter-tomcat'
}
dependencies {
    implementation("org.springframework.cloud:spring-cloud-config-
server")
    implementation("org.springframework.cloud:spring-cloud-config-
monitor")
    implementation("org.springframework.boot:spring-boot-starter-
actuator")
    implementation("org.springframework.boot:spring-boot-starter-
undertow")
}
```

对 Spring Cloud Config Server 来说，org.springframework.cloud:spring-cloud-config-server 这个依赖是必需的。在 gtm-config/src/main/java/dev/local/gtm/configserver 中建立一个新的 Application.java

文件。

```
package dev.local.gtm.configserver;

// 省略导入

/**
 * GatewayApplication
 */
@SpringBootApplication
@EnableConfigServer
@Configuration
public class Application {

    public static void main(String[] args) {
        SpringApplication.run(Application.class, args);
    }
}
```

上面的 @EnableConfigServer 注解就是让这个应用成为配置中心了。剩下的事情就是在 application.yml 中定义配置文件的存储位置，此处我们使用了 git 仓库，在调试代码时，也可以使用本地路径。但一般在生产环境中都会使用 git 仓库作为配置文件存储，因为这样可以保留版本的变更记录，在遇到问题时可以非常方便地回滚。

```
spring:
  application:
    name: configserver
  cloud:
    config:
      server:
        git:
          uri: https://gitee.com/twigcodes_group/smartoffice-config.git
          username: 请填写自己的用户名
          password: 请填写自己的密码
server:
  port: 8888
logging:
  level:
    org.springframework:
      cloud: DEBUG
---

spring:
  profiles: prod
```

如果此时将 gtm-api/src/main/resources/application.yml 放到上面配置的 git 仓库中，请注意文件名需要和 spring.application.name 一致。也就是说，在 git 仓库中 gtm-api 这个项目的配置文件应该叫作 api-service.yml，因为 gtm-api/src/main/resources/application.yml 中的 spring.application.name 是 api-service。此时，如果访问 http://localhost:8888/api-service/dev 可以看到在开发环境中的配置（见图 9-5）。

图 9-5

2. 配置中心客户端

所有希望从配置中心服务器取得配置文件的应用都是配置中心的客户端，客户端配置起来非常简单，只需在子项目的 build.gradle 中先添加依赖。

```
implementation("org.springframework.cloud:spring-cloud-starter-config")
```

然后在 src/main/resources 目录中新建一个 bootstrap.yml 文件。此处需要注意的一个地方是，默认的配置中心服务的 ServiceId 是 configserver，但这个可以在 bootstrap.yml 文件中指定 spring.cloud.config.discovery.serviceId 为 Config Server 的 spring.application.name。

```
spring:
  application:
    name: api-service
  cloud:
    config:
      uri: http://localhost:8888
---
spring:
  profiles: prod
  cloud:
    config:
      uri: http://configserver:8888
```

此时如果启动客户端，则会在 Console 中看到客户端从服务器（http://localhost:8888）拉取了配置文件。

```
  .   ____          _            __ _ _
 /\\ / ___'_ __ _ _(_)_ __  __ _ \ \ \ \
( ( )\___ | '_ | '_| | '_ \/ _` | \ \ \ \
 \\/  ___)| |_)| | | | | || (_| |  ) ) ) )
  '  |____| .__|_| |_|_| |_\__, | / / / /
 =========|_|==============|___/=/_/_/_/
 :: Spring Boot ::        (v2.0.3.RELEASE)

   2018-08-02 09:13:27.150  INFO 17027 --- [  restartedMain]
c.c.c.ConfigServicePropertySourceLocator : Fetching config from server at :
http://localhost:8888
   2018-08-02 09:13:27.197 DEBUG 17027 --- [  restartedMain]
o.s.web.client.RestTemplate              : Created GET request for
"http://localhost:8888/api-service/default"
   2018-08-02 09:13:27.342 DEBUG 17027 --- [  restartedMain]
o.s.web.client.RestTemplate              : Setting request Accept header to
[application/xml, text/xml, application/json, application/x-jackson-smile,
application/cbor, application/*+xml, application/*+json]
   2018-08-02 09:13:28.548 DEBUG 17027 --- [  restartedMain]
o.s.web.client.RestTemplate              : GET request for
"http://localhost:8888/api-service/default" resulted in 200 (OK)
   2018-08-02 09:13:28.570 DEBUG 17027 --- [  restartedMain]
o.s.web.client.RestTemplate              : Reading [class
org.springframework.cloud.config.environment.Environment] as
"application/json;charset=UTF-8" using
[org.springframework.http.converter.json.MappingJackson2HttpMessageConverter
@582fbb48]
   2018-08-02 09:13:28.638  INFO 17027 --- [  restartedMain]
c.c.c.ConfigServicePropertySourceLocator : Located environment: name=api-
service, profiles=[default], label=null,
version=ecd523ebee6f04ff11bc67c04874ae3ea006bed8, state=null
   2018-08-02 09:13:28.639  INFO 17027 --- [  restartedMain]
b.c.PropertySourceBootstrapConfiguration : Located property source:
CompositePropertySource {name='configService',
propertySources=[MapPropertySource {name='configClient'}, MapPropertySource
{name='https://gitee.com/twigcodes_group/smartoffice-config.git/api-
service.yml (document #0)'}]}
   2018-08-02 09:13:28.772  INFO 17027 --- [  restartedMain]
dev.local.gtm.api.Application : The following profiles are active: dev
```

9.2.2 发现服务

由于采用了微服务架构，所以当微服务间互相调用时，我们都要知道被调用方的 IP 和服务器端口，或者域名和服务器端口。如果是采用域名的情况，则可以利用基于 DNS 的服务发现，但因为 DNS 有缓存、无法自治等因素，这种方案就有很多局限。传统的 DNS 方式，都是通过 Nginx 或者其他代理软件来实现，物理机器的 IP 和端口都是固定的，那么 Nginx 中配置的服务 IP 和端口也是固定的，服务列表的更新只能通过手动来做，但如果后端服务很多时，手动更新容易出错，效率也很低，这在后端服务发生故障时，不可用时间就可能会加长。

在微服务中，尤其是使用了 Docker 等虚拟化技术的微服务，其 IP 都是动态分配的，服务实例数也是动态变化的，那么就需要精细而准确的服务发现机制。当微服务应用启动后，告诉一个服务发现服务器自己的 IP 和端口，这里的服务发现功能可以利用 Netflix 出品的 Eureka Server 和 Eureka Client 来配合实现。这样微服务架构内的服务都可以彼此了解对方的 IP 和端口，以及哪些服务是在线的、可用的，也方便去做负载均衡和调用。

Eureka 是 Netflix 的一个开源的服务发现框架，主要用于服务的注册发现。Eureka 由两个组件组成：Eureka Server 和 Eureka Client。Eureka Server 是一个服务注册中心，为服务实例注册管理和查询可用实例提供了 REST API，并可以用其定位、负载均衡、故障恢复后端服务的中间层服务。在服务启动后 Eureka Client 用来简化与服务器的交互，作为轮询负载均衡器，并提供服务的故障切换支持。Eureka Client 向服务注册中心注册服务同时会拉去注册中心注册表副本；在服务停止时，Eureka Client 向服务注册中心注销服务；服务注册后，Eureka Client 会定时发送心跳来刷新服务的最新状态。

客户端发现模式的优点是服务调用、负载均衡不需要和 Eureka Server 通信，直接使用本地注册表副本，因此 Eureka Server 不可用时不会影响正常的服务调用，性能也不会因为网络延迟和服务端延迟受到影响。但其缺点也很明显，在某个服务不可用时，各个 Eureka Client 不能及时知道，需要 1~3 个心跳周期才能感知，但是，由于基于 Netflix 的服务调用端都会使用 Hystrix 来容错和降级，当服务调用不可用时 Hystrix 也能及时感知到，通过熔断机制来降级服务调用，因此弥补了基于客户端服务发现的时效性的缺点。

Eureka Server 采用的是对等通信（P2P），无中心化的架构，没有 master/slave 的区分，每一个 Server 都是对等的，既是 Server 又是 Client，所以其集群方式可以自由发挥，可以各点互连，也可以链式连接。Eureka Server 通过运行多个实例及彼此之间互相注册来提高可用性，每个节点需要添加一个或多个有效的 ServiceUrl 指向另一个节点。

1. 区域与可用区

区域（region）：类似 AWS 或阿里云这种公有云服务在全球不同的地方都有数据中心，比如北美、南美、欧洲和亚洲等。与此对应，根据地理位置我们把某个地区的基础设施服务集合称为一个区域。通过区域，一方面可以使得云服务在地理位置上更加靠近我们的用户，另一方面使得用户可以选择不同的区域存储他们的数据以满足法规遵循方面的要求。

可用区（zone）：每个区域一般由多个可用区（available zone）组成，而一个可用区一般是由多个数据中心组成的。可用区设计主要是为了提升用户应用程序的高可用性。因为可用区与可用区之间在设计上是相互独立的，也就是说它们会有独立的供电、独立的网络等，这样假如一个可用区出现问题时也不会影响另外的可用区。在一个区域内，可用区与可用区之间是通过高速网络连接，从而保证有很低的延时（见图 9-6）。

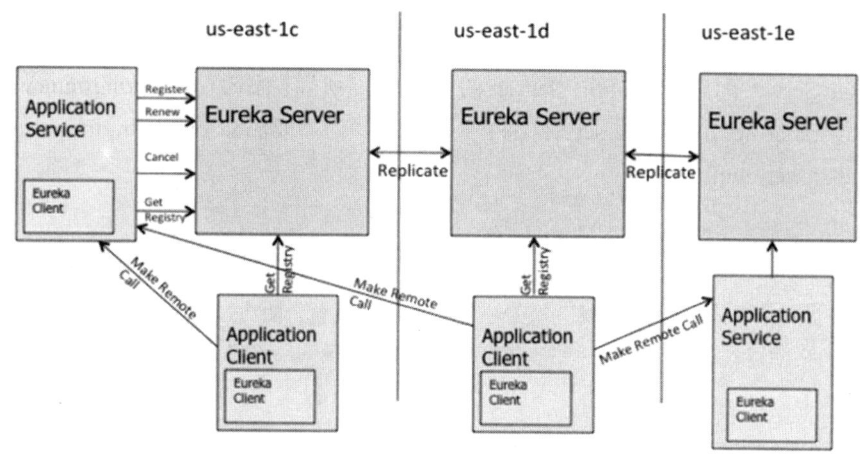

图 9-6

从图 9-5 的架构图可以看出，主要有 3 种角色。

- Eureka Server——通过 Register、Get、Renew 等接口提供注册和发现。
- Application Service（服务提供方）——把自身服务实例注册到 Eureka Server。
- Application Client（服务调用方）——通过 Eureka Server 获取服务实例，并调用 Application Service。

它们主要进行的活动如下。

- 每个区域有一个 Eureka 集群，每个区域中都至少有一个 Eureka Server。
- Service 作为一个 Eureka Client，注册到 Eureka Server，并且通过发送心跳的方式更新租约。如果 Eureka Client 到期没有更新租约，那么过一段时间后，Eureka Server 就会移除该 Service 实例。
- 当一个 Eureka Server 的数据改变以后，会把自己的数据同步到其他 Eureka Server。
- Application Client 也作为一个 Eureka Client 从 Eureka Server 中获取 Service 实例信息，然后直接调用 Service 实例。
- Application Client 调用 Service 实例时，可以跨可用区调用。

2. 建立一个 Eureka Server

首先建立一个子工程 gtm-config，然后配置 settings.gradle，在项目中添加子工程。

```
include 'gtm-api'
include 'gtm-discovery'
include 'gtm-config'
rootProject.name = 'gtm-backend'
```

接下来，需要给子工程配置依赖，新建一个 gtm-discovery/build.gradle，请注意我们这里同时将 Eureka Server 配置成 Config Server 的客户端，所以除添加 org.springframework.cloud: spring-cloud-starter-netflix-eureka-server 这个依赖之外，还需要添加 org.springframework. cloud: spring-cloud-starter-config。

```
apply plugin: 'org.springframework.boot'
configurations {
    springLoaded
    // 如果使用 undertow 或 jetty，则需要把默认包含的 tomcat 排除在外
    compile.exclude module: 'spring-boot-starter-tomcat'
}
dependencies {
    implementation("org.springframework.cloud:spring-cloud-starter-
config")
    implementation("org.springframework.cloud:spring-cloud-starter-
netflix-eureka-server")
    implementation("org.springframework.boot:spring-boot-starter-
actuator")
    implementation("org.springframework.boot:spring-boot-starter-
undertow")
}
```

然后，在 gtm-discovery/src/main/java/dev/local/gtm/discovery 中建立一个新的 Application.java 文件。注解 @EnableEurekaServer 即可配置该应用成为 Eureka Server。

```
package dev.local.gtm.discovery;

// 省略导入

/**
 * 服务发现服务器
 */
@SpringBootApplication
@EnableEurekaServer
@RefreshScope
@Configuration
```

```
public class Application {

    public static void main(String[] args) {
        SpringApplication.run(Application.class, args);
    }
}
```

最后，我们需要添加两个配置文件 applicaiton.yml 和 bootstrap.yml。其中 bootstrap.yml 中需要指定 Config Server 的 URI。

```
spring:
  application:
    name: discovery
  cloud:
    config:
      uri: http://localhost:8888

---

spring:
  profiles: prod
  cloud:
    config:
      uri: http://configserver:8888
```

而 application.yml 中需要禁用 Eureka Client 的行为，因为我们是一个 Server，此处还需要配置 eureka.instance.hostname，我们在开发环境下使用 localhost 而在生产环境使用 discovery 作为服务发现服务器的主机名。

```
server:
  port: 8761
eureka:
  instance:
    hostname: localhost
  client:
    registerWithEureka: false
    fetchRegistry: false
    serviceUrl:
      defaultZone:
http://${eureka.instance.hostname}:${server.port}/eureka/

---

spring:
  profiles: prod
```

```
eureka:
  instance:
    hostname: discovery
  client:
    registerWithEureka: false
    fetchRegistry: false
    serviceUrl:
      defaultZone:
http://${eureka.instance.hostname}:${server.port}/eureka/
```

3. 配置 Eureka Client

配置 Eureka Client 的过程更为简单一些，我们用 gtm-config 项目作为示例为其添加 Eureka Client 特性，Config Server 可以注册到服务发现上，在 build.gradle 中加上 org.springframework.cloud:spring-cloud-starter-netflix-eureka-client 这个依赖。

```
apply plugin: 'org.springframework.boot'
configurations {
    springLoaded
    // 如果使用 undertow 或 jetty，则需要把默认包含的 tomcat 排除在外
    compile.exclude module: 'spring-boot-starter-tomcat'
}
dependencies {
    implementation("org.springframework.cloud:spring-cloud-config-
server")
    implementation("org.springframework.cloud:spring-cloud-config-
monitor")
    implementation("org.springframework.cloud:spring-cloud-starter-
netflix-eureka-client")
    implementation("org.springframework.cloud:spring-cloud-starter-
stream-rabbit")
    implementation("org.springframework.boot:spring-boot-starter-
actuator")
    implementation("org.springframework.boot:spring-boot-starter-
undertow")
}
```

在 Application.java 文件中加上 @EnableDiscoveryClient 这个注解。

```
package dev.local.gtm.configserver;

// 省略导入

@SpringBootApplication
@EnableDiscoveryClient
@EnableConfigServer
```

```
@Configuration
public class Application {

    public static void main(String[] args) {
        SpringApplication.run(Application.class, args);
    }
}
```

在 application.yml 中添加 eureka.client.serviceUrl.defaultZone 指向刚才的 Eureka Server。

```
spring:
  application:
    name: configserver
  cloud:
    config:
      server:
        git:
          uri: https://gitee.com/twigcodes_group/smartoffice-config.git
          username: 请输入你的用户名
          password: 请输入你的密码
server:
  port: 8888
eureka:
  client:
    serviceUrl:
      defaultZone: http://localhost:8761/eureka/

---

spring:
  profiles: prod
eureka:
  instance:
    hostname: discovery
  client:
    registerWithEureka: false
    fetchRegistry: false
    serviceUrl:
      defaultZone: http://${eureka.instance.hostname}:8761/eureka/
```

这样配置好之后，可以启动 Config Server 和 Eureka Server，开启两个 terminal，分别输入下面两条命令。

```
./gradlew :gtm-config:bootRun
./gradlew :gtm-discovery:bootRun
```

然后，访问 http://localhost:8761/ 就可以看到 CONFIGSERVER 已经注册到 Eureka 服务上了（见图 9-7）。

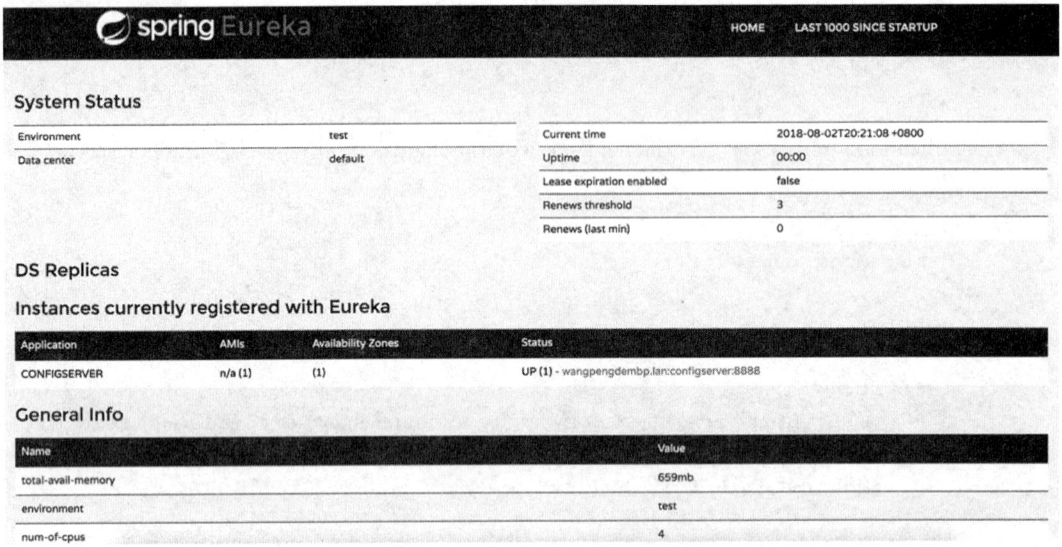

图 9-7

值得指出的是，Config Server 和 Eureka Server 是互相依赖的，Eureka Server 的配置需要到 Config Server 中获取，而 Config Server 也想要注册到 Eureka Server。但这个不会变成死循环，因为如果没有找到配置中心服务器，则会使用本地配置文件，而如果找不到服务发现服务器，则会在启动后不断以一定间隔去查找。

9.3 监控服务和路由服务

微服务是一个分布式的架构模式，这样一个结构虽然带来更灵活的特性，但也会带来单一应用中不会出现的一些问题。当系统从单个节点拓展到多节点时，如果系统的某个点出现问题，那这时的问题定位可能就会变成一个挑战，因为要定位这样的问题仅仅依赖传统的日志或者 debug 往往是不够的。

再有是当有新的业务加入或者业务更改之后，系统是否运行正常？系统的资源和性能是否正常？这样都可以通过监控手段进行衡量。出现这些问题后，监控就是一个常用的、有效的手段。

9.3.1 Spring Boot Admin

Spring Boot Admin 是一套用于监控基于 Spring Boot 的应用，它提供完整的监控 UI 界面和多种监控指标，适合中小团队用于监控微服务的状态。由于其内置了很多常用的健康指标和统计，所以属于开箱即用的一个软件包，但如果你的团队需要更多定制化的功能，就需要定制化 SpringBoot Admin，它也提供了一些方法可以自定义视图。

Spring Boot Admin 集成到系统中主要有两种方式：一种方式是建立一个 Admin Server，然后通过将每个需要被监控的微服务的配置文件指向 Admin Server（由于 SpringBoot Admin 2.0 添加了对 Spring Cloud 的支持）；另一种方式是通过 Eureka Server，将 Admin Server 注册到 Eureka，Admin Server 就可以读取到服务中心所有注册的应用以便实现监控。

我们这里要做的是第二种方式，但稍稍有点区别，因为如果单独做一个 Admin Server 则感觉有点浪费资源，我们将 Admin Server 集成到刚刚建立的 Eureka Server 上。这样的组合也比较符合逻辑，系统所有的应用都会注册到 Eureka Server 上，在这个 Server 上监控也是顺理成章的。

1. 构建 Spring Boot Admin Server

构建一个 Admin Server，首先需要添加一个依赖 de.codecentric:spring-boot-admin-starter-server。

```
apply plugin: 'org.springframework.boot'
configurations {
    springLoaded
    // 如果使用 undertow 或 jetty ，则需要把默认包含的 tomcat 排除在外
    compile.exclude module: 'spring-boot-starter-tomcat'
}
dependencies {
    implementation("de.codecentric:spring-boot-admin-starter-
server:${springbootAdminVersion}")
    implementation("org.springframework.cloud:spring-cloud-starter-
config")
    implementation("org.springframework.cloud:spring-cloud-starter-bus-
amqp")
    implementation("org.springframework.cloud:spring-cloud-starter-
netflix-eureka-server")
    implementation("org.springframework.boot:spring-boot-starter-
actuator")
    implementation("org.springframework.boot:spring-boot-starter-
undertow")
}
```

然后，给应用添加注解 @EnableAdminServer 将服务配置成 Admin Server。

```
package dev.local.gtm.discovery;

// 省略导入

/**
 * 服务发现服务器
 */
@SpringBootApplication
@EnableAdminServer
@EnableEurekaServer
@RefreshScope
@Configuration
public class Application {

    public static void main(String[] args) {
        SpringApplication.run(Application.class, args);
    }
}
```

最后，需要更改一下 application.yml，第一个要更改的是将 Admin Server 的 context-path 更改成一个除了 / 之外的路径，因为如果不更改 Eureka Server 和 Admin Server 就冲突了，都定义在 /。下面的例子中把 Admin Server 的路径定义成 /admin，也就是 http://localhost:8761/admin。

另一个要改动的配置是 eureka.instance 的续租间隔时间、健康查询路径，以及 eureka.client 的拉取间隔时间。本来这些对于 Eureka Server 是不需要的，但是 Admin Server 需要拉取注册的信息以便可以监控各服务，所以这里需要改成下面的配置。

```
server:
  port: 8761
spring.boot.admin.context-path: /admin
eureka:
  instance:
    leaseRenewalIntervalInSeconds: 10
    health-check-url-path: /actuator/health
  client:
    registryFetchIntervalSeconds: 5
    serviceUrl:
      defaultZone: ${EUREKA_SERVICE_URL:http://localhost:8761}/eureka/
management:
  endpoints:
    web:
      exposure:
```

```
        include: "*"
  endpoint:
    health:
      show-details: ALWAYS

---

spring:
  profiles: prod
eureka:
  instance:
    hostname: discovery
    leaseRenewalIntervalInSeconds: 10
    health-check-url-path: /actuator/health
  client:
    registryFetchIntervalSeconds: 5
    serviceUrl:
      defaultZone:
http://${eureka.instance.hostname}:${server.port}/eureka/
```

这样做好一个 Admin Server 之后，其他什么都不用做了，直接重新启动 discovery 和 configserver，以及 api-service，就可以在 http://localhost:8761/admin 看到如图 9-8 所示的界面列出了 3 个服务的状态。

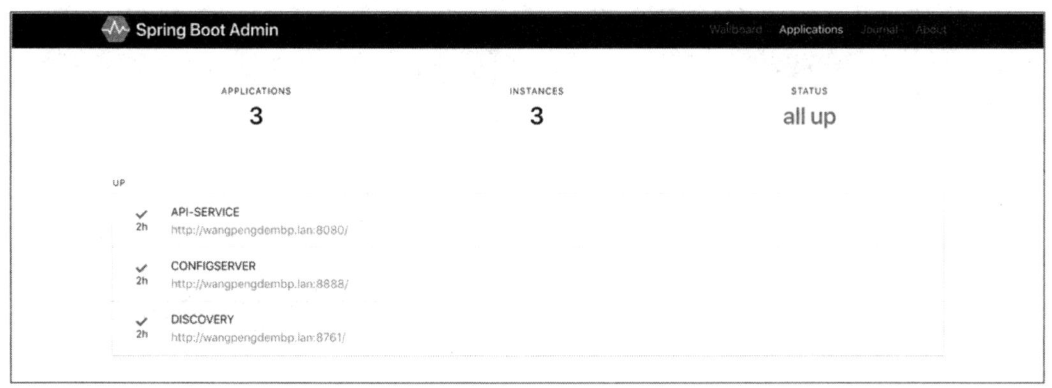

图 9-8

点击 API-SERVICE 就可以查看该微服务的详细监控界面，如图 9-9 所示。

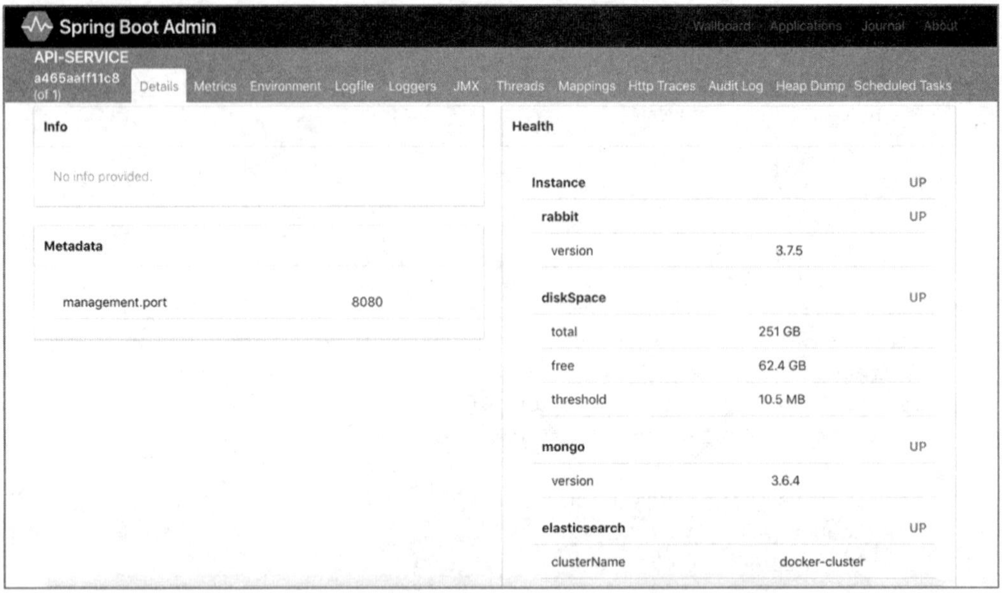

Spring Boot Admin
Wallboard Applications Journal About
API-SERVICE
a465aaff11c8
(of 1)
Details Metrics Environment Logfile Loggers JMX Threads Mappings Http Traces Audit Log Heap Dump Scheduled Tasks
Info
No info provided.
Metadata
management.port 8080
Health
Instance UP
rabbit UP
version 3.7.5
diskSpace UP
total 251 GB
free 62.4 GB
threshold 10.5 MB
mongo UP
version 3.6.4
elasticsearch UP
clusterName docker-cluster

图 9-9

在这个详情界面，除了可以看到所示的健康信息，还可以下拉看到常见的 JVM 监控指标的统计，比如进程、线程、垃圾回收、堆内存、非堆内存等（见图 9-10）。

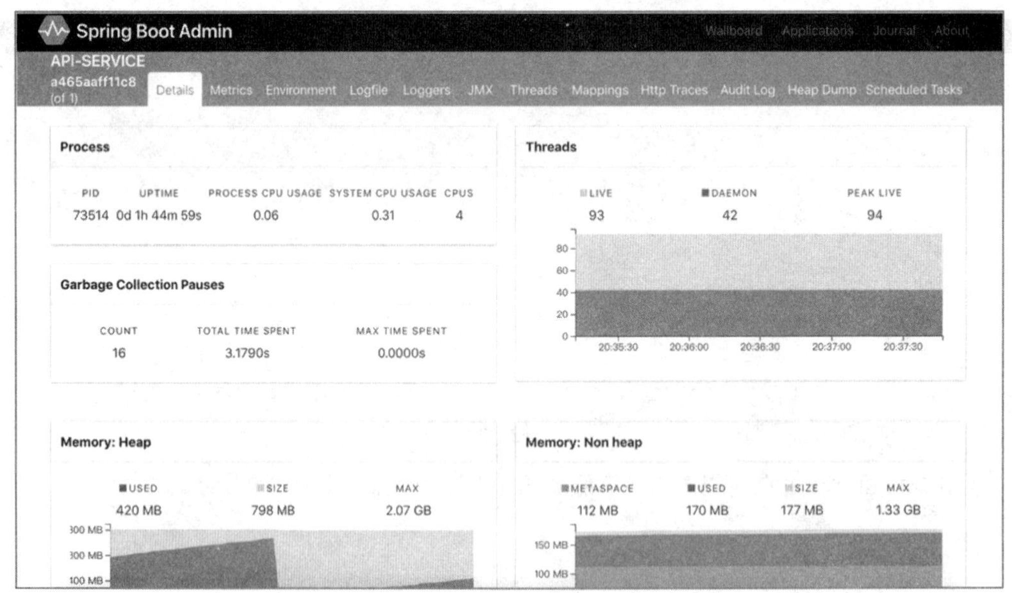

图 9-10

点击其他选项可以查看 Spring Boot Admin 提供的其他特性。总的来说，Spring Boot Admin 具有以下特性。

- 显示健康状态。
- 显示详细的指标，比如：
  - — JVM & 内存测量；
  - — micrometer.io 的度量；
  - — 数据源测量；
  - — 缓存测量。
- 显示构建版本信息。
- 跟踪和下载日志。
- 查看 JVM 的 system- 和 environment- 开头的属性。
- 支持 Spring Cloud 的 /env- 和 /refresh- 开头的路径。
- 简单的日志级别管理。
- 和 JMX-beans 进行交互操作。
- 查看线程 dump。
- 查看 HTTP 跟踪。
- 查看审计事件。
- 查看 HTTP 路径。
- 查看计划任务。
- 查看/删除会话（使用 spring-session）。
- 查看 Flyway / Liquibase 等开源数据迁移记录。
- 下载内存堆 dump。
- 状态更新通知（通过 E-mail、Slack、Hipchat 等）
- 状态变更事件的记录。

2. 监控日志

默认情况下，Spring Boot Admin 只能调节日志级别，但看不到日志本身，这个原因是 Spring Boot 如果没有设置 logging.path 或 logging.file，则是不可能通过 actuator 取得日志信息的。如果希望在 Spring Boot Admin 看到日志，则需要在对应的微服务的 applicaiton.yml 中指定 logging.path 或 logging.file。

```
logging:
  level:
    org.apache.http: ERROR
    org.springframework:
```

```
        web: ERROR
        data: ERROR
        security: ERROR
        cache: ERROR
      org.springframework.data.mongodb.core.MongoTemplate: ERROR
      dev.local.gtm.api: ERROR
    file: /Users/wangpeng/workspace/logs/gtm-api.log
    file.max-history: 20
    pattern.file: "%clr(%d{yyyy-MM-dd
HH:mm:ss.SSS}){faint} %clr(%5p) %clr(${PID}){magenta} %clr(---
){faint} %clr([%15.15t]){faint} %clr(%-
40.40logger{39}){cyan} %clr(:){faint} %m%n%wEx"
```

在上面的配置文件中，我们指定了 /Users/wangpeng/workspace/logs/gtm-api.log 作为日志文件的路径。file.max-history: 20 定义保存最近的多少天的日志文件，而 pattern.file 指定了日志的格式。这样就可以在 Spring Boot Admin 中直接看到日志了（见图 9-11）。

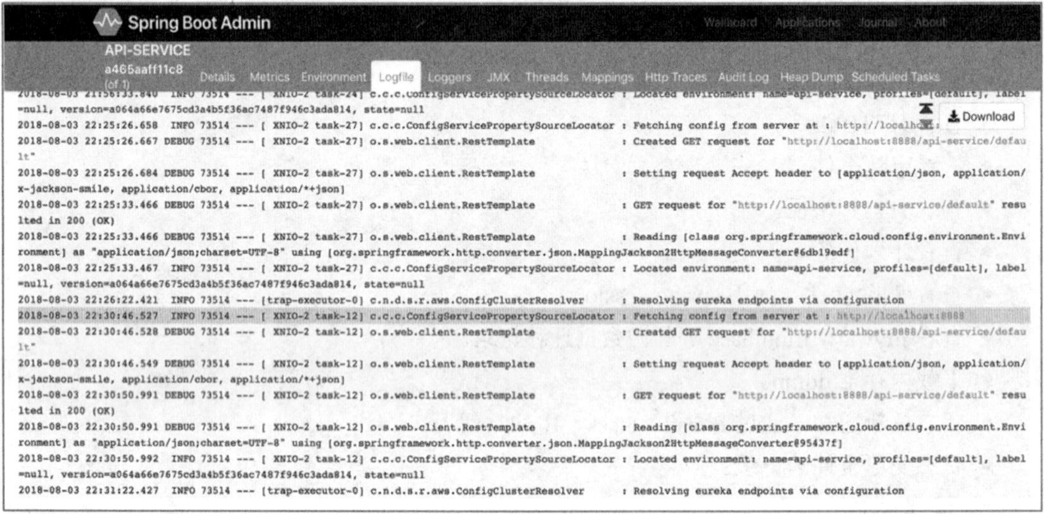

图 9-11

9.3.2 路由服务

路由服务是由于在系统中存在多个微服务的情况下，每个微服务都会有一系列的公开的 API 接口，这种分布式的状态当然对于每个微服务来说只专注于自身的业务逻辑会更清晰，但对于消费服务接口的客户端，比如前端或 App 客户端等就比较痛苦了。难道我们需要了解所有的微服务地址和其 API 才能调用吗？

如果把系统的各个微服务提供的 API 集成在一起，根据某种规则进行转发，这样消费端就可以只和这样一个集成的服务对话，而不需考虑服务架构的内部构建和部署方式了。

例如, / 可以映射到首页的前端应用，/api/users 映射到用户的微服务，/api/shop 映射到在线商铺的微服务。

要达成这样的效果，根据系统的规模，有几种可选方案。

- Nginx 作为服务路由：对于中小型规模的系统，笔者推荐这种方式，因为其配置相对简单、快速，而且效率上也没有太多损耗。
- 利用专有的路由服务：比如 Netflix 团队开源的 zuul，这种方式更专业，对于较复杂的路由规则有良好的支持，也可以方便地和 Spring Cloud 集成。

1. Nginx 作为路由服务

使用 Nginx 作为路由是非常方便的，我们只需要定义系统中需要整合的微服务为上游 Server（upstream app_server），然后在 server 那段中设计不同的 location 规则，再将对应请求路由到对应的上游 Server，或者如果是静态资源，一般就用 Nginx 处理。

下面就是一个典型的 Nginx 的配置文件。

```
upstream app_server {
  server api_server:8080;
}

server {
  listen 443;
  listen [::]:443;
  ssl on;
  ssl_certificate /etc/nginx/certs/server.crt;
  ssl_certificate_key /etc/nginx/certs/server.key;
  charset utf-8;

  location ~ ^/(api|management)/ {
    proxy_pass http://app_server;
    proxy_redirect      off;
    proxy_set_header    Host $host;
    proxy_set_header    X-Real-IP $remote_addr;
    proxy_set_header    X-Forwarded-For $proxy_add_x_forwarded_for;
    proxy_set_header    X-Forwarded-Host $server_name;
    proxy_connect_timeout   10;
    proxy_send_timeout      15;
    proxy_read_timeout      20;
  }
```

```
    location / {
      root /usr/share/nginx/html;
      index index.html index.htm;
      try_files $uri $uri/ /index.html =404;
    }
  }
```

在 Nginx 配置中的上游 Server，如果使用 IP，就不太灵活，所以我们采用了主机名的方式。但主机名如果不使用 Docker 就需要自己配置，所以我们采用容器的方式避免直接配置主机名，将需要在 Nginx 中配置的服务使用同一个网络。

```
  version: '3.2'
  services:
    elasticsearch:
      image: registry.cn-beijing.aliyuncs.com/twigcodes/elasticsearch-wo-
xpack:5.5.0
      volumes:
        - esdata:/usr/share/elasticsearch/data
        -
 ./docker/elasticsearch/config/elasticsearch.yml:/usr/share/elasticsearch/co
nfig/elasticsearch.yml:ro
      environment:
        - discovery.type=single-node
      ports:
        - "9200:9200"
        - "9300:9300"
      networks:
        - docker-app
    redis:
      image: registry.cn-beijing.aliyuncs.com/twigcodes/redis:4-alpine
      command: [ "redis-server", "--protected-mode", "no" ]
      ports:
        - "6379:6379"
      volumes:
        - redis-data:/data
      networks:
        - docker-app
    mongo:
      image: registry.cn-beijing.aliyuncs.com/twigcodes/mongo:3.6.4
      ports:
        - "27017:27017"
      volumes:
        - api_db:/data/db
      networks:
```

```
    - docker-app
  rabbitmq:
    image: rabbitmq:3-management-alpine
    ports:
      - "5672:5672" # JMS 端口
      - "15672:15672" # 管理端口 default user:pass = guest:guest
    networks:
      - docker-app
  configserver:
    image: registry.cn-beijing.aliyuncs.com/twigcodes/gtm-config
    ports:
      - 8888:8888
    depends_on:
      - rabbitmq
    networks:
      - docker-app
  discovery:
    image: registry.cn-beijing.aliyuncs.com/twigcodes/gtm-discovery
    ports:
      - 8761:8761
    depends_on:
      - rabbitmq
      - configserver
    networks:
      - docker-app
  api-server:
    image: registry.cn-beijing.aliyuncs.com/twigcodes/gtm-api-service
    environment:
      - SPRING_PROFILES_ACTIVE=prod
    ports:
      - "8080:8080"
      - "5005:5005"
    links:
      - mongo
      - redis
      - elasticsearch
    networks:
      - docker-app
  nginx:
    build:
      context: .
      dockerfile: ./frontend/docker/nginx/Dockerfile
      args:
        - env=production
```

```
      container_name: nginx
      ports:
        - 80:80
        - 443:443
  volumes:
    api_db: {}
    redis-data: {}
    esdata: {}
  networks:
    docker-app:
      driver: bridge
```

当然，这样做其实在扩展性方面还是有问题的，如果容器不在同一宿主机，就又得采用 IP 方式，但是这样就缺乏了动态配置的优点。接下来就看一下使用路由服务的方式。

2. Zuul 路由服务

Zuul 是 Netflix 开发的一个基于 JVM 的路由组件，而且可以进行服务端负载均衡。

Zuul 提供了以下特性：

- 鉴权
- 压力测试
- 动态路由
- 服务迁移
- 安全
- 静态响应处理
- 流量管控

添加一个基于 Zuul 的路由服务需要新建一个 gradle 子项目 gtm-gateway。接下来要为其添加一个依赖 org.springframework.cloud:spring-cloud-starter-netflix-zuul。

```
  apply plugin: 'org.springframework.boot'
  configurations {
      springLoaded
      // 如果使用 undertow 或 jetty，则需要把默认包含的 tomcat 排除在外
      compile.exclude module: 'spring-boot-starter-tomcat'
  }
  dependencies {
      implementation("org.springframework.cloud:spring-cloud-starter-
config")
      implementation("org.springframework.cloud:spring-cloud-starter-bus-
amqp")
      implementation("org.springframework.cloud:spring-cloud-starter-
```

```
stream-rabbit")
        implementation("org.springframework.cloud:spring-cloud-starter-
netflix-eureka-client")
        implementation("org.springframework.cloud:spring-cloud-starter-
netflix-zuul")
        implementation("org.springframework.boot:spring-boot-starter-
actuator")
        implementation("org.springframework.boot:spring-boot-starter-
undertow")
    }
```

然后，在 gtm-gateway/src/main/java/dev/local/gtm/gateway/Application.java 文件中添加 @EnableZuulProxy 这个注解。

```
package dev.local.gtm.gateway;

// 省略导入

/**
 * 路由网关服务器
 */
@SpringBootApplication
@EnableZuulProxy
@EnableDiscoveryClient
@RefreshScope
@Configuration
public class Application {

    public static void main(String[] args) {
        SpringApplication.run(Application.class, args);
    }
}
```

接下来，就可以添加 Zuul 相关的配置了，Zuul 自带强大的反向代理功能。反向代理简单的理解就是让一个 HTTP 请求被转发到对应的服务。

下面的例子中，zuul.ignored-services 这个属性是决定哪些服务 ID 是不走反向代理的，但是如果一个服务在忽略服务的表达式中匹配了，但是在 routes 中列出了，那么这个忽略就是无效的。也就是说下面的例子中虽然我们忽略的是所有服务，因为使用了通配符 *，但由于在 routes 中列出了 api 这个定义，所以 /api 开头的请求（我们在 zuul.routes.api.path 中定义了 /api/** 这个路径）都会被转发到一个 ID 为 api-service 的服务。

Zuul 代理使用 Ribbon 通过 Eureka 定位该服务。所有的请求会在一个 Hystrix 命令中执行，所以如果发生了异常，这些失败信息就会在 Hystrix 中体现出来，此时我们的反向代理就不会再去联系目标服务了。

```
zuul:
  ignoredServices: '*'
  routes:
    api:
      path: /api/**
      serviceId: api-service
      stripPrefix: true
```

stripPrefix 这个属性决定是否去掉前缀，也就是 /api，我们设置的就是去掉 api。举个例子，如果我们对路由网关发送请求 http://localhost:8090/api/actuator/info，那么转发到 api-service 时，请求就变成了 http://localhost:8080/actuator/info，去掉了 /api 这个路径前缀。

有些时候，我们转发请求时，并不需要将包含一些敏感信息的 Header 一起转发出去，这种情况下可以设置 zuul.routes.api.sensitiveHeaders，这里设置的值就不会转发出去，这个属性的默认值是 Cookie、Set-Cookie、Authorization。也就是默认情况下，这些 Header 不会被转发。但如果希望 Authorization 被保留，就可以设置 sensitiveHeaders 为 Cookie、Set-Cookie。

```
zuul:
  ignoredServices: '*'
  routes:
    api:
      path: /api/**
      serviceId: api-service
      sensitiveHeaders: Cookie,Set-Cookie
      stripPrefix: true
```

完整的 gtm-gateway 的 application.yml 现在看起来就是下面的样子。

```
server:
  port: 8090
eureka:
  client:
    serviceUrl:
      defaultZone: ${EUREKA_SERVICE_URL:http://localhost:8761}/eureka/
management:
  endpoints:
    web:
      exposure:
        include: "*"
  endpoint:
    health:
      show-details: ALWAYS
zuul:
```

```
    ignoredServices: '*'
    routes:
      api:
        path: /api/**
        serviceId: api-service
        stripPrefix: true
---

spring:
  profiles: prod
eureka:
  instance:
    hostname: discovery
    leaseRenewalIntervalInSeconds: 10
    health-check-url-path: /actuator/health
  client:
    registryFetchIntervalSeconds: 5
    serviceUrl:
      defaultZone:
http://${eureka.instance.hostname}:${server.port}/eureka/
```

Zuul 还提供很多其他功能，比如负载均衡、Filter 等，此处不再赘述。

9.4 微服务的远程调用

9.4.1 Feign Client

Feign 是 Netflix 出品的一个开源的声明式 HTTP 客户端，Feign 的目标是简化 HTTP 客户端。简单来说，开发者仅需要声明和注解一个接口，真正的接口实现会在运行时提供。我们在前面的章节中使用过 RestTemplate，那么 Feign 可以看作是一个可以感知服务注册信息的 RestTemplate。

举例来说，如果没有 Feign，使用 RestTemplate 来实现微服务之间的调用，则需要在 Controller 中注入 eurekaClient，以便可以取得要调用的微服务的地址和端口信息。

```
@Autowired

private EurekaClient eurekaClient;

public void someAPI() {
    Application application = eurekaClient.getApplication("api-service");
    InstanceInfo instanceInfo = application.getInstances().get(0);
    String hostname = instanceInfo.getHostName();
```

```
        int port = instanceInfo.getPort();
        // ...
    }
```

但显然这样做有点麻烦，Feign 可以大大地简化微服务中的这种调用。我们只需要添加一个注解 @FeignClient("api-service")，这意味着我们要调用 Eureka 中注册的 api-service 方法。其余的就可以按照标准的 RestController 去写了，这个接口的具体实现会在运行时提供：

```
    @FeignClient("api-service")
    public interface ApiServiceClient {
        @RequestMapping("/api/users")
        String getUsers();
    }
```

是的，就这么简单，而且除了对于微服务内部的调用，Feign 在实现第三方接口调用时也非常好用。尤其对于标准的 RESTful 接口来说，比 RestTemplate 要简洁很多。但是对于较老的一些接口，尤其是不太规范的接口，其支持的灵活度和可定制性比 RestTemplate 要弱一些。

1. 调用第三方 API——LeanCloud

如果同样是调用 LeanCloud 的图形验证码和短信验证服务 API，我们使用 Feign 的话，就只需定义下面的接口就可以完成 API 的接口定义。这种书写形式比 RestTemplate 要清晰和简洁。

```
    package dev.local.smartoffice.api.service.feign;

    // 省略导入

    /**
     * LeanCloudService
     */
    @FeignClient(name = "leanCloud", url = "${app.leanCloud.baseUrl}",
configuration = LeanCloudConfiguration.class)
    public interface LeanCloudFeignClient {

        @PostMapping("/requestSmsCode")
        void requestSmsCode(@RequestBody RequestSmsCodeParam
requestSmsCodeParam);

        @PostMapping("/verifySmsCode/{code}")
        void verifySmsCode(@RequestBody VerifySmsCodeParam
verifySmsCodeParam, @PathVariable String code);

        @GetMapping("/requestCaptcha")
        Captcha requestCaptcha();

        @PostMapping("/verifyCaptcha")
```

```
    String verifyCaptcha(@RequestBody VerifyCaptchaParam
verifyCaptchaParam);

    @Getter
    @Setter
    @AllArgsConstructor
    @NoArgsConstructor
    public static class RequestSmsCodeParam {
        private String mobilePhoneNumber;
        private String validate_token;
    }

    @Getter
    @Setter
    @AllArgsConstructor
    @NoArgsConstructor
    public static class VerifySmsCodeParam {
        private String mobilePhoneNumber;
    }

    @Getter
    @Setter
    @AllArgsConstructor
    @NoArgsConstructor
    public static class VerifyCaptchaParam {
        private String captcha_code;
        private String captcha_token;
    }
}
```

注意，上面代码中的 url = "${app.leanCloud.baseUrl}" 规定第三方的 API 的 URL，而这个 URL 是可以定义在 .properties 或者 .yml 文件中的。在上面例子中只需要在 applicaiton.yml 中定义 app.leanCloud.baseUrl 的值为 LeanCloud 分配的应用域名即可。

当然，我们也需要提供截断器、自定义错误处理等，这些可以都在一个 LeanCloudConfiguration 中定义，在上面的代码中只需要指定配置的类型即可 configuration = LeanCloudConfiguration.class。这个配置类中，我们可以定制化以下内容。

- Logger.Level——设置日志等级。
  - NONE——没有日志。
  - BASIC——记录请求的方法、URL 和返回的响应码以及响应时间。
  - HEADERS——除了 BASIC 提供的日志，还提供请求（Request）和响应（Response）的头（Headers）。

  — FULL——完整日志，包括请求和响应的 Header 和 Body 以及其他元数据。

- Retryer——定义重试逻辑。
- ErrorDecoder——可以进行自定义的错误处理。
- Request.Options——定义请求的连接超时和读取超时。
- Collection<RequestInterceptor>——截断器，可以对请求进行拦截，一般可以写入较通用的 Header，比如授权信息等。

```
package dev.local.smartoffice.api.config.feign;

// 省略导入

@Configuration
@RequiredArgsConstructor
public class LeanCloudConfiguration {

    private final ObjectFactory<HttpMessageConverters> messageConverters;
    private final AppProperties appProperties;

    @Bean
    public Logger.Level feignLogger() {
        return Logger.Level.FULL;
    }

    @Bean
    public ErrorDecoder errorDecoder() {
        return new LeanCloudErrorDecoder();
    }

    @Bean
    public Encoder feignEncoder() {
        return new SpringEncoder(messageConverters);
    }

    @Bean
    public LeanCloudAuthHeaderInterceptor
leanCloudAuthHeaderInterceptor() {
        return new LeanCloudAuthHeaderInterceptor(appProperties);
    }
}
```

一个典型的 Feign 拦截器，如下面的代码所示，需要实现 RequestInterceptor，在拦截中我们在所有请求的头部添加 LeanCloud 要求的 AppId 和 AppKey，以及设置 Content-Type 为 application/json。

```
package dev.local.smartoffice.api.config.feign;

// 省略导入

/**
 * LeanCloudAuthHeaderInterceptor
 */
@RequiredArgsConstructor
public class LeanCloudAuthHeaderInterceptor implements
RequestInterceptor {

    private final AppProperties appProperties;

    @Override
    public void apply(RequestTemplate template) {
        template.header("X-LC-Id",
appProperties.getLeanCloud().getAppId());
        template.header("X-LC-Key",
appProperties.getLeanCloud().getAppKey());
        template.header("Content-Type",
MediaType.APPLICATION_JSON_UTF8_VALUE);
    }
}
```

和 RestTemplate 类似，我们也可以添加自定义的错误处理机制，在连接第三方服务时，这是很有必要的。因为第三方的服务往往采用各不相同的错误处理，有必要在系统中将第三方返回的错误统一处理成系统内部规范的机制。

```
package dev.local.smartoffice.api.config.feign;

// 省略导入

@Slf4j
public class LeanCloudErrorDecoder implements ErrorDecoder {
    @Override
    public Exception decode(String methodKey, Response response) {
        if (response.status() != 200) {
            return new ErrorDecoder.Default().decode(methodKey,
response);
        }
        String json;
        try {
            json =
ErrorResponseUtil.readFully(response.body().asInputStream());
            log.debug("[LeanCloud] 解析返回错误 {}", json);
```

```
            ObjectMapper mapper = new ObjectMapper(new JsonFactory());
            JsonNode jsonNode = mapper.readValue(json, JsonNode.class);
            Integer code = jsonNode.has("code") ?
jsonNode.get("code").intValue() : null;
            String err = jsonNode.has("error") ?
jsonNode.get("error").asText() : null;

            val error = new LeanCloudError(code, err);
            log.debug("[LeanCloud] 错误: ");
            log.debug("    代码          : {}", error.getCode());
            log.debug("    信息          : {}", error.getError());
            return new OutgoingBadRequestException(error.getError());
        } catch (JsonParseException e) {
            return new OutgoingBadRequestException(e.getMessage());
        } catch (IOException e) {
            return new OutgoingBadRequestException(e.getMessage());
        }
    }

    @Data
    @AllArgsConstructor
    private final class LeanCloudError {
        private Integer code;
        private String error;
    }
}
```

利用 Feign Client 的好处是除了代码简洁流畅，还可以利用 Ribbon 进行负载均衡以及利用 Hystrix 进行熔断处理。

9.4.2 负载均衡

负载平衡自动在两台或多台计算机之间分配传入的应用程序流量。它让我们能够在应用程序中实现容错，无缝地提供路由应用程序流量所需的负载平衡容量。负载平衡旨在优化资源使用，最大化吞吐量，最小化响应时间，并避免任何单个资源的过载。使用具有负载平衡的多个组件可以通过冗余提高可靠性和可用性。

1. Ribbon

Ribbon 是一个客户端负载均衡器，可以让我们对 HTTP 和 TCP 客户端的行为进行控制。Ribbon 的客户端组件提供了很多好用的配置选项，例如连接超时、重试、重试算法（指数、边界退回）等。Ribbon 内置了可插拔和可自定义的负载平衡组件。

下面列出了一些提供的负载平衡策略：

- 简单的循环负载均衡
- 加权响应时间负载均衡
- 基于区域的轮询负载均衡
- 随机负载均衡

2. 客户端负载均衡

负载平衡的实现方法是向客户端提供服务器 IP 列表，然后让客户端从每个连接的列表中随机选择 IP。一般来说，客户端随机负载平衡往往比循环 DNS 提供更好的负载分配。利用这种方法，向客户端传送 IP 列表的方法可以有多种变化形式，甚至可以实现为 DNS 列表（在没有任何循环的情况下传递给所有客户端），或者通过将其硬编码到列表中。如果使用“智能客户端”，检测到随机选择的服务器已关闭并再次随机连接，则还提供容错功能。

Ribbon 提供以下功能：

- 负载均衡
- 容错
- 异步和反应模型中的多协议（HTTP、TCP、UDP）支持
- 缓存和批处理

Ribbon 中的一个核心概念是指定客户端的概念。每个负载均衡器都是微服务架构中多个组件的一部分，这些组件一起工作以按需联系远程服务器，并且该集合可以使用一个对开发人员友好的名称。

如果不使用 Eureka，则可以通过配置文件进行负载均衡 server 的配置。下面例子中，listOfServers 列出了用于负载均衡的 3 个服务地址和端口。

```
api-service:
  ribbon:
    eureka:
      enabled: false
    listOfServers: localhost:8091,localhost:8092,localhost:8093
    ServerListRefreshInterval: 15000
```

但如果我们使用了 Eureka，就无须任何 Ribbon 配置，当然 Eureka 的配置还是需要的。Spring Cloud 集成了 Ribbon 和 Eureka，可在使用 Feign 时提供负载均衡的 HTTP 客户端。此时默认无须指定 listOfServers，因为 Feign 会自动从 Eureka 拉取 Server 列表，并利用集成的 Ribbon 进行负载均衡。

附录 A 常见云服务使用问题汇总

1．如何使用密钥连接主机？

阿里巴巴和腾讯都是在控制台建立密钥对，注意，一般服务器不会保留私钥，请立即下载并保存好私钥。然后在客户端使用如下命令访问。以 ubuntu 镜像为例，下面这样就可以连接了。

```
ssh -i </path/to/private_key> ubuntu@xx.xxx.xx.xxx
```

第一次连接时会询问是否需要保存并信任远程主机的 footprint，输入 yes 即可。

```
The authenticity of host 'xxx.xx.xxx.xx (xxx.xx.xxx.xx)' can't be established.
ECDSA key fingerprint is SHA256:xxxxxxxx
Are you sure you want to continue connecting (yes/no)? yes
```

2．重装系统后连接失败，提示 REMOTE HOST IDENTIFICATION HAS CHANGED!，如何处理？

重装系统后，远程主机 footprint 已经改变，此时应该从客户端删除原先保存的记录。

```
ssh-keygen -R xxx.xx.xxx.xx
```

3．为何有时候会出现提示密钥权限不对的错误？

密钥的权限太广会导致系统中可以访问密钥的人过多，所以需要设置。

```
chmod 400 </path/to/private_key>
```

4．如何安装 Docker？

官方一键安装脚本，速度慢，但是比较靠谱。

```
curl -sSL https://get.docker.com/ | sh
```

当然，还有国内的镜像可用，笔者使用的是 DaoCloud 网站提供的 Docker 安装镜像。

```
curl -sSL https://get.daocloud.io/docker | sh
```

也可以通过这个网站安装 docker-compose。

```
curl -L
https://get.daocloud.io/docker/compose/releases/download/1.21.2/docker-compose-`uname -s`-`uname -m` > /usr/local/bin/docker-compose
chmod +x /usr/local/bin/docker-compose
```

5. 如何不用 sudo 使用 Docker？

```
sudo usermod -aG docker ubuntu
```

6. 设置镜像加速

腾讯云：

```
sudo mkdir -p /etc/docker
sudo tee /etc/docker/daemon.json <<-'EOF'
{
  "registry-mirrors": ["https://mirror.ccs.tencentyun.com"]
}
EOF
sudo systemctl daemon-reload
sudo systemctl restart docker
```

7. 如何修改 Ubuntu 软件仓库为国内镜像？

```
# 备份原文件
mv /etc/apt/sources.list /etc/apt/sources.list.bak

# 修改为阿里云的镜像源
cat > /etc/apt/sources.list << END
deb http://mirrors.aliyun.com/ubuntu/ trusty main restricted universe multiverse
deb http://mirrors.aliyun.com/ubuntu/ trusty-security main restricted universe multiverse
deb http://mirrors.aliyun.com/ubuntu/ trusty-updates main restricted universe multiverse
deb http://mirrors.aliyun.com/ubuntu/ trusty-proposed main restricted universe multiverse
deb http://mirrors.aliyun.com/ubuntu/ trusty-backports main restricted universe multiverse
deb-src http://mirrors.aliyun.com/ubuntu/ trusty main restricted universe multiverse
```

```
    deb-src http://mirrors.aliyun.com/ubuntu/ trusty-security main
restricted universe multiverse
    deb-src http://mirrors.aliyun.com/ubuntu/ trusty-updates main restricted
universe multiverse
    deb-src http://mirrors.aliyun.com/ubuntu/ trusty-proposed main
restricted universe multiverse
    deb-src http://mirrors.aliyun.com/ubuntu/ trusty-backports main
restricted universe multiverse
    END

    # 更新源列表信息
    apt-get update
```

8．如何修改 CentOS 软件仓库为国内镜像？

```
    # 备份原文件
    mv /etc/yum.repos.d/CentOS-Base.repo /etc/yum.repos.d/CentOS-
Base.repo.bak

    # 下载新的 CentOS-Base.repo 到/etc/yum.repos.d/

    # CentOS 5
    wget -O /etc/yum.repos.d/CentOS-Base.repo
http://mirrors.aliyun.com/repo/Centos-5.repo

    # CentOS 6
    wget -O /etc/yum.repos.d/CentOS-Base.repo
http://mirrors.aliyun.com/repo/Centos-6.repo

    # CentOS 7
    wget -O /etc/yum.repos.d/CentOS-Base.repo
http://mirrors.aliyun.com/repo/Centos-7.repo

    # 生成缓存
    yum makecache
```

9．如何高速安装 docker compose？

```
    curl -L
https://get.daocloud.io/docker/compose/releases/download/1.18.0/docker-
compose-`uname -s`-`uname -m` > /usr/local/bin/docker-compose
    chmod +x /usr/local/bin/docker-compose
```

10．安装 docker 时遇到 E: Unable to correct problems, you have held broken packages，怎么办？

```
安装 libltdl7_2.4.6
wget http://launchpadlibrarian.net/236916213/libltdl7_2.4.6-0.1_amd64.deb
```

11．如何在远程终端连接到服务器后，可以在不保持会话的条件下执行命令？

在远程主机上执行：

```
screen
```

然后，即可执行想要执行的命令，退出这个 screen 会话，可以按 Ctrl + A 组合键然后按 D。如果恢复会话，则可以使用下面的命令进行会话的恢复。

```
screen -r
```

12．在云环境部署 Elasticsearch 的容器时发生 max virtual memory areas vm.max_map_count [65530] likely too low, increase to at least [262144] 错误，怎么办？

```
sudo sysctl -w vm.max_map_count=262144
```

13．在不暴露端口的情况下，如何访问某些受限资源？

通常，我们在云服务上只会开启某些端口，但对于很多微服务来说，它们只是内网可访问，在外部无法访问。这带来了很高的安全性，但如果有问题需要我们访问内网的端口去调试，那怎么办呢？并不是所有任务都可以通过命令行方式解决，有些任务需要我们访问 Web 页面。

下面的命令会建立一个映射到服务器 8080 端口的 ssh 通道。

```
ssh -D 8080 -C -N username@example.com
```

然后，可以设置浏览器使用 SOCKS 代理 127.0.0.1:8080，这样就可以打开本机的浏览器进行访问了。

14．vscode 中打开 Java 工程显示 classpath not complete，怎么办？

在 macOS 系统中，可以到 ~/Library/Application Support/code/User/workspaceStorage 中删除对应的文件夹，然后重启 vscode，打开工程会强制重新编译。

推荐阅读 3

京东购买二维码

作者：李金洪　　书号：978-7-121-36392-4　　定价：159.00 元

完全实战的人工智能书，700 多页

这是一本非常全面的、专注于实战的 AI 图书，兼容 TensorFlow 1.x 和 2.x 版本，共 75 个实例。

全书共分为 5 篇：

第 1 篇，介绍了学习准备、搭建开发环境、使用 AI 模型来识别图像；

第 2 篇，介绍了用 TensorFlow 开发实际工程的一些基础操作，包括使用 TensorFlow 制作自己的数据集、快速训练自己的图片分类模型、编写训练模型的程序；

第 3 篇，介绍了机器学习算法相关内容，包括特征工程、卷积神经网络（CNN）、循环神经网络（RNN）；

第 4 篇，介绍了多模型的组合训练技术，包括生成式模型、模型的攻与防；

第 5 篇，介绍了深度学习在工程上的应用，侧重于提升读者的工程能力，包括 TensorFlow 模型制作、布署 TensorFlow 模型、商业实例。

本书结构清晰、案例丰富、通俗易懂、实用性强。适合对人工智能、TensorFlow 感兴趣的读者作为自学教程。